Emily Hodgins

Nontechnical Guide to

PETROLEUM Geology, Exploration, Drilling & Production

Third Edition

Nontechnical Guide to

PETROLEUM Geology, Exploration, Drilling & Production

Third Edition

Norman J. Hyne, Ph.D.

Copyright© 2012 by
PennWell Corporation
1421 South Sheridan Road
Tulsa, Oklahoma 74112–6600 USA

800.752.9764
+1.918.831.9421
sales@pennwell.com
www.pennwellbooks.com
www.pennwell.com

Marketing: Jane Green
National Account Executive: Barbara McGee

Director: Mary McGee
Managing Editor: Stephen Hill
Production Manager: Sheila Brock
Production Editor: Tony Quinn
Book Designer: Susan E. Ormston
Cover Designer: Kelly Cook

Cover photo courtesy of Helmerich & Payne. Author photo courtesy of Marshall Heim.

Library of Congress Cataloging-in-Publication Data

Hyne, Norman J.
 Nontechnical guide to petroleum geology, exploration, drilling, and production
/ Norman J. Hyne. -- 3rd ed.
    p. cm.
 Includes bibliographical references and index.
 ISBN 978-1-59370-269-4
1. Petroleum--Geology. 2. Petroleum engineering. 3. Petroleum--Prospecting.
I. Title.
 TN870.5.H9624 2011
 665.5--dc23
                                2011038698

4 5   16 15 14 13

Contents

Preface

This book contains an enormous amount of useful information on the upstream petroleum industry. It is designed for easy reading, and the information is readily accessible. The introductory chapter should be read first. It is an excellent overview that shows how everything in petroleum geology, exploration, drilling, and production is interrelated.

Each subject has its own chapter that is well illustrated with figures and plates. The rocks, minerals, and seismic examples are in color. Industry terms are defined in the text and shown in italics. All measurements are in both English and metric units. A useful index and extensive glossary are located at the back of the book, as well as an interesting list of petroleum records.

Introduction

Both crude oil and natural gas occur naturally in subsurface deposits. Crude oil is a black liquid that is sold to refineries to be refined into products such as gasoline and lubricating oil. Natural gas is a colorless, odorless gas that is sold to gas pipelines to be transported and burned for its heat content. There are many different types of crude oils and natural gases, some more valuable than others. Heavy crude oils are very thick and viscous and are difficult or impossible to produce. Light crude oils are very fluid, relatively easy to produce, rich in gasoline, and more valuable. Some natural gases burn with more heat than others and are more valuable. Some natural gases also contain almost pure liquid gasoline called condensate that separates from the gas when it is produced. Condensate is almost as valuable as crude oil. Sulfur is a bad impurity in both natural gas and crude oil. Sour crude oils contain sulfur, and sour natural gases contain hydrogen sulfide and are less valuable. Crude oil is measured by volume in barrels (bbl). Natural gas is measured by volume in thousands of cubic feet (Mcf) and by heat content in British thermal units (Btu).

In order for there to be a commercial deposit of natural gas or crude oil, three important geological conditions must be met. First, there must be a source rock in the subsurface of that area that generated the gas or oil at some time in the geological past. Second, there must be a separate, subsurface reservoir rock that holds the gas or oil. When we drill a well into that reservoir rock, the gas and oil are able to flow through the reservoir rock and into our well. Third, there must be a trap on the reservoir rock to concentrate the gas or oil into commercial quantities.

The crust of the earth in oil- and gas-producing areas is composed of sedimentary rock layers. Sedimentary rocks can be source and reservoir rocks for gas and oil. These rocks are called sedimentary rocks because they are composed of sediments that were formerly loose particles such as sand grains, mud, and seashells or salts that precipitated out of water. Sedimentary rocks are millions of years old and were deposited when the sea level rose and covered the land many times in the past. These sediments are relatively simple materials such as sands deposited along beaches, mud deposited on the sea bottom, and beds of seashells. Ancient sediments, piled layer upon layer, form the sedimentary rocks that are now sandstones composed of sand grains, shales composed of mud particles, and limestones composed of seashells. These are drilled to find and produce oil and gas.

The source of gas and oil is the organic matter—dead plant and animal material—that is buried and preserved in some ancient sedimentary rocks. The most common, organic-rich sedimentary rock and the source rock for most gas and oil is black shale. It was deposited as organic-rich mud on ancient ocean bottoms. In the subsurface, temperature and time turn organic matter into crude oil. As the source rock is covered with more sediments and buried deeper in the earth, it becomes hotter and hotter. Crude oil starts to form at about 150°F (65°C) at a depth of about 7,000 ft (2,130 m) below the surface of the land (fig. I–1). It is generated from there down to a depth of about 18,000 ft (5,500 m) at about 300°F (150°C). The reactions that change organic matter into oil are complex and take a long time. If the source rock is buried deeper where the temperatures are above 300°F (150°C), the remaining organic matter can generate natural gas.

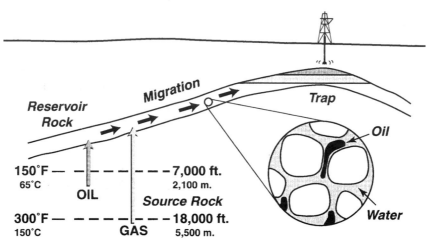

Fig. I–1. Generation and migration of gas and oil

Gas and oil are relatively light in density compared to the water that also occurs in subsurface sedimentary rocks. After oil and gas form, they rise due to buoyancy through fractures in the subsurface rocks. The rising gas and oil can intersect reservoir rock, which is a sedimentary rock layer that contains billions of tiny spaces called pores. A common reservoir rock is sandstone, composed of sand grains like those on a beach. Sand grains are like spheres—there is no way the grains will fit together perfectly. There are pore spaces between the sand grains on a beach and in a sandstone rock. Limestone, another common reservoir rock, is often deposited as shell beds or reefs, and there are pores between the shells and corals. Because limestone is soluble, there can also be solution pits in the limestone.

Porosity is the percent of reservoir rock that is pore space, and it is commonly 10 to 30%. The gas and oil flow into the pores of the reservoir rock layer. Because the reservoir rock also contains water, the gas and oil will continue to rise by flowing from pore to pore to pore up the angle of the reservoir rock layer toward the surface. The movement of gas and oil up the angle of the reservoir rock toward the surface is called migration. The ease with which the gas and oil can flow through the rock is called permeability. Because of migration, the gas and oil can end up a considerable distance, both vertically and horizontally, from where they were originally formed (fig. I–1).

As the gas and oil migrates up along the reservoir rock, it can encounter a trap. A trap is a high point in the reservoir rock where the gas or oil is stopped and concentrated. One type of trap is a natural arch in the reservoir rock (fig. I–2) called a dome or anticline. In the trap, the fluids separate according to their density. The gas is the lightest and goes to the top of the trap to fill the pores of the reservoir rock and form the free gas cap. The oil goes to the middle to fill the pores and form the oil reservoir. The saltwater, the heaviest, goes to the bottom. To complete the trap, a caprock must overlie the reservoir rock. The caprock is a seal that does not allow fluids to flow through it. Without a caprock, the oil and gas would leak up to the surface. Two common sedimentary rocks that can be caprocks are shale and salt.

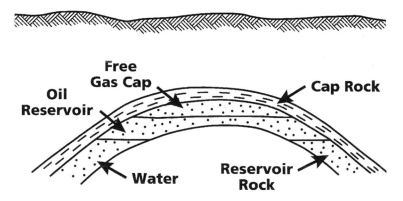

Fig. I–2. Cross section of a subsurface petroleum trap

Most gas and oil deposits are located in basins where sedimentary rocks are relatively thick. Subsurface deposits of gas and oil are found by locating traps. In some areas, the rock layers that crop out on the surface can be projected into the subsurface to discover traps (fig. I–3). Today, these surface rocks can be mapped using photographs from airplanes and satellites. In the subsurface, the rocks in different wells that have already been drilled are matched by correlation to make cross sections, and maps of the depths to the top of subsurface reservoir rocks and their thickness are drawn.

Seismic exploration is commonly used today to locate subsurface traps. The seismic method uses a source and detectors (fig. I-4). The source, such as dynamite, is located on or near the surface and gives off an impulse of sound energy into the subsurface. The sound energy bounces off sedimentary rock layers and returns to the surface to be recorded by the detectors. Sound echoes are used to image the shape of subsurface rock layers and find traps.

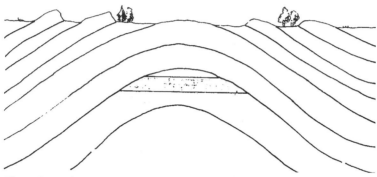

Fig. I-3. Rock outcrops on surface above a dome

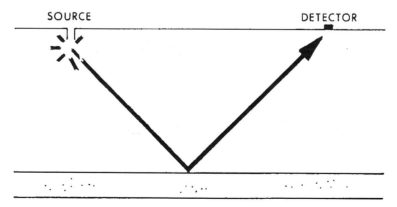

Fig. I-4. The seismic method showing sound impulse bouncing off subsurface rock layer

The only way to know for sure if a trap contains commercial amounts of gas and oil is to drill an exploratory or wildcat well. Many wildcat wells are dry holes with no commercial amounts of gas or oil. The well is drilled using a rotary drilling rig (fig. I–5). There can be thousands of feet of steel drillpipe with a bit on the end, called the drillstring, suspended in the well. By rotating the drillstring from the surface, the bit on the bottom is turned and cuts the hole. As the well is drilled deeper, every 30 ft (9.1 m) drilling is stopped and another section of drillpipe is screwed on the drillstring to make it longer. The power to the rig is supplied by diesel engines. A steel tower above the well—the derrick or mast—along with a hoisting line and pulley system, is used to raise and lower equipment in the well.

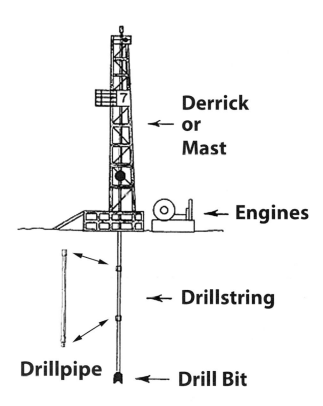

Fig. I–5. Parts of a rotary drilling rig

An important system on the rig is the circulating mud system. Drilling mud, usually made of clay and water, is pumped down the inside of the drillpipe where it jets out of nozzles on the bit and returns up the outside of the drillpipe to the surface (fig. I–6). The drilling mud removes the rock chips made by the bit, called well cuttings, from the bottom of the hole and prevents them from clogging up the bottom of the well. The well is always kept filled to the top with the heavy drilling mud as it is being drilled. The pressure of the drilling mud prevents any fluids such as water, gas, and oil from flowing out of the subsurface rocks and into the well. If gas and oil flowed up onto the floor of the drilling rig, they could catch fire, causing a blowout. Even if only water flowed out of the surrounding rock into the well, the sides of the well could cave in, and the well could be lost. As the well is being drilled, it can be drilled straight down, out at an angle as a deviated well, or out horizontally as a horizontal well through the oil and gas reservoir (fig. I–7). Horizontal wells typically produce oil and gas at a greater rate than vertical wells.

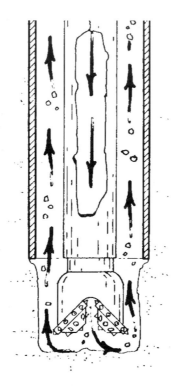

Fig. I–6. Well cutting removal by circulating drilling mud on bottom of well

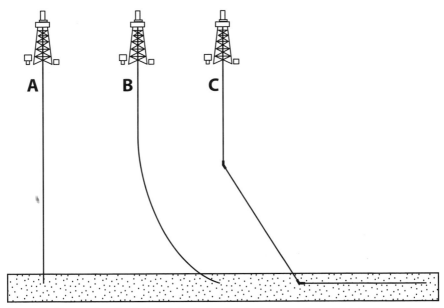

Fig. I–7. Types of wells: (a) straight hole, (b) deviated well, and (c) horizontal well

Offshore wells are drilled into sedimentary rocks on the ocean bottom the same way as on land. For offshore exploratory wells, the rig is mounted on a barge, floating platform, or ship that can be moved. Once an offshore field is located by drilling, a production platform is installed to drill the rest of the wells and produce the gas and oil. The production platform can be fixed with legs that sit on the ocean bottom or floating with anchors and cables to hold it in position.

Because drilling mud keeps gas and oil in the rocks, a subsurface deposit of gas or oil can be drilled without any indication of the gas or oil. To evaluate the well after it has been drilled, it must be logged, and well logs must be created. A well log is a record of the rocks and their fluids in the well. A mud logger is a service company that makes a mud log as the well is being drilled. The mud logger carefully analyzes both the drilling mud and well cuttings for traces of crude oil and natural gas. Another service company drives a logging truck out to the well after the well is drilled to make a wireline well log. A long cylinder containing instruments called a logging tool is unloaded from the truck and run down the well on a wireline (fig. I–8). As the logging tool is brought back up the well, the instruments remotely sense the electrical, sonic, and radioactive properties of the surrounding rocks and their fluids. These measurements are recorded on a long strip of paper called a wireline well log (fig. I–9) in the logging truck and are also digitized, encoded, and sent by radio telemetry to a data center. Well logs are used to determine the composition of each rock layer, whether the rock layer has pores and how much is pore space, and what fluid (water, gas, or oil) is in the pores. Depending on the test results, the well can be plugged and abandoned as a dry hole or completed as a producer.

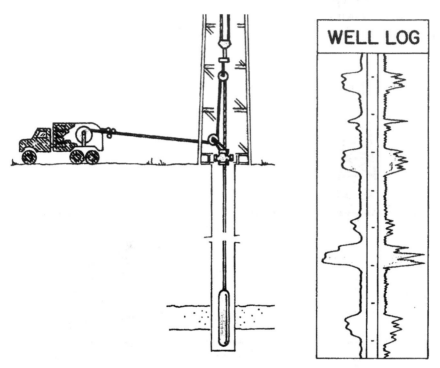

Fig. I–8. Well logging with a logging tool run down a well on a wireline

Fig. I–9. A wireline well log

To complete the well, many sections of large-diameter steel pipe called casing are screwed together to form a long length of pipe called a casing string that is lowered down the hole. Wet cement is then pumped between the casing and well walls and allowed to set (fig. I–10) during a cement job. This stabilizes the hole. The casing is done in stages called a casing program, during which the well is drilled, cased, drilled deeper, cased again, drilled deeper, and cased again (fig. 1–11). In order for the gas or oil to flow into the well, the well is either completed open-hole or with perforated casing. In an open-hole completion (fig. I–11a), the casing string is cemented down to the top of the reservoir rock and the bottom left open. In perforated casing completion (fig. I-11b), the casing is cemented through the reservoir rock and the casing is shot with explosives to form holes called perforations. A long length of narrow-diameter steel pipe called a tubing string is then suspended down the center of the well. The produced fluids (water, gas, and oil) are brought up the tubing string to the surface to prevent them from touching and corroding the casing string that is harder to repair. An expandable rubber device called a tubing packer on the bottom of the tubing string keeps the tubing string central in the well and prevents the fluids from flowing up the outside of the tubing (fig. 1–11). The tubing string is relatively easy to repair during a workover.

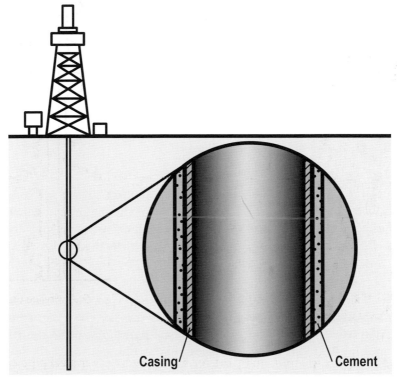

Casing

Cement

Fig. I–10. Casing cemented into a well

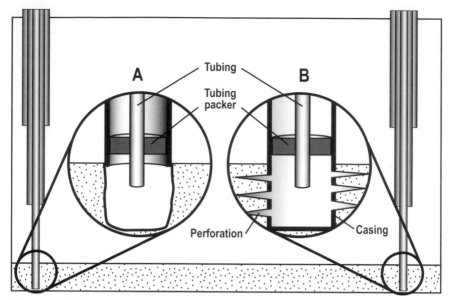

Fig. I–11. Wells with casing programs of smaller and smaller diameter casing strings. The two bottom-hole completions are (a) open-hole and (b) perforated casing. A tubing string and tubing packer are shown.

In all gas wells, gas flows to the surface by itself. There are some oil wells, usually only early in the development of the oil field, in which the oil has enough pressure to flow up to the surface. Gas wells and flowing oil wells are completed on the surface with a vertical structure of pipes, fittings, gauges, and valves called a Christmas tree, which is used to control the flow (fig. I–12a).

Most oil wells, however, do not have enough pressure for the oil to flow to the surface and the oil will fill the bottom of the well only up to a certain level. A sucker-rod pump or beam-pumping unit (fig. I–12b) is commonly used to lift the oil and water up the tubing string to the surface. A downhole pump on the bottom of the tubing string is driven by the surface beam-pumping unit. A motor causes a beam on a pivot, called the walking beam, to pivot up and down. The walking beam is connected to a downhole pump by a long, narrow, sucker-rod string that runs down the center of the tubing string (fig. 1–12b). The pump lifts the oil and water up the tubing string to the surface.

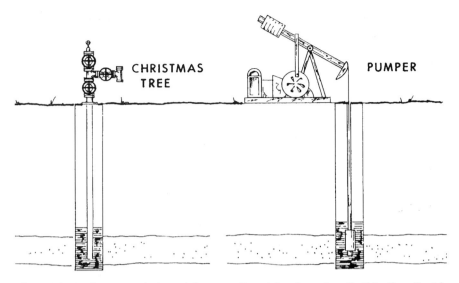

Fig. I-12. Surface completions (a) gas well and flowing oil well, (b) oil well with beam-pumping unit

On the surface, gas is prepared for delivery to a gas pipeline by gas-conditioning equipment that removes impurities such as water vapor and corrosive gases. Valuable natural gas liquids, such as condensate, are removed from the gas in a natural gas processing plant and sold separately. The gas can then be sold to a gas pipeline. For oil, a long vertical or horizontal steel tank called a separator is used to separate natural gas that bubbles out of the oil and the saltwater that settles to the bottom (plate I–1). The oil is then stored in steel stock tanks until it is sold to a refinery.

The production rate from wells can be increased by acid and frac jobs. Acid is pumped down a well to dissolve some of the reservoir rock adjacent to the wellbore during an acid job. During a frac job, the reservoir rock is hydraulically fractured with a liquid pumped under high pressure down the well. Propping agents such as sand grains are pumped down the well with the frac fluid to hold the fractures open and allow the oil and gas to readily flow into the well. Periodically, production from the well must be interrupted for repairs or to clean out the well during a workover. A service company drives out to the well with a production unit to do the workover.

As fluids are produced from the subsurface reservoir, pressure on the remaining fluids drops. The production rate of oil and gas wells and the whole field decreases with time on a decline curve. Ultimate recovery of gas from a gas reservoir is often about 80% of the gas in the reservoir. Oil reservoirs, however, are far more variable and less efficient. They range from 5 to 80% recovery but average only 35% of the oil in the reservoir. This leaves 65% of the oil remaining in the pressure-depleted reservoir. After the reservoir pressure has been depleted in an oil field, waterflood and enhanced oil recovery can be attempted to produce some of the remaining oil. During a waterflood, water is pumped under pressure down injection wells into the depleted reservoir to force some of the remaining oil through the reservoir rock toward producing wells (fig. I–13). Enhanced oil recovery involves pumping fluids such as carbon dioxide, nitrogen, or steam down injection wells to obtain more production.

Plate I–1. Oil well with separator and stock tanks

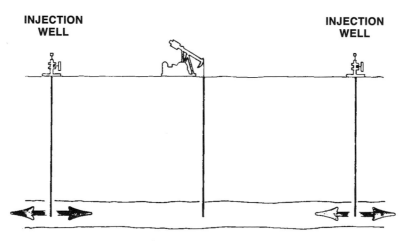

Fig. I–13. A waterflood

Recently, an enormous amount of natural gas has been produced from shales with a special technique. A shale is a very fine-grained rock. Even though many shales contain gas, shale has no permeability and the gas cannot flow through the shale into a well. Horizontal wells drilled into these shales, however, are used for a special frac job called a slickwater frac, which allows the gas to flow through the shale and into the well.

After the well has been depleted, it is required by law to be properly plugged and abandoned to prevent pollution. Cement must be poured down the well to seal the depleted reservoir and to protect any subsurface freshwater reservoirs. A steel plate is then welded to the top of the well and the well is covered with soil.

The Nature of Gas and Oil

Petroleum

The word *petroleum* comes from the Greek words *petro* for rock and *oleum* for oil. In its strictest sense, petroleum includes only crude oil. By usage, however, petroleum includes both crude oil and natural gas.

The Chemistry of Oil and Gas

The chemical composition by weight of typical crude oil and natural gas is shown in table 1–1.

Table 1–1. The chemical composition of typical crude oil and natural gas

	crude oil	natural gas
carbon	84–87%	65–80%
hydrogen	11–14%	1–25%
sulfur	0.06–2%	0–0.2%
nitrogen	0.1–2%	1–15%
oxygen	0.1–2%	0%

(Modified from Levorsen, 1967.)

The two most common elements in both crude oil and natural gas are carbon and hydrogen. Most of the molecules that make up crude oil and natural gas are composed of hydrogen and carbon atoms and are called *hydrocarbons*.

The difference between crude oil and natural gas is the size of the hydrocarbon molecules. Under surface temperature and pressure, any hydrocarbon molecule that has one, two, three, or four carbon atoms occurs as a gas. Natural gas is a mixture of the four short hydrocarbon molecules. Any hydrocarbon molecule with five or more carbon atoms occurs as a liquid. Crude oil is a mixture of several hundred hydrocarbon molecules that range in size from 5 to more than 60 carbons in length and form straight chains, chains with side branches, and circles.

Crude Oil

Hydrocarbon molecules

Four types of hydrocarbon molecules, called the *hydrocarbon series*, occur in each crude oil. The relative percentage of each hydrocarbon series molecule varies from oil to oil, controlling the chemical and physical properties of that oil. The hydrocarbons series includes paraffins, naphthenes, aromatics, and asphaltics. Hydrocarbons that have only single bonds between carbon atoms are called *saturated*. If they contain one or more double bonds, they are *unsaturated*.

A *paraffin* or *alkane molecule* is a straight chain of carbon atoms with saturated (single) bonds between the carbon atoms (fig. 1–1). The general formula is C_nH_{2n+2}. If the paraffin molecule has 18 or more carbons, it is a wax and forms a *waxy crude oil*.

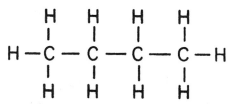

Fig. 1–1. Paraffin molecule with carbon (C) and hydrogen (H) atoms

The *naphthene* or *cycloparaffin molecule* is a closed circle with saturated bonds between the carbon atoms (fig. 1–2). The general formula is C_nH_{2n}.

These molecules are five carbon atoms and longer in length. Oils with high naphthene content tend to have a large asphalt content that reduces the value of the oil.

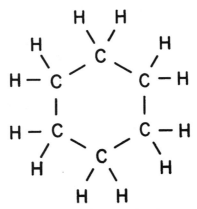

Fig. 1–2. Naphthene molecule

The *aromatic* or *benzene molecule* is a closed ring with some unsaturated (double) bonds between carbon atoms (fig. 1–3). It has a general formula of C_nH_{2n-6}. Aromatic molecules are six carbon atoms and longer in length. An aromatic-rich crude oil yields the highest-octane gasoline at the refinery and makes a valuable feedstock for the petrochemical industry. The refiner often pays a premium for aromatic-rich crude oils. Fresh from the well, normal crude oil has a pungent odor of gasoline, whereas aromatic-rich crude oil has a fruity odor.

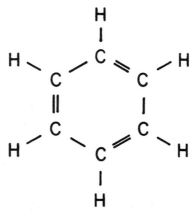

Fig. 1–3. Aromatic molecule

The *asphaltic molecule* has 40 to more than 60 carbon atoms. Asphalt is brown to black in color, solid to semisolid under surface conditions, and has a high boiling point.

Table 1–2 shows the hydrocarbon series content of crude oil.

Table 1–2. Average and range of hydrocarbon series molecules in crude oil

	weight percent	percent range
paraffins	30	15 to 60
naphthenes	49	30 to 60
aromatics	15	3 to 30
asphaltics	6	remainder

(Modified from Levorsen, 1967, and Bruce and Schmidt, 1994.)

At the refinery, there are two types of crude oils. An asphalt-based crude oil contains little or no paraffin wax and is usually black. When refined, it yields a large percentage of high-grade gasoline and asphalt. A *paraffin-based crude oil* contains little or no asphalt and is usually greenish. When refined, it yields a large percentage of paraffin wax, high-quality lubricating oil, and kerosene. A *mixed-base crude oil* is a combination of both types.

°API gravity

Crude oils are compared and described by density. The most commonly used density scale is °*API gravity* or *API gravity*. API stands for the American Petroleum Institute, based in Washington, DC. It standardizes petroleum industry equipment and procedures.

$$°API \text{ gravity is } (141.5 \div \text{specific gravity at } 60°F) - 131.5$$

Freshwater has an °API gravity of 10. The °API gravities of crude oils vary from 5 to 55. Average density (weight) crude oils are 25 to 35. Light oils are 35 to 45. They are very fluid, often transparent, rich in gasoline, and the most valuable. Heavy oils are below 25. They are very viscous and dark colored, contain considerable asphalt, and are less valuable.

Sulfur

Sulfur is an undesirable impurity in fossil fuels such as crude oil, natural gas, and coal. When sulfur is burned, it forms sulfur dioxide, a gas that pollutes the air and forms acid rain. During the refining process, the refiner must remove excessive sulfur as the crude oil is being processed. If not, the sulfur will harm some of the chemical equipment in the refinery. Crude oils are classified as sweet and sour on the basis of their sulfur content. *Sweet crudes* have less than 1% sulfur by weight, whereas *sour crudes* have more than 1% sulfur. The refiner usually pays several US dollars per barrel less for sour crude. In general, heavy oils tend to be sour, whereas light oils tend to be sweet. At a refinery, *low-sulfur crude* has 0 to 0.6% sulfur, *intermediate-sulfur crude* has 0.6 to 1.7% sulfur, and *high-sulfur crude* has above 1.7% sulfur. Most of the sulfur in crude oil occurs bonded to carbon atoms.

Benchmark crude oils

A *benchmark crude oil* is a standard for a country against which other crude oils are compared, and prices are set. In the United States, West Texas Intermediate (WTI) is 38 to 40 °API gravity and 0.3% S, and West Texas Sour, a secondary benchmark, is 33 °API gravity and 1.6% S. Brent, the benchmark crude oil for the North Sea is very similar to WTI and is 38 °API gravity and 0.3% S. Dubai is the benchmark crude oil for the Middle East at 31 °API gravity and 2% S.

Pour point

All crude oils contain some paraffin molecules. If the paraffin molecules are 18 carbon atoms or longer in length, they are waxes that are solid at surface temperature. A crude oil that contains a significant amount of wax is called a *waxy crude oil*. In the subsurface reservoir where it is very hot, waxy crude oil occurs as a liquid. As the oil comes up the well, it cools, and the waxes can solidify. This can clog the tubing in the well and flowlines on the surface. The well then has to be shut in for a workover to clean out the wax.

The amount of wax in crude oil is indicated by the pour point of the oil. A sample of the oil is heated in the laboratory. It is then poured from a container as it is being cooled. The lowest temperature at which the oil will still pour before it solidifies is called the *pour point*. Crude oil pour points range from +125° to −75°F (+52° to −59°C). Higher pour points reflect higher wax content. *Cloud point* is related to pour point. It is the temperature at which the oil first appears cloudy as wax forms when the temperature is lowered. It is 2° to 5°F (1° to 3°C) above the pour point. Very waxy crude oils are yellow in color. Slightly waxy crude oils can have a greenish color. Low- or no-wax oils are black.

Crude oil from the Altamont area in the Uinta basin of Utah are very waxy with pour points ranging from +65° to +125°F (+18° to +52°C) and are heavy (19 °API gravity) to light (54 °API gravity).

Properties

The color of crude oil ranges from colorless through greenish-yellow, reddish, and brown to black. In general, the darker the crude oil, the lower the °API gravity. The smell varies from gasoline (sweet crude) to foul (sour crude) to fruity (aromatic crude). Crude oil has a calorific heat value of 18,300 to 19,500 Btu/lb.

Crude streams

A *crude stream* is oil that can be purchased from an oil-exporting country. It can be from a single field or a blend of oils from several fields. Table 1–3 describes some crude streams.

Table 1–3. Properties of selected crude streams

crude stream	country	°API gravity	S%	pour point
Arabian light	Saudi Arabia	33.4	1.80	−30°F (−34°C)
Bachequero	Venezuela	16.8	2.40	−10°F (−23°C)
Bonny light	Nigeria	37.6	0.13	+36°F (2°C)
Brass River	Nigeria	43.0	0.08	−5°F (−21°C)
Dubai	Dubai	32.5	1.68	−5°F (−21°C)
Ekofisk	Norway	35.8	0.18	+15°F (−9°C)
Iranian light	Iran	33.5	1.40	−20°F (−29°C)
Kuwait	Kuwait	31.2	2.50	0°F (−18°C)
North Slope	USA	26.8	1.04	−5°F (−21°C)

Measurement

The English unit of crude oil measurement is a *barrel (bbl)* that holds 42 US gallons or 34.97 Imperial gallons. Oil well production is measured in barrels of oil per day (*bopd* or *b/d*). The metric units of oil measurement are *metric tons* and *cubic meters*. A metric ton of average weight crude oil (30 °API gravity) is 7.19 barrels in volume. A metric ton of heavy oil (20 °API gravity) occupies 6.75 barrels, whereas a metric ton of light oil (40 °API gravity) occupies 7.64 barrels. A cubic meter (m^3) of oil equals 6.29 barrels of oil.

Refining

First, during the refining process, various components of crude oil are separated by their boiling points. In general, the longer the hydrocarbon molecule, the higher its boiling temperature. At the refinery, crude oil is heated in a furnace until most of it is vaporized. The hot vapor is then sprayed into the bottom of a distilling column, where gases rise and any remaining liquid falls. The liquid that comes out the bottom of the distilling column is called *residuum*, or *bottom of the barrel*. It is least valuable. In the distilling column are bubble trays filled with liquid (fig. 1–4). The rising vapors bubble up through the trays and are cooled. The cooling vapors condense into liquid on the trays where they are removed by *sidedraws*. Each liquid removed by cooling is called a *cut* (fig. 1–5). Heavy cuts come out at high temperatures, whereas light cuts come out at low temperatures. In order of cooling temperatures, the cuts are heavy gas oil, light gas oil,

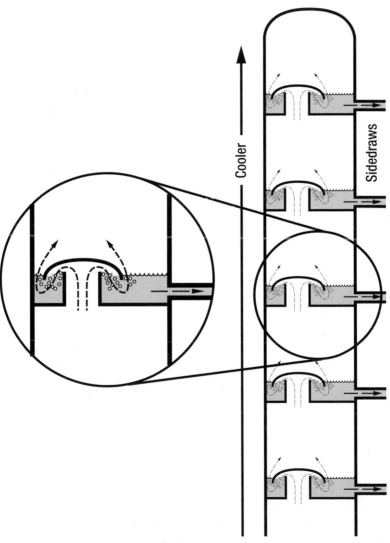

Fig. 1–4. Bubble trays in a distilling column

kerosene, naphtha, and straight run gasoline. Because gasoline is the refining product in most demand, a process called *cracking* is used to make more gasoline from the other, less valuable cuts. Gasoline is composed of short molecules with 5 to 10 carbon atoms. The longer, less valuable molecules of other cuts are used as *cracking stock*. The cracking stock is put in cracking towers at the refinery, where high temperatures and pressures and caustic chemicals split the longer molecules to form gasoline.

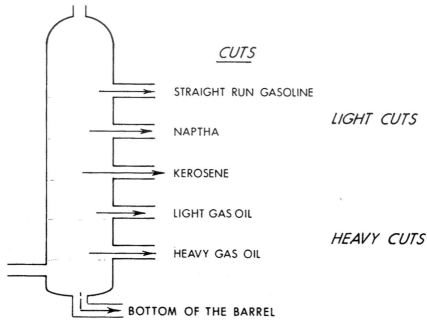

Fig. 1–5. Distilling column cuts

Refineries also produce pure chemicals called *feedstocks* from crude oil. Some common feedstocks are methane, ethylene, propylene, butylene, and naphthene. These feedstocks are sold to petrochemical industries, where the molecules are reformed and a large variety of products are made. Plastics, synthetic fibers, fertilizers, Teflon®, polystyrene, drugs, dyes, explosives, antifreeze, and synthetic rubber are examples.

The average percent yield of crude oil in a refinery is shown in table 1–4.

Table 1–4. Percent yield of crude oil

gasoline	46%
fuel oil	27%
jet fuel	9%
coke	5%
liquefied gases	4%
petrochemical feedstocks	2%
asphalt	2%
lubricants	1%
kerosene	1%

(API, 2009)

Natural Gas

Composition

Natural gas on the surface is composed of hydrocarbon molecules that range from one to four carbon atoms in length. The gas with one carbon atom in the molecule is *methane* (CH_4), two is *ethane* (C_2H_6), three is *propane* (C_3H_8), and four is *butane* (C_4H_{10}). All are paraffin-type hydrocarbon molecules. A typical natural gas composition is shown in table 1–5.

Table 1–5. Typical natural gas hydrocarbon composition

methane	70 to 98%
ethane	1 to 10%
propane	trace to 5%
butane	trace to 2%

These percentages vary from field to field, but methane gas is by far the most common, and many natural gas fields contain almost pure methane. The gas from pipelines that is burned in homes and by industry is methane gas. Propane and butane produce more heat when burned than methane, and they are often distilled from natural gas and sold separately. *Liquefied petroleum gas (LPG)* is made of pure propane gas.

Nonhydrocarbon gas impurities that do not burn in natural gas are called *inerts*. Because inerts decrease the heat content of natural gas, they decrease the value of the gas. A common inert is water vapor (steam). Another inert is carbon dioxide (CO_2), a colorless, odorless gas. In some gas reservoirs, carbon dioxide is greater than 99% of the gas. It can be used for inert gas injection, an enhanced oil recovery process, in depleted oil fields. Nitrogen (N), another inert, is also a colorless, odorless gas that can also be used for inert gas injection. Helium, a rare but important inert, is a light gas used in electronic manufacturing and filling dirigibles. Natural gas from the Hugoton gas field in western Texas, Oklahoma, and Kansas contains 0.5 to 2% helium. Amarillo, Texas, near the giant gas field, is called the "helium capital of the world."

Hydrogen sulfide (H_2S) is a gas that can occur mixed with natural gas or by itself. It is not an inert and is a very poisonous gas that is lethal in very low concentrations. The gas has the foul odor of rotten eggs and can be smelled in extremely small amounts. Hydrogen sulfide gas is also very corrosive. When it occurs mixed with natural gas, it causes corrosion of the metal tubing, fittings, and valves in the well during production. *Sweet natural gas* has no detectable hydrogen sulfide, whereas *sour natural gas* has detectable amounts of hydrogen sulfide and is less valuable. Hydrogen sulfide and most inerts must be removed in the field before natural gas can be delivered to a pipeline.

Occurrence

Because of high pressure in the subsurface reservoir, a considerable volume of natural gas occurs dissolved in crude oil. The *formation, dissolved,* or *solution gas/oil ratio* (GOR) is the number of cubic feet of natural gas dissolved in one barrel of oil in the subsurface reservoir. The measurements are reported under surface conditions of temperature and pressure (standard cubic foot per barrel, or scf/bbl). In general, as the pressure of the reservoir increases with depth, the amount of natural gas that can be dissolved in crude oil increases. When crude oil is lifted up a well to the surface, the pressure is reduced, and natural gas, called *solution gas*, bubbles out of the oil. The *producing GOR* of a well is the number of cubic feet of gas the well produces per barrel of oil.

Nonassociated natural gas is gas that is not in contact with oil in the subsurface, and it is almost pure methane. *Associated natural gas* occurs in contact with crude oil in the subsurface. It occurs both as gas in the free gas cap above the oil and gas dissolved in the crude oil. It contains other hydrocarbon gases besides methane.

Condensate

In some subsurface gas reservoirs, at high temperatures, shorter-chain liquid hydrocarbons, primarily those with five to seven carbon atoms in length, occur as a gas. When the gas is produced, the temperature decreases, and the liquid hydrocarbons condense out of the gas on the surface. This liquid, called condensate, is almost pure gasoline, is clear to yellowish to bluish in color, and has 45 to 62 °API gravities. Condensate is commonly called *casinghead gasoline, drip gasoline, white gas*, or *natural gasoline*. It is often added to crude oil in the field in a process called *spiking* to decrease the °API gravity and increase the volume and value of the oil. Condensate removed from natural gas in the field is classified as crude oil by regulatory agencies.

Refiners pay almost as much for condensate as crude oil. It does not have a high octane and must be mixed with high-octane gasoline made by cracking in the refinery. Because of the low octane, the posted price for condensate is usually slightly less than that for crude oil. Natural gas that contains condensate is called *wet gas*, whereas natural gas lacking condensate is called *dry gas*. The condensate along with butane, propane, and ethane that can be removed from natural gas are referred to as *natural gas liquids* (*NGL*).

Measurement

Natural gas is measured both by volume and heat content. The English unit of volume is a *cubic foot* (*cf*). Because gas expands and contracts with pressure and temperature changes, the measurement is made under or is converted to standard conditions defined by law. It is usually 60°F and 14.65 psi (15°C and 101.325 kPa) and is called *standard cubic feet* (*scf*). The abbreviation for 1,000 cubic feet is *Mcf*, a million cubic feet is *MMcf*, a billion cubic feet is *Bcf*, and a trillion cubic feet is *Tcf*. Condensate content is measured in barrels per million cubic feet (*BCPMM*) of gas. In the metric system, the volume of gas is measured in *cubic meters* (m^3). A cubic meter is equal to 35.315 cf.

The unit used to measure heat content of fuel such as gas in the English system is the *British thermal unit* (*Btu*). One Btu is about the amount of heat given off by burning one wooden match. Pipeline natural gas ranges from 900 to 1,200 Btus per cubic foot and is commonly 1,000 Btus. The heat content varies with the hydrocarbon composition and the amount of inerts in the natural gas. Heat content in the metric system is measured in *kilojoules*. A kilojoule of heat is equal to about 1 Btu.

Natural gas is sold to a pipeline by volume in thousands of cubic feet, by the amount of heat when burned in Btus, or by a combination of both. If the pipeline contract has a *Btu adjustment clause*, the gas is bought at a certain price per Mcf, and the price is then adjusted for the Btu content of the gas.

The amount of Btus in one average barrel of crude oil is equivalent to the Btus in 6,040 cubic feet of average natural gas and is called *barrel of oil equivalent* (*BOE*). Different companies often have a slightly different BOE numbers depending on the oil and gas composition of their production.

Reservoir Hydrocarbons

Chemists classify reservoir hydrocarbons into (1) black oil, (2) volatile oil, (3) retrograde gas, (4) wet gas, and (5) dry gas. Laboratory analysis of a sample is used to determine the type.

Both black and volatile oils are liquid in the subsurface reservoir. *Black oil* or *low-shrinkage oil* has a relatively high percentage of long, heavy, nonvolatile molecules. It is usually black but can have a greenish or brownish color. Black oil has an initial producing GOR of 2,000 scf/bbl or less. The °API gravity is below 45.

Volatile oil or high-shrinkage oil has relatively more intermediate size molecules that are shorter than black oil molecules. The color is brown, orange, or green. Volatile oil has an initial producing GOR between 2,000 and 3,300 scf/bbl. The °API gravity is 40 or above.

Retrograde gas is a gas in the reservoir under original pressure, but liquid condensate forms in the subsurface reservoir as the pressure decreases with production. The initial GOR is 3,300 scf/bbl or higher.

Wet gas occurs entirely as a gas in the reservoir, even during production, but produces a liquid condensate on the surface. It often has an initial producing GOR of 50,000 scf/bbl or higher.

Dry gas is pure methane. It does not produce condensate either in the reservoir or on the surface.

The Earth's Crust—
Where We Find It

Rocks and Minerals

The earth is composed of *rocks* that are aggregates of small grains called minerals (fig. 2–1). *Minerals* are naturally occurring, relatively pure chemical compounds. Examples of minerals are quartz, composed of SiO_2, and calcite, composed of $CaCO_3$. Rocks can be composed of numerous grains of several different minerals or numerous grains of the same mineral.

Rocks have been forming throughout the billions of years of earth's history. The same chemical and physical processes that form rocks today formed rocks throughout geological time. The molten lava flowing out a volcano in Hawaii or Italy today is forming lava rock similar to lava rock formed a long time ago. Ancient sandstone rock is composed of sand grains that were deposited the same way sand is deposited today: along beaches, in river channels, and on desert dunes. There is nothing unusual about ancient rocks—they formed the same way rocks are forming today.

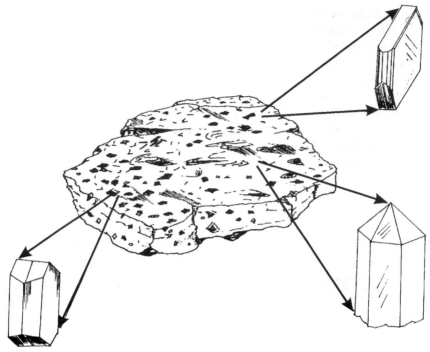

Fig. 2–1. Mineral grains in a rock

Types of Rocks

Three types of rocks that make up the earth's crust are igneous, sedimentary, and metamorphic. *Igneous rocks* have crystallized from a hot, molten liquid. *Sedimentary rocks* are composed of sediments, particles that were deposited on the surface of the ground or bottom of the ocean or salts that precipitated out of water. *Metamorphic rocks* have been recrystallized from other rocks under high temperatures and pressures.

Igneous rocks

Igneous rocks form when a molten melt is cooled. Two types of igneous rocks are plutonic and volcanic, depending on where they formed. *Plutonic* igneous rocks crystallized and solidified while still below the surface of the earth. Because the rocks that surround the cooling plutonic rocks are good insulators, plutonic rocks often take thousands of years to solidify. When given a long time to crystallize, large mineral crystals are formed. Plutonic igneous rocks are easy to identify because the mineral grains are all large enough to be seen by the naked eye. Plutonic rocks formed as hot liquids that were injected into and displaced preexisting rocks in the subsurface (fig. 2–2). Because of this, now solidified plutonic rock bodies are called *intrusions*. *Volcanic* igneous rocks crystallize on the surface of the earth as lava. As the lava flows out of a volcano, it immediately comes in contact with air or water and rapidly solidifies. The rapid crystallization forms very small crystals that are difficult to distinguish with the naked eye. In general, igneous rocks are harder to drill than sedimentary rocks. Buried lava flows and intrusions of plutonic rocks are occasionally encountered when drilling through sedimentary rocks for gas and oil.

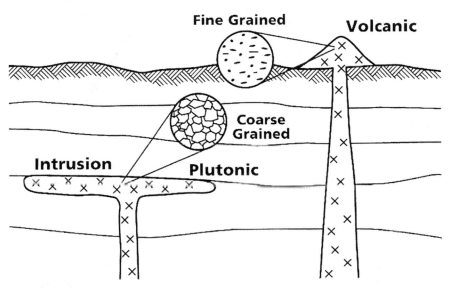

Fig. 2–2. Igneous rocks showing an intrusion of plutonic igneous rocks into sedimentary rock layers and volcanic igneous rocks on the surface; microscopic views of coarse-grained plutonic and fine-grained volcanic rocks.

Sedimentary rocks

Sedimentary rocks are composed of sediments, of which there are three types. *Clastic sediments* are whole particles formed by the breakdown of rocks and were transported and deposited as whole particles. Boulders, sand grains, and mud particles are examples. *Organic sediments* are formed biologically, such as seashells. *Crystalline sediments* are formed by precipitation of salt out of water. As sediments are buried in the subsurface, they become solid, sedimentary rocks. Sedimentary rocks are the rocks that are drilled to find gas and oil. They are the source and reservoir rocks for gas and oil.

Loose sediments (*unconsolidated sediments*) become relatively hard sedimentary rocks (*consolidated sediments*) in the subsurface by the processes of natural cementation and compaction. No matter how some sediments such as sand grains are packed together, there will be *pore spaces* between the grains (fig. 2–3). Once the grains have been buried in the subsurface, the pore spaces are filled with groundwater that can be very salty. Under the higher temperatures and pressures of the subsurface, chemicals often precipitate out of the subsurface waters to coat the grains. These coatings grow together to bridge the loose grains. This process, called *natural cementation*, bonds the loose grains into a solid sedimentary rock. The most common cement is the mineral calcite ($CaCO_3$). Also, as the sediments are buried deeper, the increasing weight of overlying rocks exerts more pressure on the grains. This compacts the sediments that also solidify the rock.

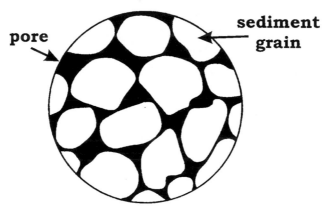

Fig. 2–3. Microscopic view of pores between sediment grains

Clastic sedimentary rocks often consist of three parts when examined under a microscope (fig. 2–4). First, there are sediment grains. Second, there are natural cements that coat and bond the grains together. Third, there are spaces called *pores*. In the subsurface, these pores are filled with fluids (water, gas, or oil), usually water.

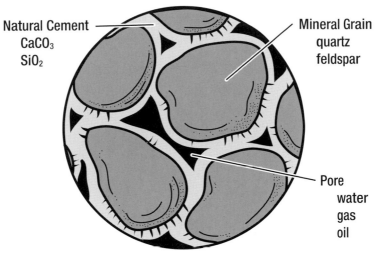

Natural Cement
$CaCO_3$
SiO_2

Mineral Grain
quartz
feldspar

Pore
water
gas
oil

Fig. 2–4. Clastic sedimentary rock under a microscope

There is an enormous amount of water below the surface, called *groundwater*, in sedimentary rock pores (fig. 2–5). Groundwater is described by salt content in parts per thousand (*ppt*). *Freshwater* contains so little salt (0–1 ppt) that it can be used for drinking water. *Brines* are subsurface waters that contain more salt than seawater (35–300 ppt). *Brackish waters* are mixtures of freshwaters and brines (1–35 ppt). Below the surface is a boundary called the *water table* between the dry pores above and pores filled with groundwater below. The water table can be on the surface or very deep depending on how much rain falls in that area. Just below the water table, the groundwater is usually fresh because of rainwater that percolates down from the surface. Deep waters, however, are usually brines. When a well is drilled, completed, and producing, near-surface freshwaters that are or can be used for drinking or irrigation are protected from pollution by law.

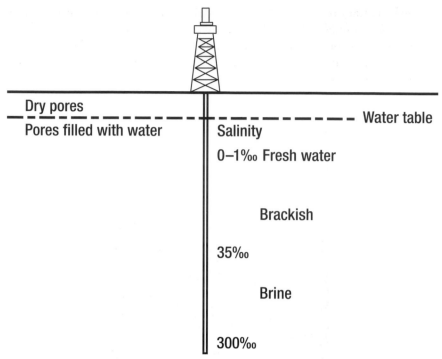

Fig. 2–5. Groundwater

The sizes of the clastic grains that make up an ancient sedimentary rock are important. The rock is often classified according to the grain size. Sandstones are composed of sand-sized grains, whereas shales are composed of fine-grained, clay-sized particles. Also, the size of the grains controls the size of the pore spaces and the quality of the oil or gas reservoir. Larger grains have larger pores between them. It is easier for fluids, such as gas and oil, to flow through larger pores and into a well. Clastic grains in sedimentary rocks are classified by their diameters in millimeters (fig. 2–6). They are called *boulder, cobble, pebble, granule, sand, silt,* and *clay*-sized particles. The finest grains (sand, silt, and clay-sized) are the most common.

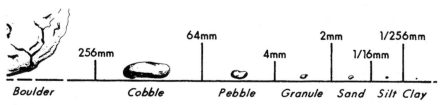

Fig. 2–6. Grain sizes in millimeters (1 mm = 1/25 in.)

Sedimentary rocks are identified by their characteristic layering, called *stratification* or *bedding* (plate 2–1). As the sediments are deposited, there are frequent variations in the amount and composition of sediment supply and sea level that cause the layering. These sediment layers are originally deposited horizontally in water.

Plate 2–1. Layering in sedimentary rocks, Dolomite Mountains, Italy. (Courtesy of Robert Laffi, Paola Ronchi, and American Association of Petroleum Geologists.)

Geologists can interpret how ancient sedimentary rocks were deposited by looking for clues. *Lithology* (rock composition) is an important clue as to how a sedimentary rock was formed. Sand grains, mud particles, and shell beds each form different sedimentary rocks, and each is originally deposited in a very different environment. *Sedimentary structures* such as ripple marks, mud cracks, and flow marks help to visualize the environment in which the rock was deposited. Another aid to interpretation consists of *fossils*, preserved remains of plants and animals.

Metamorphic rocks

Metamorphic rocks are any rocks that have been altered by high heat and pressure. Marble ($CaCO_3$), a metamorphic rock, is formed by recrystallization from the sedimentary rock limestone ($CaCO_3$). Since temperatures and pressures become greater with depth, a rock often becomes metamorphosed when buried very deep in the earth.

Structure of the Earth's Crust

The earth is estimated to be about 4.5 billion years old. Even the sedimentary rocks that generated and hold gas and oil are millions to hundreds of millions of years old. Where did these sedimentary rock layers come from? During that vast expanse of geological time, sea level has not been constant but has been rising and falling. During the rise and fall of sea level, sediments were deposited in layers. Sands were deposited along the ancient beaches, mud was deposited in the shallow seas offshore, and seashells were deposited in shell beds. These ancient sediments form the sedimentary rocks that are drilled to find gas and oil. The rise and fall of sea level has occurred in numerous cycles (fig. 2–7). The largest cycles occurred every few hundreds of millions of years. There are shorter cycles within the large cycles and even shorter cycles within them. At least five orders of sea level cycles have occurred, with the shortest occurring every few tens of thousands of years. The shorter cycles are caused by the freezing and melting of glaciers.

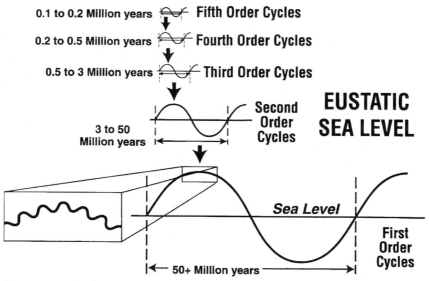

Fig. 2–7. Sea-level cycles (Hyne, 1995)

In Tulsa, Oklahoma, a typical section of the earth's crust, about 5,000 ft (1,500 m) of well-layered sedimentary rocks are underlain by very old metamorphic or igneous rocks (fig. 2–8). There are about 100 layers of sedimentary rocks. Sands form the rock sandstone, mud forms the rock shale, and seashells form the rock limestone. The unproductive rocks for gas and oil, usually igneous and metamorphic rocks underlying the sedimentary rocks, are called *basement rocks*. When drilling encounters basement rock, the drilling is usually stopped.

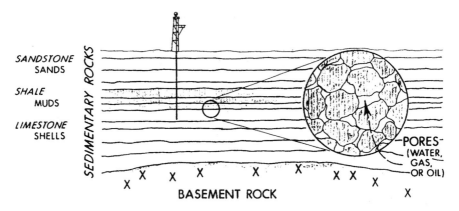

Fig. 2–8. Cross section of the earth's crust

In some areas of the earth, there are no, or very few, sedimentary rocks, and the basement rock is on or near the surface. These areas are called *shields*, and there is no gas or oil. Every continent in the world has at least one shield area (fig. 2–9). A shield, such as the Canadian shield in eastern Canada, tends to be a large, low-lying area. Ore minerals such as iron, copper, lead, zinc, gold, and silver are mined from the basement rock in shield areas. All the gas and oil deposits in Canada on land are located to the west of the Canadian shield where there are sedimentary rocks. Offshore eastern Canada, where there are sedimentary rocks, contains gas and oil fields. The southwest portion of Saudi Arabia is a shield (fig. 2–10). All the Saudi Arabian oil fields are located in sedimentary rocks to the northeast of the Arabian shield.

Fig. 2–9. Map of the world showing the location of shields in black where unproductive rocks for gas and oil occur on or near the surface

In other areas, called *basins*, the sedimentary rocks are very thick. Most basins have been filled in with sedimentary rocks and are dry land today. Some basins, however, are partially filled with sedimentary rocks and parts are still covered with water such as the Gulf of Mexico basin. The Caspian basin (Caspian Sea) has about 85,000 ft (26,000 m) of sedimentary rock cover. However, 20,000 to 40,000 ft (6,000 to 12,000 m) of sedimentary rocks is typical of many basins. Basins such as the Gulf of Mexico and the Anadarko basin of southwestern Oklahoma are large areas that are often more than 100 miles (160 km) across. It is in the sedimentary rock basins where most of the gas and oil is found and produced. Because of the thick sedimentary rock, most basins have source rocks that have been buried deep enough in the geological past to generate gas and oil (fig. 2–11). The deep part of the basin where the organic matter is cooked to form gas and oil is called the *kitchen*. After the gas and oil are generated, a lot of it migrates upward into the overlying rocks where it can be trapped. The trap, such as an anticline, is a relatively small feature compared to the basin. Numerous traps can occur along the flanks of the basin. There are about 600 sedimentary rock basins in the world. Of the basins that have been explored and drilled, about 40% are productive. About 90% of the world's oil occurs in only 30 of those basins. The other 60% of the explored basins are relatively barren. The unproductive basins either have no source rocks, the source rocks have never been buried deep enough to generate gas and oil, or the basin was overheated and the oil was destroyed.

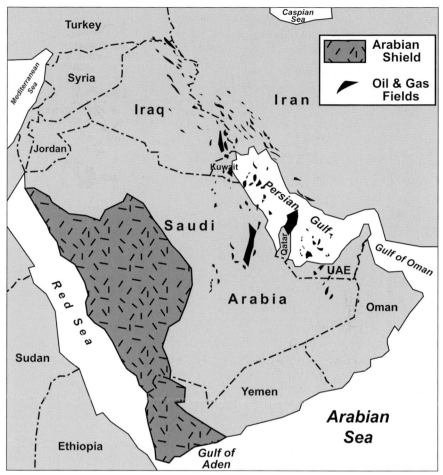

Fig. 2–10. Map of the Arabian shield and oil fields. (Modified from Beydown, 1991.)

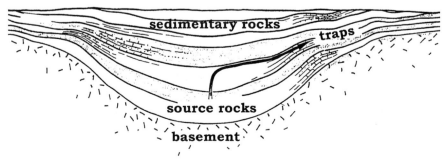

Fig. 2–11. Cross section of a sedimentary rock basin

Identification of Common Rocks and Minerals

Just a few common rocks and minerals make up the bulk of the earth's crust. All of these are readily identified by simple tests that can be made in the field such as at the drillsite without elaborate equipment.

Identification of Minerals

Minerals occur as crystals and grains in rocks. Color is the first property observed in a mineral. Many colors, such as the brassy yellow of pyrite, are diagnostic. Some transparent minerals, such as quartz, can be misleading, as they are tinted by slight impurities such as gas bubbles, iron, or titanium. Rose quartz, milky quartz, and smoky quartz are examples. *Luster* is the appearance of light reflected from the surface of a mineral. Two common lusters are metallic and nonmetallic. Nonmetallic lusters have descriptive names as greasy, glassy, and earthy. A few minerals are transparent in thin sheets, and others are translucent (they transmit light but not an image), but most are opaque and do not transmit light.

The form that a mineral crystal takes, such as cube or pyramid, can also be diagnostic. Other minerals have no crystal form and are called *amorphous*. The tendency for some minerals to break along smooth surfaces is called *cleavage*. Cleavage is described by three properties (fig. 3-1). The first is the number of cleavage surfaces of different directions. The second is the quality of the surfaces, such as poor or excellent. The third is the angle between the surfaces. *Fracture* is the breakage of the mineral along an irregular surface.

The hardness of a mineral is quantified by *Moh's scale*, which ranges from 1 to 10. The mineral talc is the softest (1), and diamond is the hardest (10). A mineral that is higher on Moh's scale can scratch a mineral that is lower on the scale. Some common objects that are used for hardness comparisons are a fingernail (2.5), a copper penny (3.5), a knife or steel key (6), and glass (7).

Specific gravity is the relative weight of a mineral compared to the weight of an equal volume of water. A specific gravity of 2.5 means the mineral weighs 2.5 times an equal volume of water. The specific gravity of an average rock or mineral is about 2.5. Ore minerals mined for metals, such as iron, copper, or nickel, are heavy and have specific gravities of 3.5 and above.

Certain minerals have unique characteristics that can be used to identify them. The mineral halite (common table salt) can be identified by its taste. A very important test is the application of cold, dilute acid to a sample, which causes only the mineral calcite to bubble.

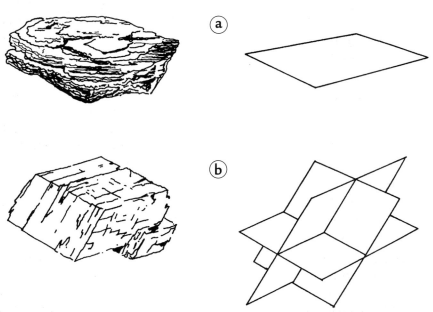

Fig. 3–1. Cleavage: (a) one perfect cleavage and (b) three perfect cleavages, not at right angles

Minerals

Mica is a common mineral that breaks along one perfect cleavage plane, forming very thin, elastic flakes. Two types of mica are white mica and black mica. *White mica (muscovite)* is composed of $KAl_3Si_3O_{10}(OH)_2$. It is colorless and is transparent in thin flakes (plate 3–1a). *Black mica (biotite)* is composed of $K(Mg,Fe)_3AlSi_3O_{10}(OH)_2$. It is brown to black in color (plate 3–1b).

Quartz (SiO_2) is a very common mineral (plate 3–1c) that is colorless when pure. It is often, however, tinted by impurities such as iron or gas bubbles. Common varieties include rose, cloudy, milky, and smoky quartz. Quartz is the hardest of the common minerals (7 on the Moh's scale). It will scratch all other common minerals and cannot be scratched by a knife. It forms six-sided, prismatic crystals but can occur as amorphous grains. Most sand grains on a beach or in sandstone rock are composed of quartz.

Calcite ($CaCO_3$) is a common mineral that is either colorless or white. Calcite breaks along three perfect cleavage planes that are not at right angles to form rhombs (plate 3–1d). Calcite is relatively soft (3 on Moh's scale) and can be scratched by a knife. Calcite will bubble in cold, dilute acid. Most seashells are composed of calcite.

Halite ($NaCl$) is common table salt. It is colorless to white (plate 3–1e). Halite forms a granular mass or crystallizes in cubes. It breaks along three perfect cleavage planes at right angles, forming rectangles and cubes. Halite tastes salty. The mineral halite forms from the evaporation of seawater. It is very common in ancient salt deposits.

Gypsum ($CaSO_4 \cdot 2H_2O$) is colorless to white (plate 3–1f). It forms tabular crystals and has one perfect cleavage plane. Gypsum is very soft (2 on Moh's scale) and can be scratched by a fingernail. It has a specific gravity of 2.3. Gypsum is also called *selenite* or *alabaster*. Gypsum and a similar mineral, *anhydrite* ($CaSO_4$), form by the evaporation of seawater.

Pyrite (FeS_2) is known as fool's gold. It has a brassy yellow color and a metallic luster (plate 3–1g). Pyrite forms either cubes or an earthy mass and is relatively heavy, with a specific gravity of 5. Pyrite is an iron ore and can sometimes be found as grains in river sands.

Table 3–1 lists the properties of these minerals. Each of these minerals has one or two characteristic tests that readily distinguish it from other minerals. For example, quartz is the hardest of the common minerals and cannot be scratched by a knife. Calcite is relatively soft, can be scratched with a knife, and will bubble in cold, dilute acid.

Table 3–1. Mineral properties

name	hardness	specific gravity	luster
white mica	2 to 3	3	pearly to vitreous
black mica	2.5 to 3	3	pearly to vitreous
quartz	7	2.65	vitreous to greasy
calcite	3	2.72	vitreous to dull
halite	2 to 2.5	2.1	vitreous
gypsum	2	2.3	vitreous, pearly, oily
pyrite	6 to 6.5	5	metallic

Identification of Rocks

Rocks are classified and identified by their textures and mineral compositions. Igneous rock textures are based on the size of the mineral crystals that range from large enough to see with the naked eye to glassy with no crystals. Metamorphic rock textures are based on the size and orientation of the mineral crystals. A *foliated* metamorphic rock has parallel, platy crystals (fig. 3–2a). *Nonfoliated* metamorphic rock has either uniform-sized crystals or a nonparallel orientation of platy crystals (fig. 3–2b). Sedimentary rock textures are based on the nature, size, and shape of the grains and how they are bound together.

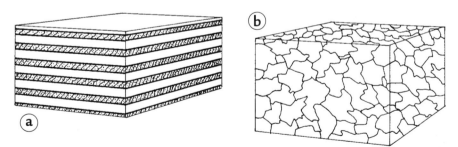

Fig. 3–2. Metamorphic rock textures: (a) foliated and (b) nonfoliated

Rocks

Igneous rocks

Granite is the most common plutonic igneous rock. It has the coarse-grained texture characteristic of all plutonic rocks. Granite is composed of the minerals quartz, feldspar, biotite, and hornblende. Quartz grains are the most common, giving granite a light color (plate 3–2a). The dark-colored mineral grains give it a speckled texture. Some granites are reddish or pinkish from iron impurities. Granite is commonly used for building stone.

Basalt is the most common volcanic rock. It has the fine-grained texture characteristic of lava. Basalt is black to gray in color (plate 3–2b). In some instances, the basalt coming out of a volcano cooled so rapidly that gas bubbles were frozen in the basalt. Fragments of basalt with numerous gas bubbles are called *scoria*.

Metamorphic rocks

Gneiss is a product of intense metamorphism. It is easily identified by its foliated texture of alternating, wide bands of light- and dark-colored coarse mineral grains (plate 3–3a).

Marble is the result of heat and pressure on the sedimentary rock limestone. It is composed of large, sparkling crystals of calcite ($CaCO_3$) (plate 3–3b).

Slate is slightly metamorphosed shale, a sedimentary rock. It is harder than shale, usually dark in color, and readily breaks along flat, parallel planes (plate 3–3c).

Sedimentary rocks

Conglomerate is a clastic rock with a wide range of pebble- to clay-sized grains (plate 3–4a). The coarse grains distinguish it from other clastic sedimentary rocks. The particles are all well-rounded. A conglomerate is commonly deposited in a river channel or on an alluvial fan formed where a mountain stream empties into the desert. If the particles are angular, the rock is called *breccia*.

Sandstone is composed primarily of sand grains (plate 3–4b) that have been naturally cemented together. Sandstone is rough like sandpaper to the touch. The sand grains can be broken off the rock if they are loosely cemented. The rock can be white to buff to dark in color. Sandstones are commonly deposited on beaches, river channels, or dunes. It is a common reservoir rock for gas and oil and is the most important oil reservoir rock in the United States.

Shale is composed of fine-grained, clay-sized particles (plate 3–4c) and is the most common sedimentary rock. It is usually well layered and relatively soft. Shale breaks down into mud when exposed to water. The color of shale commonly ranges from gray to black, depending on the organic content. The darker the shale, the higher the organic content. Shale is commonly deposited on river floodplains and on the bottom of oceans, lakes, or lagoons. Black shales are common source rocks for gas and oil. A gray shale can be a caprock on a reservoir rock in a petroleum trap. *Mudstone* is similar to shale but is composed of both silt- and clay-sized grains.

Limestone is composed of calcite mineral grains that range in size from very fine to large, sparkling crystals (plate 3–4d). The rock is commonly white or light gray in color. The calcite mineral grains are soft enough to be scratched by a knife and will bubble in cold, dilute acid. Limestones often have fossil fragments that are also usually composed of calcite. Limestone is a common reservoir rock and is the most important reservoir rock in the Middle East oil and gas fields. An organic-rich, dark-colored limestone can also be a source rock for gas and oil.

Coal is brown to black in color and very brittle (plate 3–4e). It usually has no layers. Coal is composed of woody plant remains that were buried in the subsurface and transformed by heat and time. *Lignite, bituminous,* and *anthracite* are varieties of coal formed by increasing heat that causes the coal to become harder and change in texture and composition.

Chert or *flint* is amorphous quartz (plate 3–4f). It is very hard and cannot be scratched by a knife. Being amorphous (without crystals), chert breaks along smooth, curved surfaces, forming sharp edges and points. Native Americans used chert to make arrowheads. Colored varieties of chert include jasper, chalcedony, and agate. Chert can be formed by precipitation directly out of groundwater or by recrystallization of fossil shells composed of SiO_2 by heat and pressure. Chert is the hardest of all sedimentary rock to drill.

Of the sedimentary rocks that make up the earth's crust, 99% are shales, sandstones, and limestones. Many sedimentary rocks are a combination of these three types and are described as sandy, shaly, and limey or calcareous (fig. 3–3).

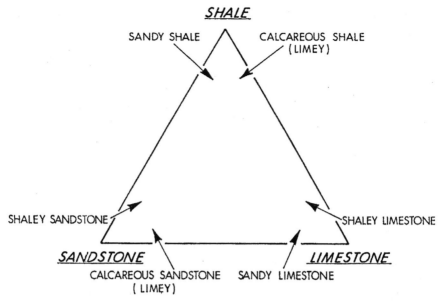

Fig. 3–3. Common sedimentary rocks

Plate 3–1. Minerals:
(a) white mica,
(b) black mica,
(c) quartz,
(d) calcite,
(e) halite,
(f) gypsum, and
(g) pyrite

Plate 3–2. Igneous rocks: (a) granite and (b) basalt

Plate 3–3. Metamorphic rocks: (a) gneiss, (b) marble, and (c) slate

Plate 3–4. Sedimentary rocks: (a) conglomerate, (b) sandstone, (c) gray shale and black shale, (d) limestone, (e) coal, and (f) chert

Geological Time

Two methods used for dating the formation of rocks and events in the earth's crust are absolute and relative age dating. *Absolute age dating* puts an exact time (e.g., 253 million years ago) on the formation of a rock or an event. *Relative age dating* arranges the rocks and events into a sequence of older to younger.

Absolute Age Dating

Exact dates for the formation of rocks are made by radioactive analysis. *Radioactivity* is the spontaneous decay of radioactive atoms that occur naturally in rocks (fig. 4-1). Radioactive atoms decay by giving off atomic particles and energy. For example, uranium-238 (^{238}U) decays by giving off atomic particles to form lead (^{206}Pb). The original radioactive atom, uranium (^{238}U), is called the *parent*. The product of radioactive decay, lead (^{206}Pb), is the *daughter*. Four relatively abundant radioactive atoms occur in rocks: two isotopes of uranium (^{238}U and ^{235}U), potassium (^{40}K), and rubidium (^{87}Rb). Each decays at a different rate. The rate of radioactive decay is measured in *half-lives*. A half-life is the time in years that it takes one-half of the parent atoms to decay into daughter atoms (fig. 4-2). The half-lives are shown in table 4-1.

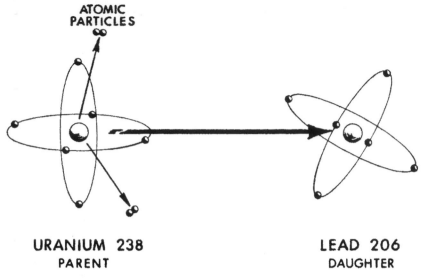

URANIUM 238
PARENT

LEAD 206
DAUGHTER

Fig. 4–1. Radioactive decay

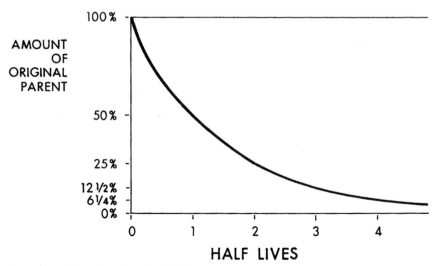

Fig. 4–2. Radioactive decay half-life

Table 4–1. Half-lives of common radioactive atoms

atom	years
^{238}U	4.5×10^9
^{235}U	0.7×10^9
^{87}Rb	4.7×10^{10}
^{40}K	1.3×10^9

One daughter atom is formed by the decay of each parent atom. As time goes on, the amount of radioactive parent atoms decreases, and the amount of daughter atoms increases. By measuring the amount of parent atoms left and daughter atoms created, the age of the mineral grains in a rock can be determined. This parent-daughter technique is used in the potassium-argon method. Potassium is a common element. A potassium isotope (^{40}K) decays into argon (^{40}Ar) with a half-life of 1.3 billion years. The assumption is made that when a mineral crystal forms, only potassium is accepted into the crystal structure, never argon because it is an inert gas. Any argon that is found in the crystal today could have come only from radioactive decay of potassium. By measuring the amount of ^{40}K and ^{40}Ar in the mineral, the ratio can be applied to the radioactive decay curve, and the age of the mineral determined. For example, if the ratio of ^{40}K to ^{40}Ar is 1 to 3, the age of the mineral grain is 2 half-lives, or 2.6 billion years (fig. 4–3).

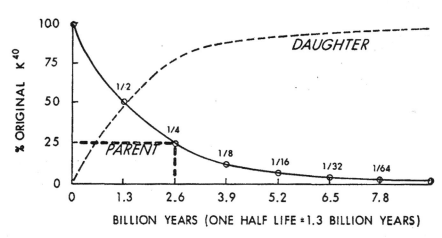

Fig. 4–3. Using half-lives to determine age

Carbon (^{14}C) is a radioactive isotope that decays very quickly and is not useful for most rocks. The half-life is only 5,710 years. After about 10 half-lives or about 60 thousand years, there is not enough parent left to date the material. It is used for archeology where ages are much younger but not for rocks that are millions of years old.

Radioactive age dating is used primarily on igneous and metamorphic rocks and cannot be used directly on sedimentary rocks. Sedimentary rocks are derived from the erosion of preexisting rocks. Absolute age dating of sedimentary rock grains will tell the age of the formation of the mineral grains in the preexisting rocks but not the time the sediments were deposited.

Relative Age Dating

In sedimentary rock sequences, relative age dating is used. Sedimentary rocks and events are put in order from oldest to youngest. In a sequence of undisturbed sedimentary rock layers, the youngest rock layer is on top, and the oldest layer is on the bottom. Events such as faulting, folding, intrusions, and erosion can also be relative age dated. If any one of those events affected a sedimentary rock layer, the event must be younger than the affected rock. The sequence in figure 4–4, from oldest to youngest is (a) deposition of sedimentary rocks 1, 2, and 3; (b) faulting; (c) erosion (unconformity); and (d) deposition of sedimentary rocks 4 and 5.

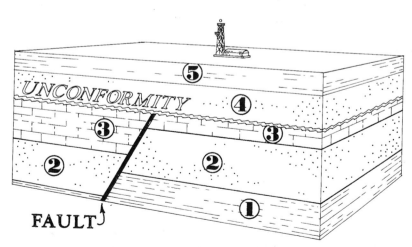

Fig. 4–4. Relative age dating

Fossils

An important tool in relative age dating of sedimentary rocks is fossils. *Fossils* are the preserved remains of plants and animals (fig. 4–5). There are several ways in which fossils are preserved. Although the soft parts of animals decay and are not preserved, the hard parts such as shells and bones can be preserved. Plants can be preserved as films of carbon in mud, which becomes shale. Some organisms are preserved when the original matter is completely replaced by another mineral in the subsurface. Petrified wood is formed by the replacement of wood by silicon dioxide, which preserves the grain structure of the wood. *Trace fossils*, such as burrows, tracks, or trails, are indirect evidence of ancient life.

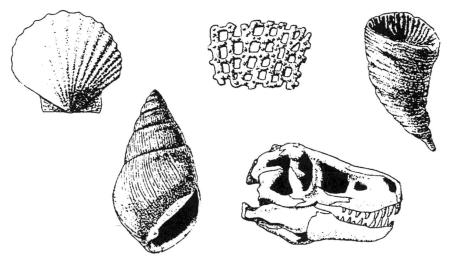

Fig. 4–5. Fossils

Certain species of plants and animals lived during certain geologic times. They eventually became *extinct* (disappeared from the earth) and were replaced by newer plants and animals. This continuous succession of organisms throughout geologic time is known as *evolution*. Vertical sequences of sedimentary rock layers that have been relative age dated can be used to determine the relative ages of the fossils in those rock layers (fig. 4–6). Geologists have collected and established the relative ages of most fossils. The evolutionary sequence of the fossils can be used to relative age date any sedimentary rocks that contain those fossils. In figure 4–7, the rocks labeled A are older than those labeled B.

SEDIMENTARY ROCKS

FOSSILS

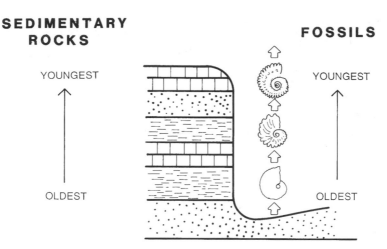

Fig. 4–6. Relative age dating of fossils

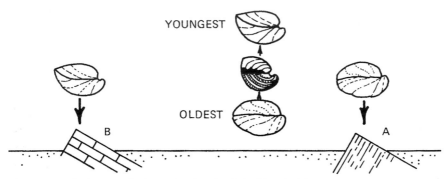

Fig. 4–7. Relative age dating using fossils: (A) older and (B) younger

A *guide* or *index fossil* is a distinctive plant or animal that lived during a relatively short span of geologic time. This fossil species identifies the age of any sedimentary rock in which it occurs. A *fossil assemblage* is a group of fossils found in the same sedimentary rocks. It identifies that zone of rocks and the geologic time during which those rocks and fossils were deposited.

Fossils can also be used to determine the environment in which the sediments were deposited. Different plants and animals live in different environments such as beach, marsh, and deep ocean.

Fossils can be indirectly dated by radioactivity (fig. 4–8) using volcanic ash layers. If the volcanic ash layers that occur above and below the fossil are dated, the estimated age of the fossil can be determined: The fossil must be younger than the underlying ash layer and older than the overlying ash layer.

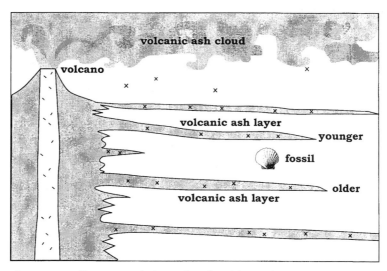

Fig. 4–8. Indirect age dating of a fossil by radioactive age dating volcanic ash layers

Microfossils

Wells are rarely drilled down to basement rock. That would be too expensive. Instead a well is drilled down to a *drilling target*, a known potential reservoir rock in that area. But as the well is drilled, it penetrates hundreds of sedimentary rock layers consisting of shales, sandstones, and limestones that look very similar. How can each sedimentary rock layer and the drilling target be identified while drilling through them? Each sedimentary rock layer was deposited during a different time and has different fossils that help identify the subsurface rock layers. Large fossils such as clams and corals, however, are broken into small chips (well cuttings) by the drill bit. It is almost impossible to identify the pieces when they are finally flushed to the surface by the drilling mud. The sedimentary rocks, however, also contain abundant microfossils.

Microfossils are fossils that are so small that they can be identified only with a microscope (plate 4–1). They are often undamaged by the drill bit and are flushed unbroken in the well cuttings up to the surface. Microfossils can used to identify subsurface rocks and their ages. Rock layers that contain a characteristic species of microfossil are often named after that microfossil and called a *zone* or *biozone*. The Siphonina davisi zone is composed of sedimentary rock layers that contain the microfossil *Siphonina davisi* (fig. 4–9). A horizon in a well that is identified by the first appearance, most abundant occurrence, or last appearance of a specific microfossil species when drilling is called a *paleo pick*. Paleo picks can also be used to determine if the sedimentary rock layers in a well are higher or lower in elevation than those in a well that has already been drilled (fig. 4–10).

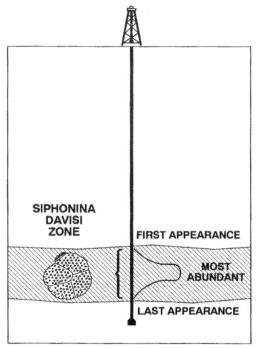

Fig. 4–9. Microfossil zone

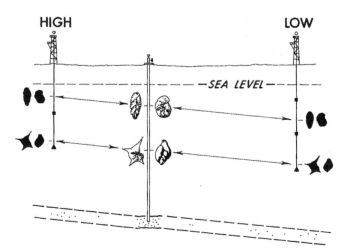

Fig. 4–10. High and low wells

Many microfossils are shells of single-celled plants and animals that live in the ocean.

Foraminifera (*forams*) are single-celled animals with shells (plate 4–1a) composed primarily of $CaCO_3$. They live by either floating in the ocean or growing attached to the bottom. More than 30,000 species of forams have existed throughout geologic time.

Radiolaria are single-celled, floating animals that live in the ocean and have shells (plate 4–1c) of SiO_2. Some chert layers were formed by alteration of radiolarian shell deposits.

Coccoliths are plates from the spherical $CaCO_3$ shell (coccolithophore) of algae that float in the ocean. They are so small that they can be identified only with a scanning electron microscope. Ancient, relatively pure $CaCO_3$ deposits of coccoliths or foraminifera microfossils form a limestone rock called *chalk*.

Diatoms are single-celled plants that float in water and have shells of SiO_2 (plate 4–1b). Ancient deposits of relatively pure diatom microfossils are called *diatomaceous earth*. Some chert layers formed from ancient diatom shell deposits.

Spores and pollens given off by plants to reproduce also are good microfossils. Scientists who study fossils are called *paleontologists*, and those who specialize in microfossils are *micropaleontologists*. Because they pick microfossils (bugs) from well cuttings to examine them under a microscope, they are often called *bug pickers*. Micropaleontologists specializing in spores and pollens are called *palynologists* or *weed and seed people*.

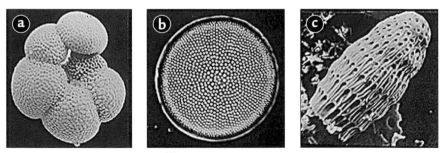

Plate 4–1. Magnified microfossils (a) foraminifera, (b) diatom, and (c) radiolaria

The Geologic Time Scale

The geologic time scale was developed during the early 1800s by relative age dating sedimentary rocks and fossils in Europe. Large divisions of geologic time are called *eras*. Eras are subdivided into *periods*, and periods into *epochs*. The geologic time scale is presented in table 4–2. In Europe, the Mississippian and Pennsylvanian periods are combined into the Carboniferous period.

Table 4–2. Geologic time scale

Era	Period	Epoch	Absolute age (years)
			0
		Holocene	
	Quaternary		10,000
		Pleistocene	
Cenozoic			1.8 million
		Pliocene	
			5.3 million
		Miocene	
	Tertiary		23 million
		Oligocene	
			33.9 million
		Eocene	
			55.8 million
		Paleocene	
			65.5 million
	Cretaceous		
			145.5 million
Mesozoic	Jurassic		
			201.6 million
	Triassic		
			251 million
	Permian		
			299 million
	Pennsylvanian		
			318 million
	Mississippian		
			359 million
Paleozoic	Devonian		
			416 million
	Silurian		
			444 million
	Ordovician		
			488 million
	Cambrian		
			542 million
Precambrian			
			4.5 billion

Earth History

It is known by radioactive age dating that the earth is about 4.5 billion years old. For the first part of the Precambrian era, there is no fossil evidence that life existed on earth. The first life, probably bacteria followed later by algae floating in the ocean, appeared approximately 3.5 billion years ago. The fossil record throughout the later part of the Precambrian is sparse.

At the start of the Paleozoic era (542 million years ago), a great abundance of diverse plants and animals were living in the ocean. All the major animal phyla that we know today in the oceans, except the vertebrates, were present. Nothing, however, existed on the land. During the Ordovician period, fish, the first vertebrates, came into existence. Plants and animals finally adapted to life on the land in the next period, the Silurian. During the Pennsylvanian period, swamps covered large areas of the land. Primitive plants, such a ferns and horsetail rushes, grew to great heights. These Pennsylvanian swamp deposits formed many of the world's coal deposits. During the last period of the Paleozoic, the Permian, the climate was very dry and warm, and the lands were covered with deserts. During the Permian period, egg-laying reptiles appeared. At the end of the Permian, the greatest extinction of plants and animal species in the history of the earth occurred when 95% of all species of marine organisms and 70% of all species of land organisms suddenly disappeared. There have been four great extinctions during geological time, but this was the most extensive. It is called the *great dying* and is thought to have been caused by enormous volcanic eruptions in Siberia giving off gases that poisoned the air and the oceans.

The Mesozoic era, starting about 251 million years ago, is known as the age of reptiles. These animals, which include the dinosaurs, dominated the earth. They filled a great diversity of ecological niches for more than 150 million years. Most dinosaurs were plant eaters, but some were carnivores. During this time, great reptiles lived in the oceans, while others flew through the air with wingspans of more than 50 ft (16 m). During the Jurassic period, the middle period of the Mesozoic, mammals appeared. They were small and were dominated by the reptiles throughout the remainder of the Mesozoic.

At the very end of the Mesozoic era (65.5 million years ago), another sudden extinction of life occurred. All the dinosaurs died out, along with the flying reptiles, most swimming reptiles, and 60% of all species of plants and animals. This extinction was remarkable because reptiles apparently dominated their environment until the very end. Most then disappeared in an instant of geological time called the *great killing*. This extinction was caused when an asteroid made of rock hit the earth. It was at least 6 miles (10 km) in diameter and hit the earth head on with a speed of about 60,000 miles per hour (100,000 km/h). In every location throughout the world, the thin clay layer separating the Mesozoic and Cenozoic sedimentary rocks has an abnormally high concentration of iridium-121 (^{121}Ir). This rare, radioactive isotope is also found in abundance in meteorites. The ^{121}Ir layer is thickest in the Caribbean, indicating that the impact occurred in that area. A crater, now filled with sediments and buried 3,000 ft (1,000 m) deep, is under the fishing village of Chicxulub on the Yucatan peninsula of Mexico (fig. 4–11). It is 110 miles (180 km) in diameter and is the same age as the great killing. This is where the asteroid hit.

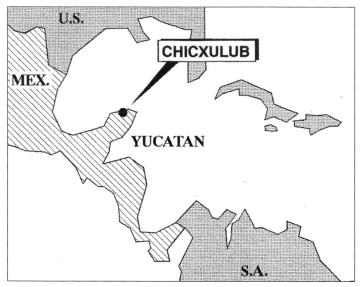

Fig. 4–11. Cretaceous-Tertiary impact site

Soot is common in the iridium-121 layer, indicating that a firestorm was caused by the heat of the impact during the collision. The soot and rock dust caused by the collision and ejected into the atmosphere must have thrown the earth into total darkness for several decades. During this time the world's climate cooled. After the soot and dust settled out of the air, greenhouse gases caused by the collision made the world's climate warmer. Large amounts of sulfur dioxide were vaporized during the impact. The vaporized sulfur dioxide mixed with rainwater in the atmosphere to form sulfuric acid, which rained down upon the earth. In addition, evidence in Texas and other areas indicates that the waters of the Gulf of Mexico were thrown up into a great wave hundreds of feet high that swept across North America. There is also evidence that an enormous earthquake felt throughout the earth was caused by the asteroid when it hit.

Some mammals survived the great killing and flourished after their reptilian competition was eliminated. The Cenozoic era is known as the age of mammals. Grasses evolved during the Cenozoic and became an important food source for the mammals. Late in the Cenozoic, during the Pleistocene epoch or the ice ages, the climate was colder than it is today. Extensive ice sheets, thousands of feet thick, covered approximately one-third of the land. During four separate times, the glaciers advanced across the land and then retreated. The last ice sheet did not retreat until the end of the Pleistocene, just 10,000 years ago.

Comets made of ice and asteroids made of stone have hit the earth many times in the geologic past. The impact fractures and breaks up rocks to form potential reservoir rock. The crater that is formed is often filled with organic-rich lake sediments that are source rocks, and if the crater is covered with a caprock, it can trap gas and oil. About 50% of the buried craters that have been drilled contain commercial amounts of gas and oil. The oil and gas fields surrounding the Chicxulub crater of Mexico have an estimated 30 billion barrels of oil and 15 billion cubic feet of gas reserves. The Cantarell field is the largest oil field in Mexico and one of the largest in the world. It will eventually produce about 18 billion barrels of oil. The field is located in the Gulf of Mexico, and the trap is a large anticline. Seventy percent of the production from the field comes from a 950-ft (290-m) thick reservoir rock that was formed by the ejected debris from the Chicxulub crater during the impact.

A comet or asteroid hit western Oklahoma during the Ordovician period when it was covered with shallow, tropical seas. The impact crater was discovered by seismic exploration at about 9,000 ft (2,750 m) in the subsurface (fig. 4–12). It is 8 miles (13 km) in diameter and has an uplifted rim and a central uplift. The Ames oil field, with 25 million bbl of recoverable oil and 15 Bcf of recoverable gas, produces from highly fractured granite that forms a granite breccia reservoir under the crater rim and central uplift. Both the source rock and caprock are an organic-rich black shale layer that was deposited when ocean water filled the crater and overlies the reservoir rock. One well in the crater, producing out of 285 ft (87 m) of granite breccia, set the modern record in Oklahoma for the highest calculated open flow rate of over 10,000 bbl (1,590 m³) of oil per day.

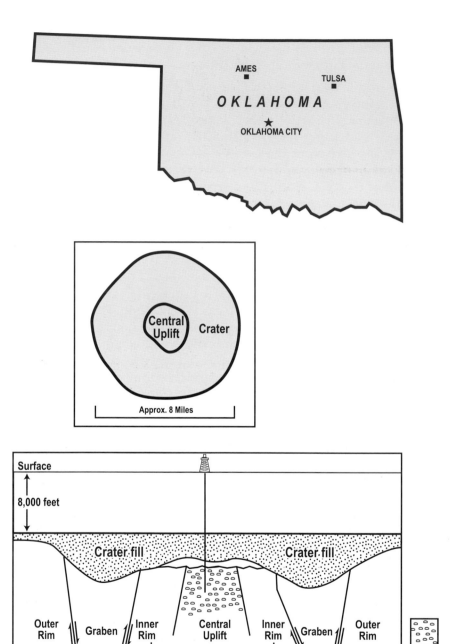

Fig. 4–12. Ames crater field location, map, and cross section. (Modified from Koeberl, Reimold, and Powell, 1994.)

Deformation
of Sedimentary Rocks

Sedimentary rocks are originally deposited in horizontal layers. One type of oil and gas trap, a *structural trap*, is formed by the deformation of these rock layers.

Weathering, Erosion, and Unconformities

Weathering is the breakdown of solid rock. Once a rock is exposed on the surface of the earth, either to the atmosphere or ocean bottom, it will eventually be mechanically broken into particles or chemically dissolved by the forces of weathering. Some sedimentary rocks, such as sandstones, are more resistant to weathering, and others, such as shales, readily break down. *Erosional processes* are those that transport and deposit sediments. These processes include rivers, wind, waves, gravity (landslides), and glaciers.

Sea level has been rising and falling throughout geological time. Whenever sea level was lower, the land was exposed to erosion, and some of the sedimentary rocks were stripped off the surface of the land. Buried, ancient erosional surfaces that were formed during these times are called *unconformities*. Two types are disconformities and angular unconformities.

A *disconformity* is an erosional channel in which the sedimentary rock layers above and below the erosional surface are parallel (fig. 5–1). It is an ancient river channel usually filled with sand that has become sandstone.

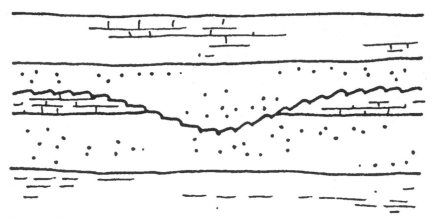

Fig. 5–1. Disconformity

An *angular unconformity* is an ancient erosional surface in which the sedimentary rock layers below the unconformity are tilted at an angle to the layers above the unconformity (plate 5–1). An angular unconformity represents a time of mountain building followed by erosion. It often covers a large, subsurface area. The formation of an angular unconformity started with the deposition of horizontal sediment layers as ancient seas covered the earth. After the seas retreated, exposing the earth, the sedimentary rocks were tilted to form hills and mountains. The hills and mountains were then eroded down, leaving an erosional surface. The seas again covered the land, depositing horizontal sedimentary rock layers on the erosional surface, burying it in the subsurface.

Angular unconformities can form gas and oil traps (fig. 5–2). One of the sedimentary rock layers tilted at an angle below it must be a reservoir rock that can store gas and oil, usually a sandstone or limestone. The sedimentary rock layer above it must be a caprock that acts as a seal, usually a shale or salt layer. The gas and oil form below the unconformity in a source rock such as black shale. They migrate up into and then through the pore spaces of the reservoir rock until they reach the angular unconformity surface where they are trapped below the caprock. Because angular unconformities can cover large subsurface areas, they can form giant gas and oil fields. The two largest oil fields in the United States, the East Texas field and the Prudhoe Bay field in Alaska, are both in angular unconformity traps. In both fields, the horizontal rock layers on the surface of the ground do not give any indication of the subsurface angular unconformities and their giant oil accumulations.

Plate 5-1. Angular unconformity in a sea cliff in England showing flat sedimentary rocks above and sedimentary rocks tilted at an angle below.

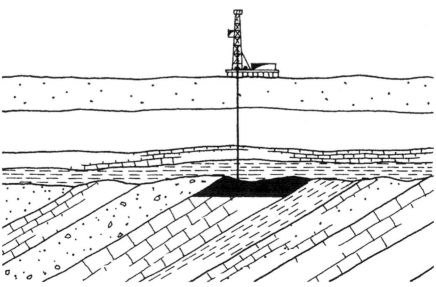

Fig. 5-2. Angular unconformity trap

The East Texas field originally contained more than 7 billion bbl (1.1 billion m³) of oil. The oil is located in the Woodbine Sandstone below an angular unconformity (fig. 5–3). The Austin Chalk, a very fine-grained limestone that forms the caprock, directly overlies the angular unconformity. The Woodbine Sandstone was originally deposited as a horizontal layer of sand when shallow seas covered East Texas about 100 million years ago (fig.5–4a). The sandstone was then buried in the subsurface as it was covered with other sediments (fig. 5–4b). Later, the Sabine uplift, along the Texas-Louisiana border, arched up and exposed the Woodbine Sandstone (fig. 5–4c). Erosion removed the Woodbine Sandstone from the top of the arch (fig. 5–4d). After that, the seas invaded the area, depositing the Austin Chalk and other sediments, covering the angular unconformity (fig. 5–4e). The oil formed in the Eagle Ford Shale source rock below and migrated up into the Woodbine Sandstone. It then flowed along the porous sandstone toward the east until it was trapped under the angular unconformity, unable to flow into the Austin Chalk.

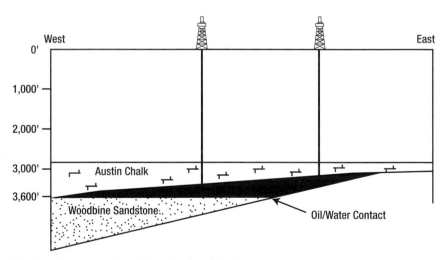

Fig. 5–3. Cross section of East Texas oil field

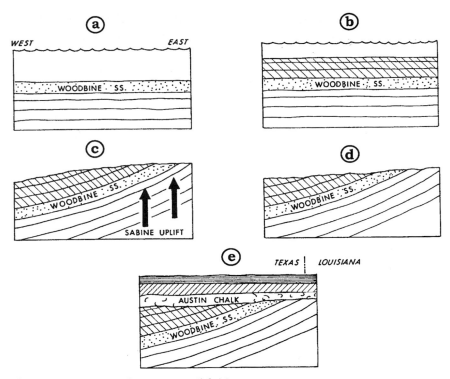

Fig. 5–4. Formation of East Texas oil field trap

The discovery of the East Texas oil field in 1930 is a classic example of petroleum history. Oil companies explored this area in the early 1900s. There were no oil seeps in the area, and the companies became discouraged after drilling some dry holes. The companies abandoned this area by the mid-1920s and instead drilled in the newly discovered West Texas oil fields. Because of this, Columbus Marion "Dad" Joiner, a driller and promoter, was able to obtain leases for drilling in a large area of eastern Texas by promising the cotton farmers who owned the land a share of any oil revenue if he found oil. He started to drill in the area in the late 1920s using a method best described as "random drilling." His only geological help was from a veterinarian named Dr. A. D. Lloyd. The local farmers would often volunteer to help drill the well. Because Dad Joiner had little money, he traded shares in the well for room and board, equipment repair, supplies, and hired help.

After two wells had caved in, the No. 3 Daisy Bradford well finally reached below the angular unconformity at 3,725 ft (1,135 m) after 16 months of drilling and blew in the East Texas oil field on October 5, 1930. The well initially tested 6,800 bbl/day (1,080 m³/day) and was completed to produce 300 b/d (48 m³/day). Unfortunately, Dad Joiner, in his financial need during drilling, had sold 300% of the Daisy Bradford well and was in legal trouble. The investors had filed a lawsuit to take all the leases away from him. Dad had no money to hire lawyers and defend himself. H. L. Hunt negotiated a deal with Dad to settle his legal problems and pay him $1,335,000, mostly in future oil production for 5,000 acres of prime leases in the field.

The East Texas oil field is 45 miles (72 km) long and 5 miles (8 km) wide (fig. 5-5). More than 30,000 wells were drilled in the field that has now produced more than 5 billion bbl (800 million m³) of light, sweet crude oil. Hundreds of poor farming families that had land in the field became Texas millionaires, and H. L. Hunt became a billionaire. Dad Joiner went to another area to drill a series of dry holes and died penniless.

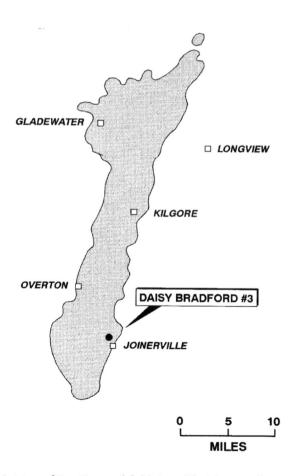

Fig. 5-5. Map of East Texas oil field. (Modified from Halbouty, 1991.)

Anticlines and Synclines

An *anticline* is a large, upward arch of sedimentary rocks (fig. 5–6), whereas a *syncline* is a large, downward arch of rocks. Anticlines, but not synclines, form high areas in reservoir rocks and can be gas and oil traps. Folds such as anticlines expose the rocks to erosion. If the anticlines are relatively young, they have not been very eroded and appear as topographic ridges on the surface. A series of young, rising anticlines that are also prolific petroleum producers occur as a line of hills that cross the Los Angeles basin (fig. 5–7). These trend from Beverly Hills in the north, through the Inglewood (Baldwin Hills) and Dominguez fields, southward to Long Beach and the Wilmington field, and offshore into the Huntington Beach field.

Most anticlines and synclines are not level and are tilted with respect to the surface of the earth. These are called *plunging anticlines* (fig. 5–8) and *plunging synclines*. When a plunging anticline or syncline is eroded down, it leaves a characteristic lobate-shaped pattern on the surface (fig. 5–9).

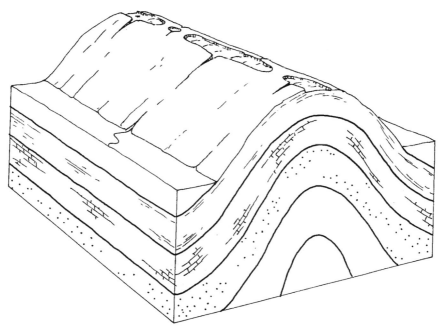

Fig. 5–6. Anticline

Fig. 5–7. Los Angeles basin showing trend of anticline oil fields

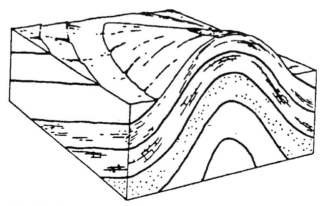

Fig. 5–8. Plunging anticline

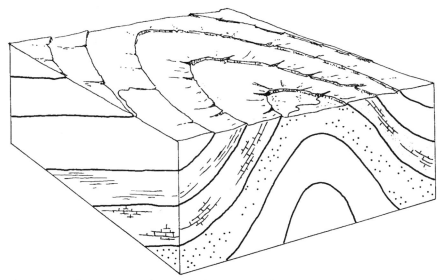

Fig. 5–9. Surface pattern of eroded, plunging anticline

The formation of anticlines and synclines results in shortening of the earth's crust (fig. 5–10). Forces that shorten the earth's crust are compressional. If an area of the earth's crust is compressed, the rocks will be folded into anticlines and synclines. If folds are present in the rocks of the earth's crust, that area probably has been compressed some time in the past.

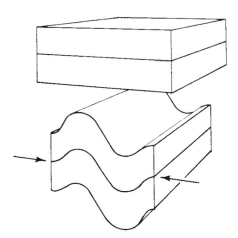

Fig. 5–10. Formation of anticlines and synclines by compression

• 55 •

Domes

A *dome* is a circular or elliptical uplift. Domes also form gas and oil traps. Before a dome is eroded down, it forms a hill. Oil was first discovered in the Middle East in Bahrain, an island in the Persian Gulf, in 1932. The traps in Bahrain were domes with a low hill on the surface above each of them. A similar low hill above a dome in Saudi Arabia was drilled to find the first oil field there in 1937. If the dome is eroded, it leaves a characteristic bull's eye pattern (fig. 5–11) on the surface.

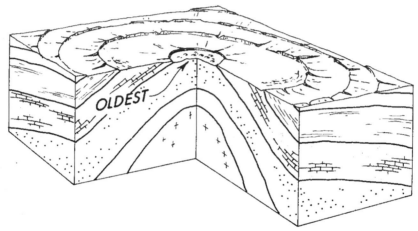

Fig. 5–11. Surface pattern of eroded dome

Anticlines and domes were the first type of petroleum trap recognized. They form many of the giant oil and gas fields of the world. Most of the Middle East oil fields are in anticline and dome traps. The Cushing oil field of Oklahoma, discovered in 1912, is located southwest of Tulsa. The trap is an anticline with three domes superimposed on it (fig. 5–12). The major reservoir rock is the Bartlesville sandstone. The best producing wells are on domes. The Cushing oil field will produce 450 million bbl (72 million m^3) of oil. It was the largest oil field in the world during World War I.

Homoclines

Sedimentary rocks dipping uniformly in one direction are known as a *homocline* (fig. 5–13). Although homoclines are common, they do not form gas and oil traps.

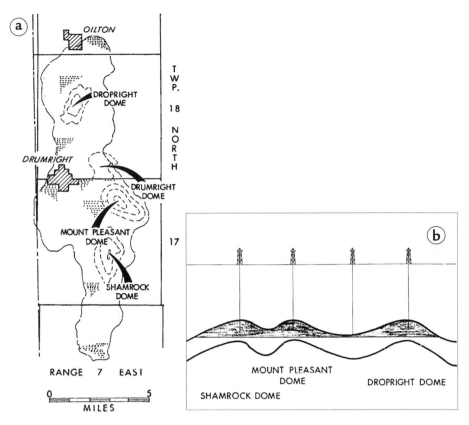

Fig. 5-12. Cushing oil field, Oklahoma: (a) map and (b) cross section

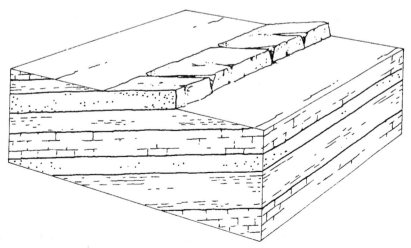

Fig. 5-13. Homocline

Fractures

Two types of natural fractures in rocks are joints and faults.

Joints

A *joint* is a fracture in the rocks with no movement of one side relative to the other (plate 5–2). Joints are common in sedimentary rocks and are oriented perpendicular (90°) to the bedding planes. There are usually two sets of joints oriented at right angles (90°) to each other. They were formed when erosion removed sedimentary rocks that were located above and stress on the rock was relieved. Joints in sedimentary rocks improve the reservoir quality of the rock. They slightly increase the fluid storage capacity of the rock (porosity) and greatly increase the ability of the fluid to flow through the rock (permeability). Any naturally fractured rock is a potential reservoir rock.

Plate 5–2. Joints in a sandstone (Winding Stair Mountains, Oklahoma)

Faults

Faults are breaks in the rocks along which one side has moved relative to the other (plate 5–3). The relative movement of each side is used to classify faults (fig. 5–14). *Dip-slip* faults move primarily up and down, whereas *strike-slip* faults move primarily horizontally. *Oblique-slip* faults have roughly equal dip-slip and strike-slip displacements. The side of a fault that extends under the fault plane is called the *footwall* (fig. 5–15), and the side that protrudes above the fault plane is the *hanging wall*. *Throw* (fig. 5–16) is the vertical displacement on a dip-slip fault. The side of the dip-slip fault that goes down is called the *downthrown* side, and the side that goes up is called the *upthrown* side.

Plate 5–3. Fault showing displacement of sedimentary rock layers (Austin Chalk, Texas)

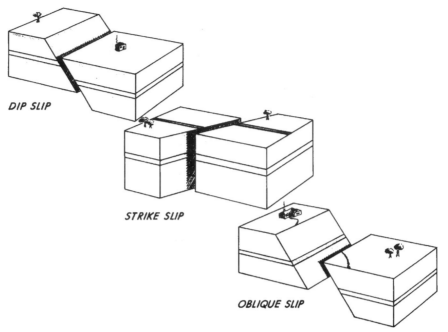

Fig. 5–14. Types of faults

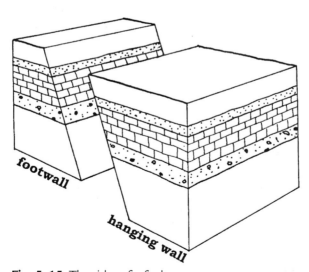

Fig. 5–15. The sides of a fault

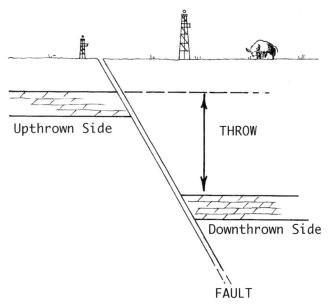

Fig. 5–16. Dip-slip fault terminology (upthrown and downthrown)

Two types of dip-slip faults are normal and reverse. If the hanging wall has moved down relative to the footwall, it is a *normal dip-slip fault* (fig. 5–17). In a normal dip-slip fault, the beds are separated and pulled apart. A normal dip-slip fault is identified in the subsurface by a *lost section*, a missing layer or layers of rocks when a well is drilled through the fault (fig. 5–18).

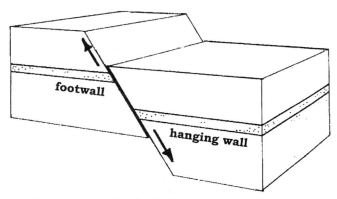

Fig. 5–17. Normal dip-slip fault

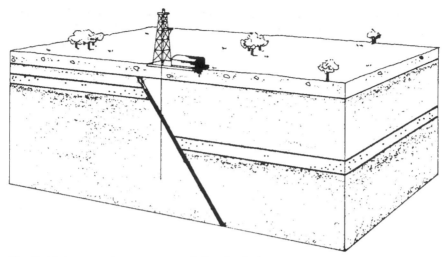

Fig. 5–18. Lost section on a normal dip-slip fault

A series of parallel, normal dip-slip faults forms a structure called horst and graben (fig. 5–19). A *graben* is the down-dropped block between two normal faults. A *horst* is the ridge left standing between two grabens. These can range in size from inches to tens of miles across.

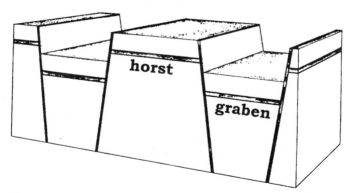

Fig. 5–19. Horst and graben

If the hanging wall has moved up relative to the footwall, it is a *reverse dip-slip fault*. In a reverse dip-slip fault, some subsurface beds overlap. It is possible to drill through this fault and encounter the same rock layers twice in a *double section* (fig. 5–20). A *thrust fault* is a reverse fault with a fault plane less than 45° from horizontal (fig. 5–21). On a thrust fault, the upper hanging wall has been thrust up and over the lower footwall. There are some thrust faults in the earth's crust where the hanging wall has been thrust horizontally tens of miles over the footwall. Several large thrust faults, called the Rocky Mountain overthrust belt, occur in a band along the Rocky Mountains. A series of large gas and oil traps are located in the overthrust belt.

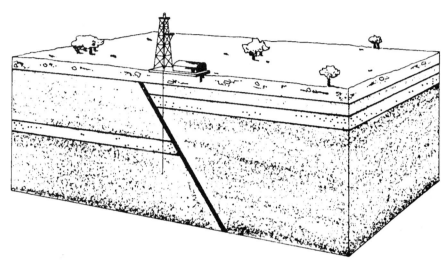

Fig. 5–20. Double section on a reverse dip-slip fault

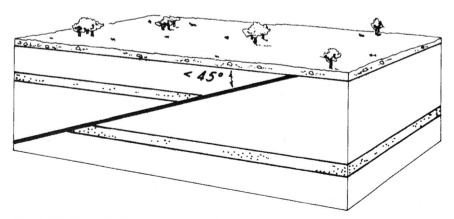

Fig. 5–21. Thrust fault

A normal dip-slip fault is formed when the rocks are pulled apart by tensional forces. A reverse dip-slip fault is formed by shortening the rocks with compressional forces (fig. 5–22). When the earth's crust is pulled apart, normal dip-slip faults with horsts and grabens are formed. When the earth's crust is squeezed, reverse dip-slip and thrust faults and folds, such as anticlines and synclines, are formed.

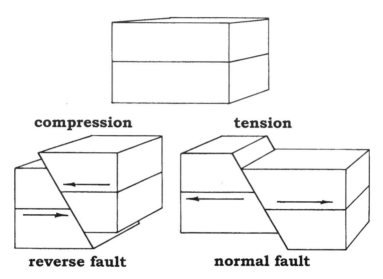

Fig. 5–22. Forces that form a normal dip-slip fault and a reverse dip-slip fault

Faults can be both active and inactive. When a fault moves, it can produce shock waves called an *earthquake*. Many faults, however, moved a long time ago and are inactive today. Two very large faults occur in Oklahoma, the Seneca and Nemaha faults. Both were active hundreds of millions of years ago but are inactive today.

Dip-slip faults form traps by displacing the reservoir rock (fig. 5–23). The fault must be a *sealing fault*, which means it prevents fluid flow across or along the fault. Any gas and oil migrating up a reservoir rock will be trapped under the sealing fault. The largest oil field on land in England is the Wytch Farm field, located southwest of London on the South Dorset coast. There are natural oil seeps along the coast, and the field was discovered in 1973. The trap was formed by a fault cutting the Sherwood Sandstone reservoir rock (fig. 5–24). It contains 286 million bbl (45 million m^3) of recoverable oil.

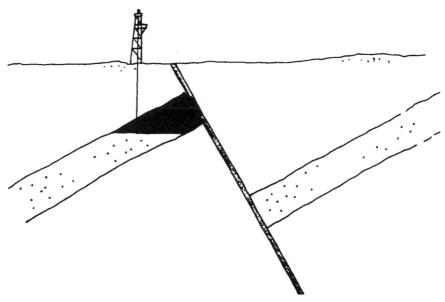

Fig. 5–23. Fault trap

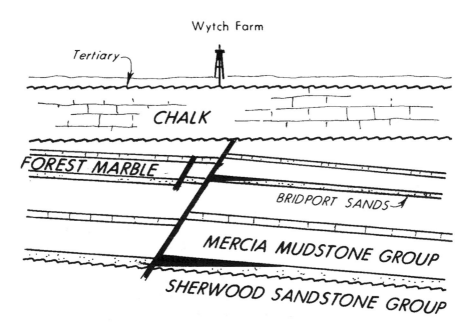

Fig. 5–24. Cross section of Wytch Farm oil field, England. (Modified from Colter and Harvard, 1981.)

A strike-slip fault is described by the horizontal movement of one side relative to the other (fig. 5–25). If the opposite side of the fault as you face it moves to the right, it is a *right-lateral strike-slip fault*. If it moves to the left, it is a *left-lateral strike-slip fault*. The San Andreas fault of California is an active right-lateral strike-slip fault. It is hundreds of miles long and has moved many tens of miles over a long time. The Potrero oil field in California (fig. 5–26) is formed by an anticline on sandstone reservoir rocks. The crest of the anticline is displaced 1,200 ft (365 m) by the Potrero fault, a right-lateral strike-slip fault.

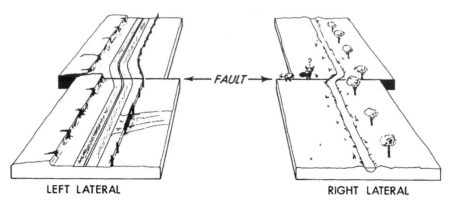

LEFT LATERAL   RIGHT LATERAL

Fig. 5–25. Strike-slip faults

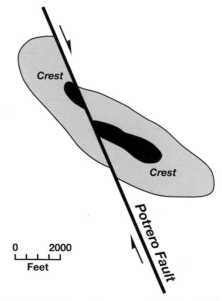

Fig. 5–26. Map of Potrero oil field, California

Ocean Environment
and Plate Tectonics

Continental Margins

All continents are surrounded by a shallow, almost flat platform, the *continental shelf* (fig. 6–1). It extends from the shoreline out with a slope of less than 1° to the shelf break. At the *shelf break*, the ocean bottom sharply increases in slope. The shelf break is located in an average water depth of 450 ft (137 m). The width of the continental shelf varies from ½ mile (0.8 km) to more than 500 miles (805 km), with an average width of 50 miles (80 km).

The continental shelf is geologically part of the continents. Sedimentary rocks that are encountered in drilling along the beach extend out under the continental shelf. Many large structures such as faults and folds continue from land onto the continental shelf. The giant Wilmington oil field is formed by an anticline that lies partially under land (Long Beach, California) and partially offshore. The San Andreas Fault extends offshore onto the continental shelf in northern California. Throughout geologic time, sea level has been rising and falling, and the seas are now covering this part of land, the continental shelves. The continental shelf is a very active petroleum exploration and production area. The same source rocks, reservoir rocks, and traps that occur on land are found on the continental shelves.

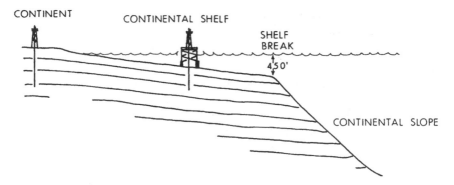

Fig. 6–1. Cross section of a continental shelf

Seaward of the continental shelf and slope break is the *continental slope* that extends down to the bottom of the ocean. It has a slope of about 3° and is the geological edge of the continents.

Eroded into the continental shelves and slopes in many areas are *submarine canyons* (fig. 6–2). They often extend from shallow depths off the shoreline down to the bottom of the continental slope, thousands of feet deep. Submarine canyons are relatively common throughout the world and often occur offshore from rivers. The Mississippi, Amazon, Ganges, Niger, Nile, and many other rivers have submarine canyons offshore.

Submarine canyons are eroded, and sediments are transported down submarine canyons by turbidity currents. *Turbidity currents* are masses of water with suspended sediments such as sand, silt, and clay. The turbidity current is denser than the surrounding seawater and is pulled by gravity down the submarine canyon similar to river water being pulled by gravity down a river channel on land. Turbidity currents can originate from rivers with a large sediment load flowing into the ocean. Like rivers on land, they erode submarine canyons offshore.

A turbidity current will continue to flow down a submarine canyon as long as a slope exists. When the turbidity current flows onto the relatively flat ocean bottom, it stops, and the sediments settle out of the water. The coarsest sediments (usually sand) settle out first and the finest sediments (silt and then clay) settle out last. This deposits a *graded bed* (fig. 6–3) with the coarsest sediments on the bottom and the finest on the top. Accumulations of turbidity current sediments at the base of a submarine canyon form a large sedimentary deposit called a *submarine fan* (fig. 6–2). A channel usually leads out of the submarine canyon and divides into smaller distributary channels on the submarine fan.

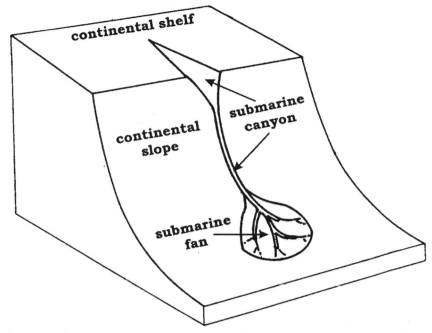

Fig. 6–2. Submarine canyon and fan

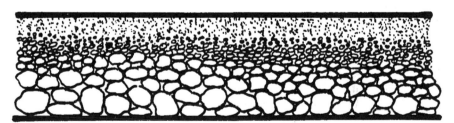

Fig. 6–3. Graded bed

The sands in turbidity current deposits are called *turbidites* and can be reservoirs (fig. 6–4) for gas and oil. Relatively thin sandstones separated by shales are characteristic of submarine fan reservoirs deposited as graded beds. Relatively thick sandstones were deposited in submarine canyons and channels where each turbidity current flowing down the channel eroded away the finer-grained sediments on top of the underlying graded bed to deposit sand on top of sand. Deep-water production on the continental slopes of the Gulf of Mexico, western Africa, and Brazil are from turbidite sands deposited in submarine canyons, submarine fan distributary channels, and submarine fans now buried in the ocean bottom. The sands are relatively young and have not been buried too deep. Because of this, they often have very high porosities and permeabilities and are excellent reservoir rocks. They yield oil and gas at a high rate to justify the high cost of drilling in deep water.

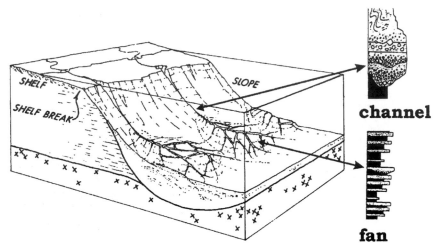

Fig. 6–4. Sandstone reservoirs in submarine canyons and fans

Many areas on land today used to be deep water in the geological past. The Los Angeles basin of Southern California is a graben basin that was originally occupied by the Pacific Ocean. It has been filled in with sediments and is dry land where the city of Los Angeles is located today. The Santa Fe Springs oil field in the Los Angeles basin is a symmetrical dome (fig. 6–5). It produces gas and oil from 25 sandstones, each located at the base of Miocene and Pliocene age graded beds that were deposited when the basin was filled in. The field has 622 million bbl (99 million m^3) of recoverable oil and 0.87 Tcf (28 million m^3) of recoverable gas. There is more known crude oil in the Los Angeles basin per cubic mile of sedimentary rocks than any other basin in the world, including the Middle East.

The largest producing offshore gas field in the North Sea is the Frigg field, located in both the United Kingdom and Norwegian sectors (fig. 6–6a). It was formed by a stratigraphic trap, a type of petroleum trap formed by a reservoir rock completely encased in shale that is both the source rock for the gas and the caprock for the reservoir. Stratigraphic trap fields often take the shape of the reservoir rock. A map of the Frigg field reservoir rock (fig. 6-6b) shows it produces from an ancient submarine canyon and fan distributary channel sandstone (the Frigg Sandstone of Paleocene age) at a depth of about 6,000 ft (1,800 m) below the bottom of the North Sea (fig. 6-7). The field will eventually produce 7 Tcf (200 million m^3) of natural gas.

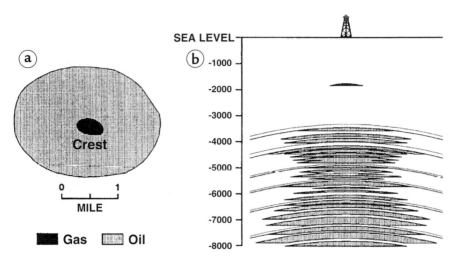

Fig. 6-5. Santa Fe Springs field, Los Angeles basin, California, (a) map (b) cross section

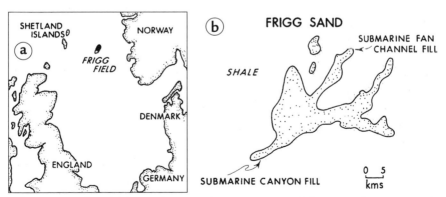

Fig. 6-6. Frigg gas field, North Sea (a) location map and (b) map of reservoir rock. (Modified from Heritier, Lassel, and Waltne, 1980.)

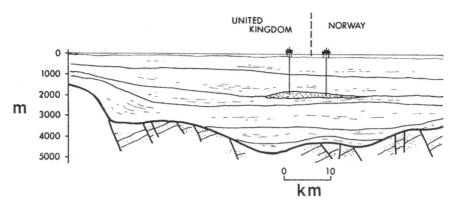

Fig. 6–7. Cross section of Frigg gas field, North Sea. (Modified from Heritier, Lassel, and Waltne, 1980.)

Deep Water

The deepest parts of the seafloor are *ocean trenches* that are long, narrow depressions usually located along the margins of the oceans. Adjacent to many deep ocean trenches are active volcanic islands.

Located almost in the very center of the Atlantic Ocean is a segment of *mid-ocean ridge*. It is the world's longest mountain chain and can be traced for 40,000 miles (64,000 km). The ridge extends down the Atlantic Ocean, around south Africa, into the Indian Ocean and continues between Australia and Antarctica and up into the eastern Pacific Ocean (fig. 6–8). It bifurcates in several locations. In the Indian Ocean, one segment extends into the Gulf of Aden and the Red Sea. The ridge is very wide, averaging 1,000 miles (1,600 km) and rises about 1 to 2 miles (1.6 to 3 km) above the adjacent ocean floor. The center of the ridge typically has a rift valley (graben). It protrudes through the surface of the ocean at Iceland and the Azores, both islands formed by active, basalt volcanoes. Observations from submarines have shown that active basalt volcanoes occur all along the floor of the submerged graben on the crest of the mid-ocean ridge.

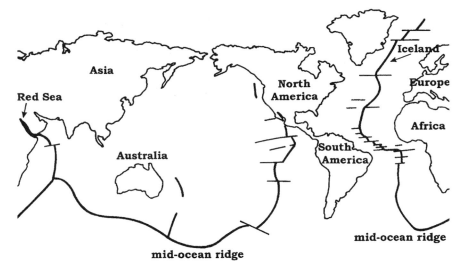

Fig. 6–8. Map of the mid-ocean ridge

Ocean Sediments

The average sediment thickness on the ocean bottom is about ½ mile (0.8 km). This is unevenly distributed. Sediments are very thick along the edge of the oceans in continental shelves and slopes and thin or absent on the mid-ocean ridge. The oldest sedimentary rocks in the ocean basins anywhere in the world are only Jurassic in age, about 150 million years old. Many sedimentary rocks found on land are considerably older than the oldest ocean-bottom sediments. Because of lack of sedimentary rocks, no one will ever drill a well for gas or oil near the center of the ocean.

Earth's Interior

The crust of the earth under continents is granite in composition and is 20 to 45 miles (32 to 72 km) thick. The crust under oceans is basalt lava rock in composition and is only about 3 miles (5 km) thick. Continents are high in elevation because they are composed primarily of granite that is relatively light in density compared to basalt, and it floats higher on the rocks in the interior of the earth. Because no one has drilled or mined very deep into basement rock, there is little direct evidence of what occurs below the crust of the earth. However, temperature and pressure both increase with depth. Because of the high temperatures, rocks below a certain depth are partially melted and act as thick, viscous liquids. The solid rock above the partially melted rocks is called the lithosphere (fig. 6–9). It is about 90 miles (145 km) thick below continents and about 60 miles (97 km) thick below oceans.

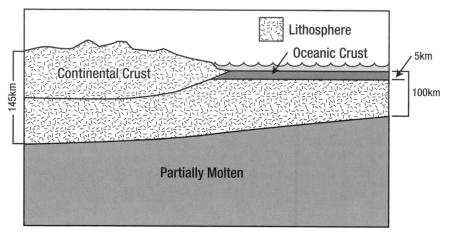

Fig. 6–9. Cross section of the earth's crust and lithosphere

Continental Drift

The theory of *continental drift* dates back to the early 1900s. It suggests that all the present-day continents were previously joined into one supercontinent, Pangaea (fig. 6–10). During the early Jurassic, about 200 million years ago, Pangaea broke up. The fragments of the supercontinent drifted (moved) across the face of the earth into their present positions to form the modern continents. The theory was not widely accepted at first. It was not known what process would cause Pangaea to break up and the continents to move.

Fig. 6–10. Map of Pangaea, the supercontinent

Seafloor Spreading

A new theory, *seafloor spreading*, was presented in the early 1960s. This theory provided the processes for the breakup of Pangaea and the drifting of the continents. Seafloor spreading postulates that large, slow-moving convection currents occur in the interior of the earth (fig. 6–11) where rocks act as viscous liquids. A convection current is a cell of flowing liquid caused by heating and cooling. Where the liquid is heated, it becomes less dense and rises. Where the liquid is cooled, it becomes more dense and sinks. Convection currents cause the interior of the earth to be constantly moving. A rising hot current from the interior of the earth cannot penetrate the crust of the earth. It arches the crust up to form the mid-ocean ridge (fig. 6–12). The hot, molten current then divides and flows to either side of the mid-ocean ridge. This splits the solid crust of the earth (lithosphere) at the ridge crest and drags it to either side of the ridge. The term *seafloor spreading* comes from the seafloor being spread out at right angles from the crest of the mid-ocean ridge. The existence of a graben, a tensional feature that runs along the center of the mid-ocean ridge, supports this idea. The graben contains erupting volcanoes that produce basalt lava. This new basalt crust of the earth is split and spread out from the ridge crest.

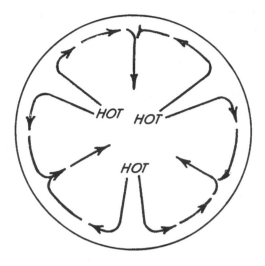

Fig. 6–11. Convection currents in the interior of the earth

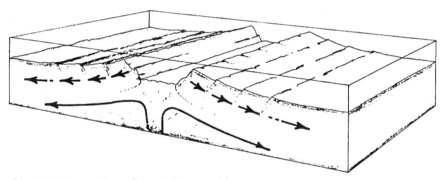

Fig. 6–12. Formation of the mid-ocean ridge

The seafloor is spreading out from several mid-ocean ridges in different oceans. Areas where seafloors from two different mid-ocean ridges collide are called *subduction zones*. There are three types of subduction zones. First, if two seafloors from different mid-ocean ridges meet (fig. 6–13), one seafloor is thrust below the other. This forms a long, narrow depression, an ocean trench. The deeper the seafloor is thrust into the interior of the earth, the hotter it becomes. When the subducted seafloor becomes too hot, it melts, and the light, molten rock rises to the surface to form a series of volcanoes adjacent to the ocean trench. The Aleutian Trench and Aleutian Islands off Alaska are an example of this. Second, where one seafloor meets another seafloor with a continent riding on it (fig. 6–14), the seafloor without the continent is thrust under the one with the continent. This forms an ocean trench off the coast of the continent. The edge of the continent is compressed to form a coastal mountain range. Molten rock from the subducted seafloor under the edge of the continent rises to form volcanoes in the coastal mountains. The west coast of South America is an example of this. The Peru-Chile Trench occurs just offshore, and the volcanic Andes Mountains occur along the coast. Third, when two seafloors meet, both carrying continents (fig. 6–15), neither continent is subducted into the interior because both are composed of relatively light granite. One seafloor is thrust under the other seafloor, and the colliding continents are compressed to form a mountain range between the continents. The Himalayan Mountains between India and Asia are an example of this.

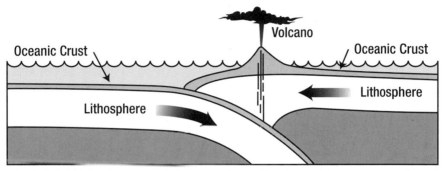

Fig. 6–13. Subduction zone with two seafloors forming a deep ocean trench and adjacent volcanic islands. (Modified from Kious, W. J. and R. J. Tilling, 1996.)

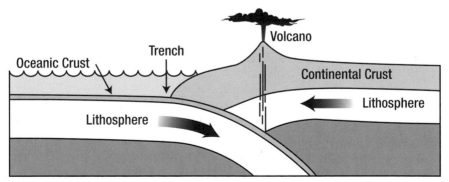

Fig. 6–14. Subduction zone with two seafloors, one with a continent, forming a deep ocean trench offshore from a coastal mountain range with volcanoes. (Modified from Kious, W. J. and R. J. Tilling, 1996.)

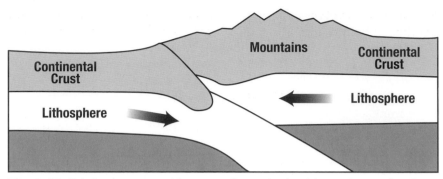

Fig. 6–15. Subduction zone with two seafloors, both with continents, forming a mountain range. (Modified from modified from Kious, W. J. and R. J. Tilling, 1996.)

Since the early 1960s, considerable evidence has accumulated to support the theory of seafloor spreading. By dividing the distance of the basalt crust from the mid-ocean ridge where it formed into the basalt seafloor's age, the seafloor spreading rate can be calculated. Spreading rates vary with the location of the ridge and range from 7 in/yr (18 cm/yr) to 0.5 in/yr (1 cm/yr). These are very fast rates for geologic processes. The mid-ocean ridge in the North Atlantic Ocean is spreading at the rate of 1 in/yr (2.5 cm/yr). Because of this, the North Atlantic Ocean is getting wider by the rate of 2 in/yr (5 cm/yr). The North American continent is moving at the rate of 1 in/yr (2.5 cm/yr) to the west, whereas the European continent is moving at the rate of 1 in/yr (2.5 cm/yr) to the east.

Seafloor spreading and continental drift are compatible theories. A mid-ocean ridge formed under Pangaea during the Jurassic time and caused it to break up. The continents, riding on the spreading seafloor, would have been carried to their present positions as the Atlantic Ocean became wider. There are modern examples of a newly formed ocean and a continent that is breaking up. A segment of the mid-ocean ridge from the Indian Ocean enters the Gulf of Aden and bifurcates into two sections. One section is located on the bottom of the Red Sea (fig. 6–8). The Red Sea is a long, narrow arm of the ocean that separates Egypt and Sudan, in Africa, from Saudi Arabia. Africa and Saudi Arabia were joined millions of years ago. A mid-ocean ridge rose beneath them about 20 million years ago, split them apart, and created the Red Sea. The Red Sea is growing wider by inches each year. It is similar to the Atlantic Ocean when Pangaea first broke up.

Another section of the mid-ocean ridge underlies the Great Rift Valley of East Africa. The valley is a series of large, long grabens with active volcanoes, earthquakes, and deep lakes. East Africa is breaking up today. A long, narrow arm of the ocean, similar to the Red Sea, will eventually occupy the rift valley in the next few thousands of years, forming two Africas.

Plate Tectonics

The modern day theory of *plate tectonics*, suggested in 1967, combines the ideas of seafloor spreading and continental drift. Plate tectonics postulates that the solid lithosphere of the earth is divided into large, moving plates (fig. 6–16). Every location on the earth's surface, whether a continent or a seafloor, is on a moving plate that is sliding across the partially molten rocks below it. Each plate originates at a mid-ocean ridge where new seafloor is being formed. The plate is moving at right angles away from the crest of the ridge at the spreading rate of that ridge. At the opposite side of the plate from the mid-ocean ridge is a subduction zone, an ocean trench, and/or a mountain range. Large strike-slip faults occur where different plates scrape against each other. Continents ride along on the moving plates.

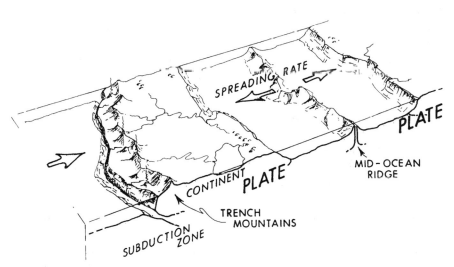

Fig. 6–16. Cross section of plates

At present, there are eight large plates and many smaller ones (fig. 6–17). The North American Plate is moving to the west at 1 in/yr (2.5 cm/yr). Below California and off the west coast of the United States and Canada, the North American Plate is obliquely colliding with the Pacific Plate, which is moving to the northwest. The subduction zone has formed the San Andreas fault (fig. 6–18), earthquakes, volcanoes such as Mount St. Helens, and a mountain range (Coastal Ranges). The major features of the earth's surface, both modern and ancient, can be explained by moving plates.

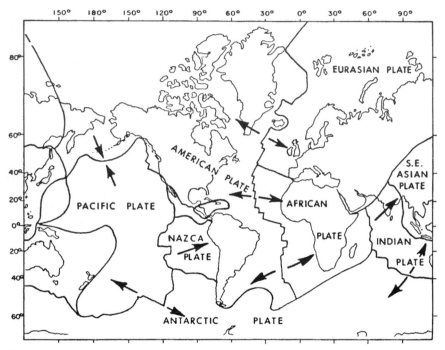

Fig. 6–17. Plate tectonics

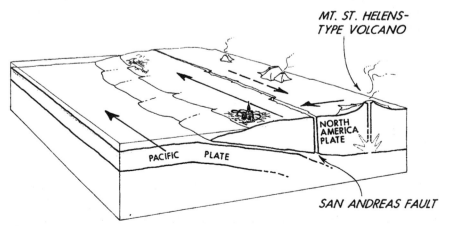

Fig. 6–18. Cross section of California showing underlying subduction zone

In the geologic past, the number and size of the plates have varied along with their rates and directions. Mountains are formed by the collision of plates. For hundreds of millions of years, thick sediments accumulated along continental margins on two different seafloor plates (fig. 6–15). The continents eventually collided, and the sediments were compressed, forming mountain ranges such as the Himalayan Mountains. Mount Everest is composed of sedimentary rocks deposited in the seas between India and Asia before they collided. The collision between one seafloor plate with a continent and another seafloor plate (fig. 6–14) forms a coastal mountain range. The Andes Mountains along the west coast of South America are an example.

Failed Arm Basins

The initial breakup of a continent by plate tectonics can take the form of a *triple junction*. A triple junction has three rifts (arms) that join in the center (fig. 6–19). Usually, two of the arms unite and continue rifting to form an ocean. The other arm stops spreading and is called a *failed arm*. A failed arm is a graben that can be filled with sediments.

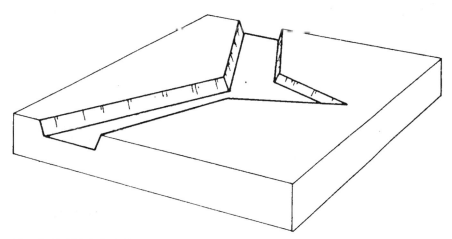

Fig. 6–19. Triple junction

Several failed arms are oil and gas producers. During the Mesozoic breakup of the supercontinent Pangaea, several triple-junctions with failed arms formed. As North America separated from Europe, a triple junction formed near the present-day North Sea. Two arms joined to become the

North Atlantic Ocean. The failed arm became the central graben that runs down the center of the North Sea (fig. 6–20). The central graben and other related grabens are now filled with sedimentary rocks that are 3,000 to 6,000 ft (900 to 1,800 m) thicker than the sedimentary rocks in adjacent areas on the North Sea bottom. Many of the North Sea gas and oil fields are in the sedimentary rocks filling this graben. As South America pulled away from Africa, a triple junction formed under Nigeria (fig. 6–21). Two arms joined to become the South Atlantic Ocean. The other arm, the Benue Trough, failed and is located under Nigeria. The Niger River deposited its delta along the length of the Benue Trough out into the Atlantic Ocean. Most of the Nigerian oil production comes from the sedimentary rocks in the Niger River Delta filling the Benue Trough.

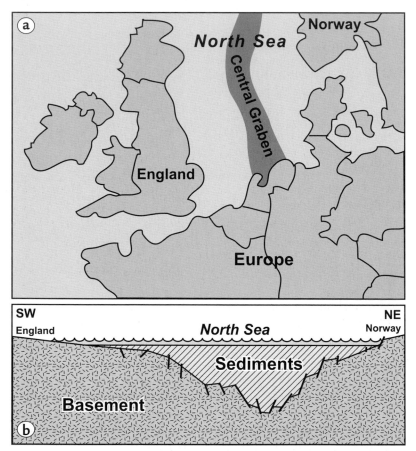

Fig. 6–20. (a) Map of central graben in North Sea and (b) cross section of North Sea from England to Norway showing thick sediments filling the central graben. (Modified from Graversen, 2005, and Thorne and Watts, 1989.)

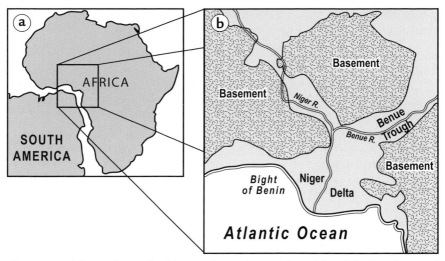

Fig. 6–21. (a) Breakup of Africa and South America with triple junction and (b) present-day Nigeria showing Benue Trough, a failed rift, and the Niger River and Niger Delta. (Modified from Short and Stauble, 1967.)

Middle East Oil Fields

The large petroleum traps of the Middle East (fig. 6–22) were formed by plate tectonics. Almost all the Middle East oil field traps are anticlines and domes. The mid-ocean ridge in the Red Sea (fig. 6–8) is causing Saudi Arabia on the Arabian Plate to move northeastward and collide with the Eurasian Plate. The Persian Gulf area is being compressed between the two plates, forming the Middle East Field traps. The deformation becomes more intense in a northeastward direction from Saudi Arabia toward Iran and Iraq and forms the Zargos Mountains. Saudi Arabia alone has 85 fields with 325 reservoirs containing 25% of the world's oil reserves.

The Ghawar oil field in Saudi Arabia is the largest conventional oil field on earth. The trap is an anticline 174 miles (280 km) long and up to 18.6 miles (30 km) wide (fig. 6–23a). The reservoir rock is fractured and dissolved limestone, the Jurassic age Arab D Limestone with an average oil pay zone of 200 ft (60 m). The reservoir rock is not very deep (–1,500 ft or –457 m below sea level). Salt layers in the overlying Hith Formation are the seal (fig. 6–23b). The source rock is a Jurassic age black limestone. The average production of a Ghawar well is 11,400 bbl (1,800 m^3) of oil per day, and the field will eventually produce 82 billion bbl (13 billion m³) of oil. The oil is 32 to 36 °API gravity and has 1.7 to 2% sulfur.

The Gashsaran oil field in Iran is a deformed anticline (fig. 6–24). Large thrust and reverse faults associated with the anticline are also the result of the compression between the moving plates. The reservoir rock, the Asmari Limestone of Oligocene-Miocene age, is 1,000 to 1,500 ft (305 to 457 m) thick. The limestone is overlain by salt that forms the seal. On the steep side of the anticline, the thick, steep-dipping limestone reservoir rock forms a 6,000-ft (1,800-m) net oil pay zone. Ultimate production from this field will be 8.5 billion bbl (1.35 billion m³) of oil.

Fig. 6–22. Map of Middle East oil fields. (Modified from Beydown, 1991.)

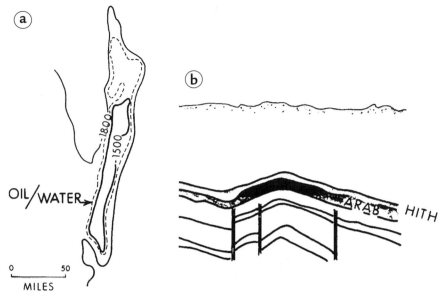

Fig. 6–23. Ghawar oil field, Saudi Arabia, (a) structural map on Arab D Limestone and (b) cross section. (Modified from Arabian American Oil Company Staff, 1959.)

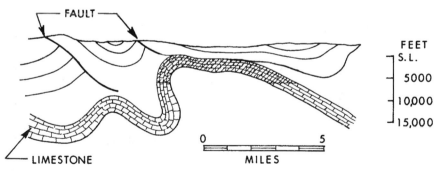

Fig. 6–24. Cross section of Gashsaran oil field, Iran. (Modified from Hull and Warman, 1970.)

Sedimentary Rock Distribution

Petroleum is not evenly distributed throughout the world. Sweden has none, whereas the Middle East has more than one-third of the world's known oil supply. Sedimentary rocks are both source and reservoir rocks for petroleum. Where the sedimentary rocks are thick, a large amount of petroleum can occur. Where there are no sedimentary rocks, there is no petroleum.

Basin Formation

A *basin* is a large area with relatively thick sedimentary rocks. It is where most gas and oil are found. There are several ways to form a basin.

A basin can be formed by subsidence of the basement rock. The depression is originally filled with ocean water and, eventually, sediments. The Michigan basin, with a maximum thickness of about 14,000 ft (4,267 m) of sedimentary rocks, is an example of this type of basin (fig. 7–1). Subsidence occurred during the Paleozoic era, and very thick carbonates and evaporates accumulated during the Silurian period. Petroleum occurs in reefs that surrounded the basin during that time.

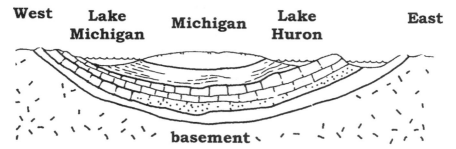

Fig. 7–1. Cross section of Michigan basin

The southern California oil basins were formed by grabens. Some basins are located on land (Los Angeles and Ventura basins), whereas others are offshore (Santa Barbara and San Pedro basins). An east-to-west cross section (fig. 7–2) of the basins shows that the area was subjected to tensional forces that formed a series of parallel normal dip-slip faults with north-south horsts and grabens in the basement rock. The grabens were originally filled with ocean water and many of the horsts stood above sea level to form islands. Sediments filling the grabens came primarily from the east, where land was being eroded. Because of the easterly source of sediments, the eastern grabens (Los Angeles and Ventura basins) were the first to be filled and are now dry land. The city of Los Angeles is located on the Los Angeles basin. To the northwest is the Ventura basin, where the city of Ventura is located. Presently, the sediments are filling the next basins to the west (San Pedro and Santa Barbara basins). The basins furthest to the west will be the last to be filled. The sedimentary rocks, rich in source rocks and reservoir rocks, are tens of thousands of feet thick in the eastern basins. Anticlines, domes, and faults form numerous traps.

The Los Angeles basin (fig. 7–3) is the most prolific oil basin on earth. There are 61 oil fields that have produced over 9 billion bbl (1.4 billion m^3) of oil. Almost all the reservoir rocks are Miocene to Pleistocene age sandstones. There is more known oil per cubic mile of sedimentary rocks here than anywhere else in the world.

Half-graben basins are formed by subsidence along one side of a normal fault (fig. 7–4). These basins are common and are productive in the North Sea, offshore western Africa, and offshore Brazil.

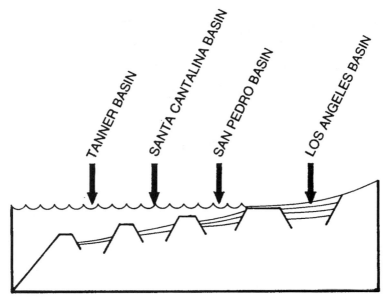

Fig. 7-2. East-to-west cross section of Southern California graben basins

Fig. 7-3. Oil fields of the Los Angeles basin

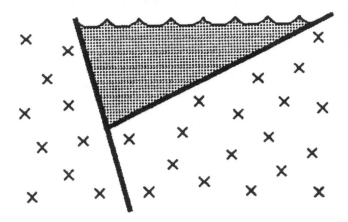

Fig. 7–4. Half-graben basin

Most mountain ranges on land were formed by the compression of sedimentary rocks when plates collided (see figs. 6–14 and 6–15 in chapter 6). They display large compressional features such as anticlines, synclines, reverse faults, and thrust faults. Mountain ranges, however, are relatively unproductive because most petroleum reservoirs have been breached by erosion, and the oil and gas have leaked out. In many areas, the sedimentary rocks have been eroded away, exposing the basement rock. Many of the remaining sedimentary rocks in the mountains have been metamorphosed by the high heat and pressure that occurred during the compression. Only intermontane basins, which form between mountain peaks, are good areas to explore for oil and gas.

Intermontane basins form when mountain ranges are created. The basin is located between the mountain peaks and is often occupied by a lake. Algae growing in the lake contribute organic matter to the bottom sediments for source rocks. Streams, eroding the surrounding mountains, deposit numerous channel and beach sandstone reservoir rocks in the basin. When the Rocky Mountains were uplifted during the Cretaceous time, several intermontane basins were formed. Many of these basins, such as the Big Horn, Powder River, Green River, and Uinta basins, are good petroleum producers today (fig. 7–5).

Basins also form along the edges of mountains. As the mountains are eroded by streams, sediments fill in the areas adjacent to the mountains. The Alberta, Denver-Julesburg, and Raton basins formed in this manner.

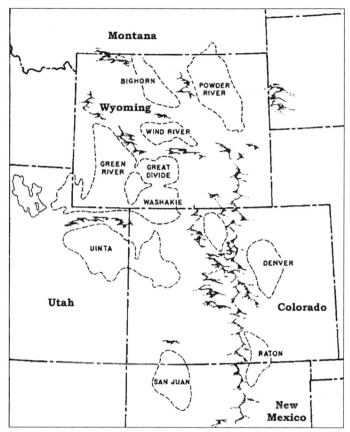

Fig. 7–5. Rocky Mountain intermontane basins

Coastal plains are formed by thick sediments deposited adjacent to an ocean (fig. 7–6). They originate when mountains are uplifted adjacent to a coast. As erosion lowers the mountains, streams deposit sands along the beaches, and waves carry the silts and clays offshore. The sandy beaches are deposited out into the ocean, forming the coastal plain. The Gulf of Mexico coastal plain was created by this process. Because the sediments on the surface are young and have never been buried, they are loose and uncemented (unconsolidated). The Atlantic and Gulf coastal plains of the United States were deposited by sediments eroded from the Appalachian and Ouachita mountains, which rose during the Pennsylvanian period. Underlying the Gulf of Mexico coastal plain is 40,000 to 60,000 ft (12,000 to 18,000 m) of sedimentary rocks. This area is one of the most prolific petroleum-producing areas of the world. The Atlantic coastal plain, however, is barren for gas and oil, possibly due to the lack of source rocks.

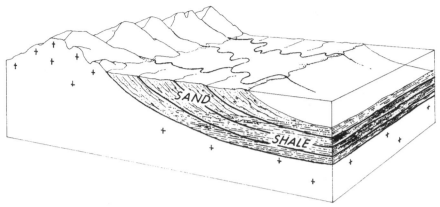

Fig. 7–6. Coastal plain

Sedimentary Rock Facies

Sedimentary rock layers are not always deposited uniformly. A single layer can be composed of two or more different rock types (fig. 7–7). Each is a *facies*, a distinctive portion of the rock layer. The change between rock types is called a *facies change*. Modern sediments being deposited along the margin of an ocean are forming both a sandstone and shale facies. As waves wash ashore, silt and clay are suspended in the water. The mud is carried offshore and deposited in deep water to become shale. Sand, too heavy to be suspended in the water by the waves, is deposited on the beach to become sandstone (fig. 7–8). This forms a layer of sediments with a shale (deep-water) and sandstone (beach) facies.

The Jurassic age Smackover Limestone occurs in the subsurface of the northern Gulf of Mexico coastal plain. It is 15,000 to 20,000 ft (4,600 to 6,000 m) deep and in some areas is a good reservoir rock. A facies map of the Upper Smackover (fig. 7–9) illustrates the different depositional environments and textures of the limestone. The best reservoir rock is the oolite facies that was deposited on a tropical, shallow-water shelf and has 20 to 25% porosity in Arkansas. In Texas, parts of the oolite facies have been dolomitized, increasing the porosity to 30%. The salt facies, in contrast, is impermeable and acts as a caprock.

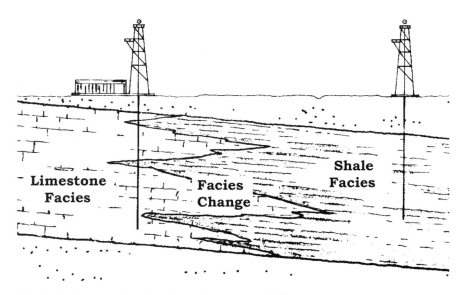

Limestone Facies   **Facies Change**   **Shale Facies**

Fig. 7–7. Cross section showing sedimentary rock facies

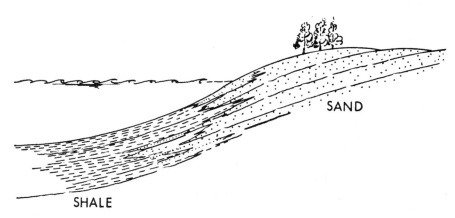

SAND

SHALE

Fig. 7–8. Deposition of sand on the beach and mud offshore to form a future facies change

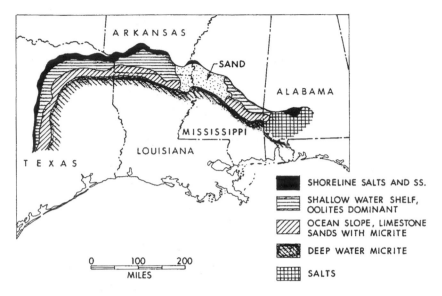

Fig. 7–9. Facies of the Smackover Limestone in the Gulf Coast. (Modified from Bishop, 1968.)

The largest conventional gas field on the North American continent, the Hugoton field of Texas, Oklahoma, and Kansas (fig. 7–10a), was formed by a facies change. The field covers an enormous area that is 275 miles (443 km) long and 8 to 57 miles (13 to 92 km) wide. It will eventually produce 81 Tcf (2 trillion m³) of natural gas. In Texas there is also oil production along the eastern margin of the field. This is called the Panhandle field and will eventually produce 1.4 billion bbl (230 million m³) of oil. The reservoir rocks are primarily limestones and dolomites known as the Chase Group of Permian age. The Wichita Formation, containing salt layers, lies directly over the reservoir rocks and forms the seal. The Chase Group was deposited as limestone reservoir rocks to the east and impermeable, red-colored shales and sands to the west in a facies change (fig. 7–10b–1). A later uplift to the west, the Stratford arch, completed the trap (fig. 7–10b–2). Enormous volumes of gas formed in the deep Anadarko basin to the southeast. The gas was trapped as it migrated updip to the west by the change from permeable into impermeable Chase Group limestones.

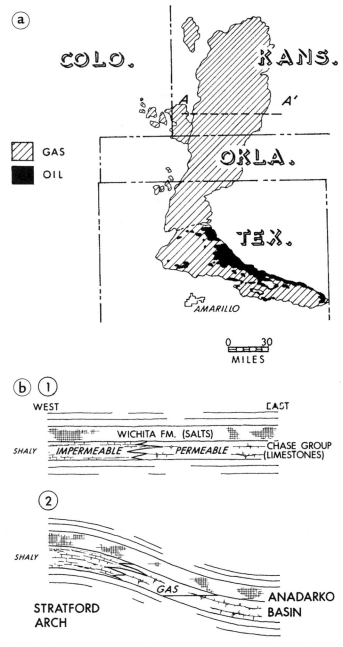

Fig. 7–10. Hugoton gas field and Panhandle oil field, Kansas, Oklahoma, and Texas: (a) map and (b) formation. (Modified from Pippen, 1970.)

Subsurface Rock Layers

Few sedimentary rock layers are uniform in thickness and rock type. A rock layer often thins in one direction and thickens in another. It can grade from one facies into another facies. Sometimes the boundary between two facies is sharp, and the rocks *interfinger* or *wedge out* into each other, and sometimes the boundary is gradational (fig. 7–11). Often a single rock layer will *pinch* or *wedge out* (fig. 7–12) in another rock layer. Sandstone wedges, deposited as the edges of beaches or river channels, are common in shale layers.

An updip pinch-out of a reservoir rock in a shale or salt layer can form a petroleum trap. This type of trap is very common in coastal plains with buried beach sandstone reservoir rocks, such as the south Texas coastal plain. The Glenn Pool oil field, located just south of Tulsa, Oklahoma, is formed by an updip pinch-out of a sandstone wedge in a shale layer (fig. 7–13). The Pennsylvanian age reservoir rock, the Bartlesville Sandstone, is a river channel tilted up to the east at 1°. The wells are only about 1,500 ft (500 m) deep. The field has already produced 327 million bbl (53 million m³) of sweet, 36 to 41 °API gravity oil. The Glenn Pool was discovered in 1905 and started the Oklahoma oil boom.

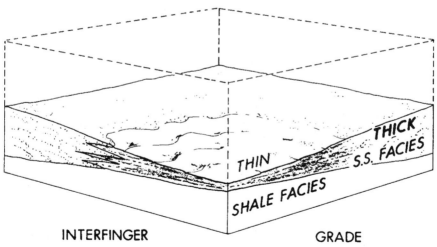

Fig. 7–11. Variations in a sedimentary rock layer

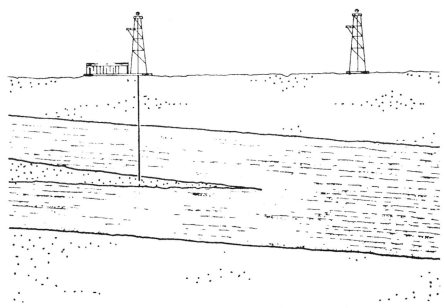

Fig. 7–12. Sandstone pinch or wedge out in a shale layer

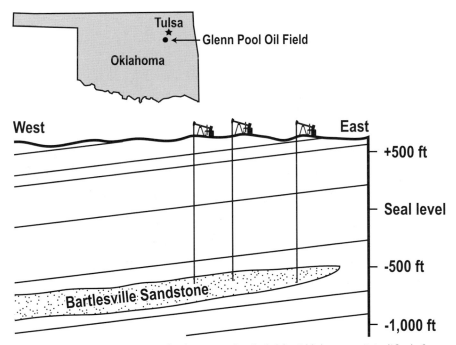

Fig. 7–13. Cross section of Glenn Pool oil field, Oklahoma. (Modified from Kuykendall and Matson, 1992.)

Mapping

Surface and subsurface maps are important tools that geologists use to find gas and oil. All maps are oriented with north to the top, south to the bottom, east to the right, and west to the left.

Topographic Maps

A *topographic map* shows the elevation of the earth's surface (fig. 8–1). To illustrate the third dimension (elevation) on a flat, two-dimensional map, contour lines are used. A *contour line* is a line of equal value on a map, and a contour line on a topographic map is a line of equal elevation. A contour line is always labeled with an elevation that is above or below sea level. All along that contour line, the elevation is exactly the same. For example, anywhere along the +400 ft contour line on a topographic map, the elevation is exactly 400 ft above sea level. The *contour interval* of a topographic map is the difference in elevation between two adjacent contour lines. The contour interval of the topographic map in figure 8–1 is 100 ft. If the elevations on contour lines increase in a direction, the slope is rising (fig. 8–2). If the contours are spaced relatively close together, the elevation is changing rapidly, and the slope is steep. If the contours are relatively far apart, the slope is gentle.

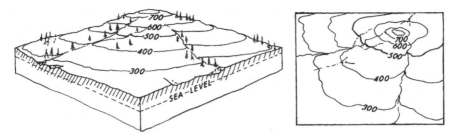

Fig. 8–1. Land and a topographic map of the land

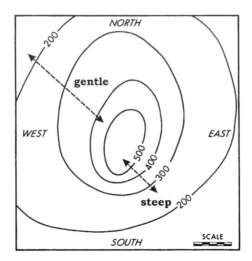

Fig. 8–2. Contoured topographic map
showing steep and gentle slopes

There are some important characteristics of contours on a topographic map. Contour lines never cross. Contour lines are single lines; they never branch. Contour lines are continuous; they always close or run off the map and never end on the map.

Elevations can be accurately estimated from a topographic map. If a point is on the +300 ft contour, it must be, by definition, exactly 300 ft above sea level. If the point is about halfway between the +300 and +400 ft contour, an elevation of +350 ft is a good estimate. The shape of the contours is characteristic for many topographic features such as hills, ridges, and canyons.

A topographic map (or any contoured map) cannot be drawn without some accurately surveyed points. After the elevations or values are located on a map (*spotted*), contours can be drawn between the points. Contouring of any map can be done either by hand or computer. The position of a contour line between two data points can be accurately located by using proportions. For example, the 400 contour line must run between data points of 402 and 399 (fig. 8–3). A straight line is drawn between the two data points. Because there is a difference of 3 between the data points (402 and 399), the line is divided into three equal segments. The 400 contour line is located one segment from the 399 point and two segments from the 402 point. Anything that can be expressed by mathematics can be programmed into a computer, and computer-generated contour maps can be made.

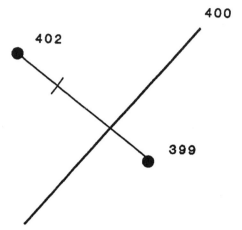

Fig. 8–3. Locating a contour using proportions

Geologic Maps

A *geologic map* (fig. 8–4) shows where each rock layer crops out on the surface of the earth. Each rock layer is given a different pattern, color, and symbol on the map. The basic sedimentary rock layer used for geologic mapping is called a *formation*. A formation is a mappable rock layer with a definite top and bottom. Geologists have divided all sedimentary rocks into formations. Each formation has a two-part name. The first part is a town where the layer crops out on the surface. The second part is the dominant rock type, such as sandstone or limestone. San Andreas Limestone, Bartlesville Sandstone, and Barnett Shale are formation names. If the sedimentary rock layer is a mixture of rock types, such as alternating thin sandstones and shales, the word *formation* is used, for example, the Coffeyville Formation. Formations can be subdivided into smaller units called *members*. A member is a distinctive but local bed in a formation (fig. 8–5). It is also given a formal, two-part name. For example, the Layton Sandstone Member is part of the Coffeyville Formation. Adjacent formations of similar rocks can be joined to form a *group* and given a geographic name (*i.e.*, the Chase Group). If a rock layer occurs deep in the subsurface and does not appear to crop out on the surface or if it is located offshore, it is given a letter and number designation such as the H5 sands.

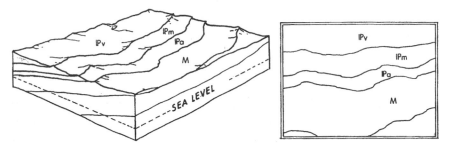

Fig. 8–4. Geologic map

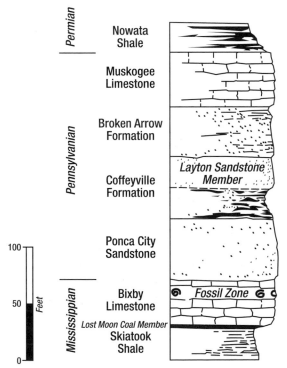

Fig. 8–5. Stratigraphic column showing formations and members

A geologic map is a flat, two-dimensional representation of the earth's surface. The orientation of rock layers, the third dimension, is shown with a strike-and-dip symbol. *Strike* is the horizontal orientation of a plane (fig. 8–6a), such as a sedimentary rock layer or a fault. It is measured with a compass orientation, such as north 30° east. Strike is shown as a short line on the geological map (fig. 8–6b) that is oriented in the measured compass direction. *Dip* is the direction and vertical angle of the plane. It is measured perpendicular (90°) to the strike (fig. 8–6a). The dip symbol on the map is a small bar attached to the middle of the strike line (fig. 8–6b). It points in the direction that the plane goes down into the earth. The angle in degrees is often on the dip symbol. The dip of a rock layer is the angle and direction it goes into the subsurface. Drilling *updip* means that the drillsite will be up the angle (dip) of the rock layer from the last drillsite. Updip in a reservoir is usually a favorable position from a dry hole (fig. 8-7). You may assume that any reservoir rock is filled with water. Gas and oil are lighter than water and will flow (migrate) updip in the reservoir rock to a high area. One would almost never want to drill downdip from a dry hole; one would want to drill updip.

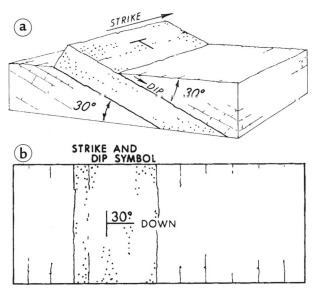

Fig. 8–6. (a) Strike and dip of a sedimentary rock layer and (b) strike-and-dip symbol on a geologic map

Fig. 8–7. Updip from a dry hole

A stratigraphic column (fig. 8–5) is a convenient method for presenting the vertical sequence of rocks on a geologic map or in a basin. Any deformation of the rocks, such as faulting or tilting, has been removed. The youngest formation is at the top of the column, and the oldest is located at the bottom. The column is drawn as a cliff of weathered rocks with the weaker rock types (e.g., shales) indented. Stronger rock types (e.g., sandstones) protrude outward as they would weather in nature.

Common geological symbols (fig. 8–8) are used for rocks, structures, and wells on a geological map.

Base Maps

A *base map* is a map that shows the location of all the wells that have been drilled in an area. *Spotting a well* involves locating a wellsite and placing the well symbol (fig. 8–8) on a base map. Base maps can also include seismic lines and other data.

Fig. 8–8. Common geological symbols

Global Positioning System

Accurate positioning is very important to geologists, geophysicists, and petroleum engineers. They need to know the exact location of proposed drillsites, existing wells, and seismic lines. These sites used to be located with considerable time and expense using surveying tools. Since the 1980s, accurate location in all weather and anywhere on the earth with no cost has been determined by the *Global Positioning System* (GPS). GPS involves the use of satellites and a receiver. There are 24 solar-powered satellites very precisely orbiting the earth twice a day at an altitude of 12,550 miles (20,200 km) in six planes with four satellites each. Each satellite transmits extremely accurate time signals and the satellite's orbital information. The receiver at the location has an antenna tuned to each satellite's frequency, a processor, and a very stable clock. It compares the satellite time signal with the same time on the receiver to determine how much time it took the satellite signal to reach the receiver. It then uses that information to compute the distance from the receiver to the satellite. By using the computed distance from three, or more accurately from four satellites, the location, usually in latitude and longitude, and the altitude of the receiver are calculated and displayed on the receiver. Very precise receivers can calculate positions to less than 10 ft (3 m) on average.

Subsurface Maps

Three important types of subsurface maps are structural, isopach, and percentage. All three maps use contour lines to describe a subsurface rock layer.

Structural map

A *structural map* uses contour lines to show the elevation of the top of a subsurface sedimentary rock layer (fig. 8–9). The contour lines are usually in minus feet below sea level, as most rocks are located below sea level. An important structural map would be one contoured on the top of a potential reservoir rock or drilling target.

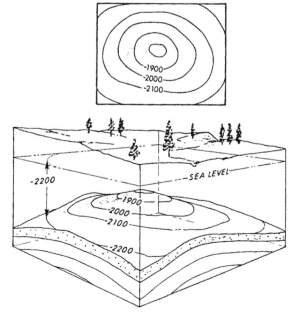

Fig. 8–9. Structural map

Domes, anticlines, and faults can be identified on structural maps. Both a hill on a topographic map and a dome on a structural map have a bull's eye pattern (fig. 8–10) with the highest elevation in the center. Both a ridge on a topographic map and an anticline on a structural map have a concentric but oblong pattern (fig. 8–11) with the highest elevation in the center. Dip-slip faults are characterized by a rapid change in elevation along a relatively straight line (fig. 8–12). A normal dip-slip fault that causes a lost section in the rock layer being mapped (see fig. 5–18 in chapter 5) is seen on the map as two lines separating the contour lines (fig. 8–13).

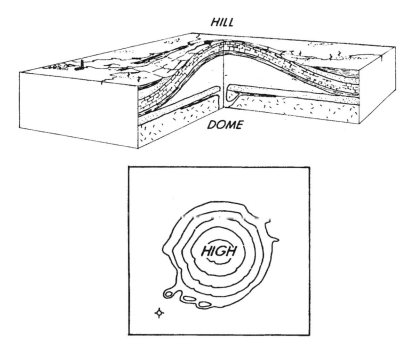

Fig. 8–10. Topographic map of a hill and structural map of a dome

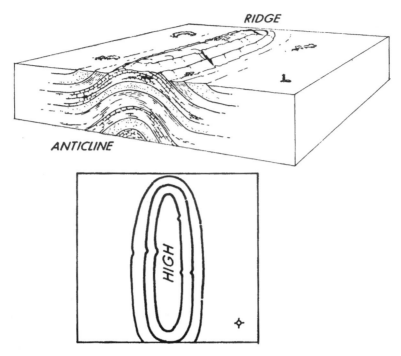

Fig. 8–11. Topographic map of a ridge and structural map of an anticline

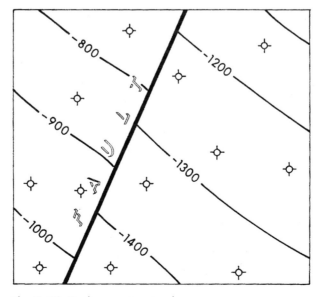

Fig. 8–12. Fault on a structural map

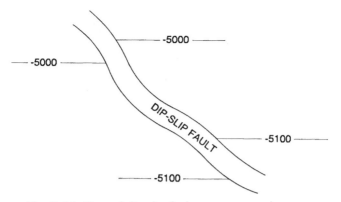

Fig. 8–13. Normal dip-slip fault on a structural map

Isopach map

An *isopach map* (fig. 8–14) uses contour lines to show the thickness of a subsurface layer. If an oil or gas field has been drilled, an isopach map can be made of the reservoir rock pay zone. The *pay zone* is the vertical distance in a well that produces gas and/or oil. *Gross pay* contours the entire reservoir thickness including nonproductive water-bearing and shaly zones. *Net pay* contours only the productive thickness of the reservoir. A net pay isopach map of a reservoir is used to calculate the oil and gas volume and reserves.

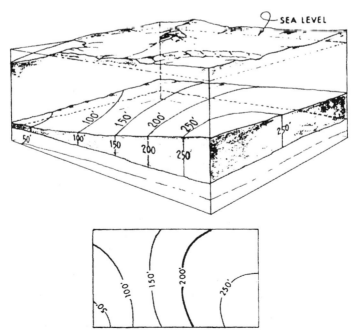

Fig. 8–14. Isopach map

An isopach map can be used in exploration to delineate a sandstone pinch-out (fig. 8–15a) where the isopach contour line becomes zero. The aerial patterns of beach and river channel sandstones are seen on an isopach map (fig. 8–15b).

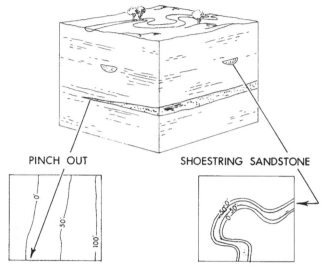

PINCH OUT   SHOESTRING SANDSTONE

Fig. 8–15. Isopach map of (a) a sandstone pinch-out and (b) a beach or river channel sandstone

An isopach map of a limestone layer can also be used to locate a reef. A reef is a mound and is shown by thick contour lines (fig. 8–16). Barrier reefs that are long (fig. 8–16a) can be distinguished from pinnacle reefs that are circular (fig. 8–16b).

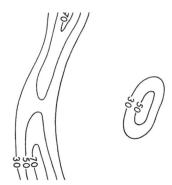

Fig. 8–16. Isopach map of (a) a barrier reef and (b) a pinnacle reef

Percentage map

A *percentage map* (fig. 8–17) plots the percentage of a specific rock type such as sandstone in a formation. Higher percentages of reservoir quality rocks, such as sandstones and carbonates, imply a better reservoir quality.

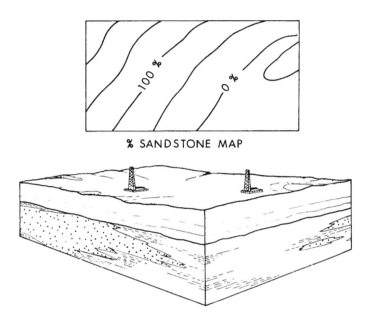

% SANDSTONE MAP

Fig. 8–17. Sandstone percentage map of a formation composed of sandstone and shale

Source Rocks, Generation, Migration, and Accumulation of Petroleum

Source Rocks

A *source rock* is a rock that can generate natural gas and/or crude oil. Gas and oil form from ancient organic matter preserved in sedimentary rocks. As sediments are deposited, both inorganic mineral grains, such as sands and mud, and organic matter (dead plants and animals) are mixed. Most organic matter is lost on the surface by decay, a process of oxidation. The decaying organic matter on land gets oxygen from the air, and the decaying organic matter on the ocean bottom gets the oxygen from out of the water. Some organic matter, however, is preserved. It was either rapidly buried by other sediments before it decayed or was deposited on the bottom of a sea with stagnant, oxygen-free waters. The black color in sedimentary rocks comes primarily from its organic content. Black-colored, organic-rich sedimentary rocks include coal, shale, and some limestones.

When woody plant material is buried, it is transformed into coal and methane gas (CH_4) by temperature and time. This is why coal mines are dangerous; they contain methane gas and sometimes explode. Coal deposits are drilled to produce *coal seam* or *coal bed gas*, which is pure methane gas.

Shale is the most common sedimentary rock, and many are black. A black shale commonly has 1 to 3% organic matter by weight and can have up to 20%. Green or gray shale has only about 0.5% organic matter. Black shales contain a large variety of organic matter that includes single-celled plants and animals that live floating in the ocean, algae, spores, pollen and bacteria. They have the right chemical composition to generate both natural gas and crude oil. In some areas, such as North Africa and the Middle East, organic-rich dark limestones are also source rocks.

Some sedimentary basins are gas-prone and produce primarily natural gas. Examples of these are the Sacramento basin of northern California, the Arkoma basin of southern Oklahoma and Arkansas, and the southern portion of the North Sea. This is because the only effective source rock is coal.

Generation

The most important factor in the generation of crude oil from organic matter in sedimentary rocks is temperature. A minimum temperature of about 150°F (65°C) is necessary for oil generation under typical sedimentary basin conditions (fig. 9–1). This temperature is obtained by burying the organic-rich source rocks. The deeper the depth, the higher the temperature.

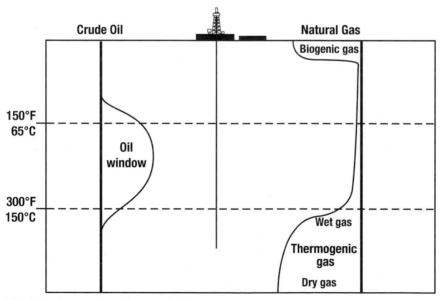

Fig. 9–1. Generation of gas and oil

At relatively shallow depths, the temperature is not sufficient to generate oil. There, just a few feet below the surface, bacterial action on the organic matter forms large volumes of *biogenic* or *microbial gas*. It is generated very fast and is almost pure methane gas. This gas is commonly known as swamp or marsh gas. Biogenic gas is not commonly trapped and usually leaks into the atmosphere in enormous volumes. However, one of the largest gas fields in the world, Urengoy in Siberia, is believed to be filled with biogenic gas. The gas is trapped below the permanently frozen ground (permafrost). The field contains 285 Tcf (8 trillion m^3) of gas. Generation of biogenic gas decreases with depth as bacterial action decreases with increasing temperature.

In a typical sedimentary basin, oil generation starts at about 150°F (65°C) and ends at about 300°F (150°C). If the source rock is buried deeper, where temperatures are above 300°F (150°C), *thermogenic gas* is generated. It is the gas that is often trapped. The zone in the earth's crust where the oil is generated is called the *oil window*. It occurs from about 7,000 to 18,000 ft (2,100 to 5,500 m) deep on land. Crude oil generated in the oil window is originally good oil with °API gravities between 30 and 40. Where does heavy oil come from? It is formed later when bacteria, along with chemical and physical processes, degrade the good oil to form heavy oil. Below the oil window, thermogenic gas is generated from thermal cracking of crude oil and organic matter left in the source rock. Wet gas is formed under more shallow depths and cooler temperatures. Under higher temperatures at deeper depths, dry gas is formed.

At temperatures higher than about 300°F (150°C), crude oil is irreversibly transformed into graphite (carbon) and natural gas. The process is similar to thermal cracking in a refinery. This temperature occurs at a depth of about 18,000 ft (5,500 m) in a sedimentary rock basin and is a floor below which only gas can occur in the reservoir. Deep wells are drilled for natural gas. In several instances, a deep well has discovered a gas reservoir, and the sand grains in the sandstone reservoir rock are coated with carbon. Apparently, there was oil originally in the reservoir, but it was buried too deep and was thermally cracked. In some areas of the Gulf of Mexico, the geothermal gradient is very low, the floor for oil is very deep, and crude oil has been found down to 35,000 ft (10,670 m) below the seafloor.

Many sedimentary basins are unproductive. An unproductive basin might not have an organic-rich source rock that could generate petroleum. Even if an unproductive basin has a source rock, it might never have been buried into the oil window. *Maturity* is the degree to which petroleum generation has occurred in a source rock. A mature source rock has experienced the temperature and time to generate petroleum in contrast to an immature source rock. In sedimentary rock basins, between 30 to 70%

of the organic matter in the source rock that has been buried deep enough generates gas and oil.

Migration

After gas and oil are generated in shale source rock, some is expelled from the impermeable shale. The generation of a liquid (crude oil) or gas (natural gas) from a solid (organic matter) causes a large increase in volume. This stresses the source rock and fractures the shale. The hydrocarbons escape through the fractures. After the pressure is released, the fractures close, and the shale becomes impermeable again.

Because gas and oil are light in density compared to the water that also occurs in the pores of the subsurface rocks, petroleum rises (fig. 9–2). Oil and gas can flow upward along faults and fractures. It can also flow laterally and upward along unconformities and through carrier beds. *Carrier beds* are rock layers that are very permeable and transmit fluids. The vertical and lateral flow of the petroleum from the source rock is called *migration*. If there is no trap on the migration route, the gas and oil will flow out onto the surface as a gas or oil seep. If there is a trap along the migration route, the gas and oil can accumulate in the trap. Of all the gas and oil that form in sedimentary rock basins, only from 0.3 to 36% is ever trapped. On average, only 10% of the gas and oil is trapped. The rest of the gas and oil did not get out of the source rock, was lost during migration, or seeped into the earth's surface.

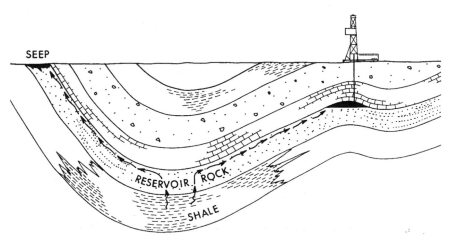

Fig. 9–2. Migration of gas and oil in a sedimentary rock basin

Because of migration, where the petroleum formed in the deep basin and where it ends up in the trap are different both vertically and horizontally. This is why a reservoir of thermogenic gas that originally formed deeply can be found at shallow depths. In the Williston basin of Montana, the oil has migrated more than 200 miles (320 km) horizontally out from the source area in the deep basin to the traps on the flanks of the basin.

Accumulation

The trap must be in position before the gas and oil migrate through the area. If the trap forms after the migration, no gas or oil will occur in the trap (fig. 9–3).

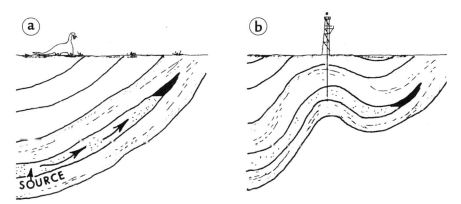

Fig. 9–3. (a) Generation, migration, and accumulation of oil in a sandstone pinch-out and (b) later formation of an anticline that is barren of oil

Once the gas and oil migrate into the trap, they separate according to density. The gas, being lightest, goes to the top of the trap to form the *free gas cap*, where the pores of the reservoir rock are occupied by gas. The oil goes to the middle of the trap, the *oil reservoir*. Saltwater, the heaviest, goes to the bottom.

The most common trap is a *saturated pool*, which always has a free gas cap on top of the oil reservoir (fig. 9–4a). The oil in the reservoir has dissolved all the natural gas it can hold and is saturated. An *unsaturated pool* lacks a free gas cap (fig. 9–4b). The oil has some dissolved gas, but it can hold more and is unsaturated. Sometimes there is only a gas reservoir on water (fig. 9–4c).

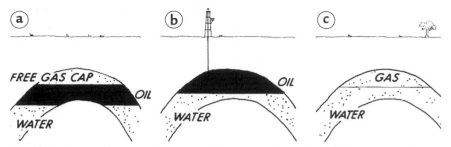

Fig. 9–4. Types of hydrocarbon traps: (a) saturated pool and (b) gas reservoir

The boundary in the reservoir between the free gas cap and the oil is the *gas-oil contact* (fig. 9–5). The boundary between the oil and water reservoir is the *oil-water contact*. The gas-oil and oil-water contacts are either relatively sharp or gradational and are usually level.

The top of the trap is called the *crest*. The first exploratory well is usually drilled *on-structure*, on the crest of the structure where the probability is highest that petroleum will be encountered (fig. 9–5). When a well is drilled to the side of the crest, it is drilled *off-structure*. If a well is drilled too far off-structure, it might not encounter commercial amounts of oil or gas and is called a *dry hole, duster*, or *wet well*.

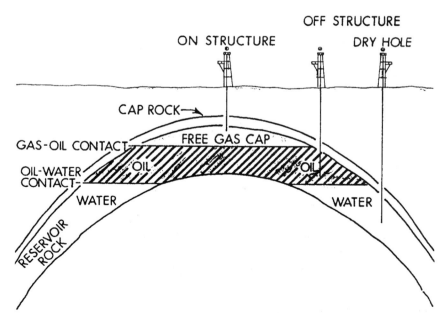

Fig. 9–5. Details of a saturated pool (oil-water and gas-oil contact)

In a trap, the reservoir rock must be overlain by a *caprock* or *seal* (fig. 9–5), an impermeable rock layer that does not allow fluids to flow through it. Without a caprock or seal, the gas and oil would leak onto the surface. Two common caprocks are shales and salt layers. Well-cemented or shaly rocks, very-fine-grained limestone (micrite or chalk), and permafrost can also be caprocks.

A *field* is the surface area directly above one or more producing reservoirs on the same trap, such as an anticline (fig. 9–6). Fields are commonly named after a geographic area such as a town, hill, or creek, for example, the Tulsa oil field. A *reservoir* is a subsurface zone that produces oil and gas but does not communicate with other reservoirs. Fluids cannot flow from one reservoir to another. The oil or gas in a single reservoir has the same characteristics throughout the reservoir but can be very different between reservoirs in the same field. A pressure system that is common to several adjacent reservoir rocks indicates a single reservoir. The Wilmington oil field of California produces from several separate reservoirs. The oil is the same in each reservoir but varies between reservoirs in the field from 12 to 34 °API gravity and is both sweet and sour. Reservoirs are usually named after their reservoir rock, for example, the Bartlesville Sandstone reservoir.

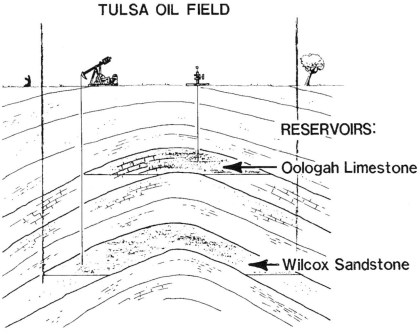

TULSA OIL FIELD

RESERVOIRS:

Oologah Limestone

Wilcox Sandstone

Fig. 9–6. Field and reservoir names

A giant oil field has at least 500 million bbl of recoverable oil, and a super giant has at least 5 billion bbl of recoverable oil. A giant gas field has at least 3 Tcf of recoverable gas, and a supergiant gas field has at least 20 Tcf of recoverable gas.

Age

The time in which the crude oil was generated, migrated through the rocks, and accumulated in the trap can be very old to very recent. Oil in Oklahoma was generated and trapped during the Pennsylvanian time. In the North Slope of Alaska, it was during the Cretaceous time. In the Gulf of Mexico coastal plain and offshore, however, crude oils are very young, just several million years old, and some oil is probably still forming, migrating, and accumulating today.

Reservoir Rocks

A *reservoir rock* is a rock that can both store and transmit fluids. A reservoir rock must have both porosity and permeability. *Porosity* is the percent volume of the rock that is not occupied by solids. These spaces are called *pores*. In the subsurface, the pores are filled with fluids such as water, gas, and oil. Porosity measures the fluid storage capacity of a reservoir rock.

There are several accurate methods to measure porosity using well cuttings, cores, and wireline well logs. First, when a well is drilled, the rock chips (well cuttings) made by the drill bit are flushed up the well by the drilling mud. The well cuttings are sampled at regular intervals. A geologist examines them under a binocular microscope to identify the rock types and see the pores. The geologist can often visually estimate the porosity of the rock to within 1 to 2%.

Second, a *core* is a cylinder of rock that is drilled from the well. A small plug in the shape of a cylinder, 1 to 1½ in. (2.5 to 3.8 cm) in diameter and 1 to 3 in. (2.5 to 7.6 cm) long, is cut from the core. The plug is then dried to remove the fluids from the pores. An instrument called a *porosimeter* is used to measure the porosity of the plug.

Third, accurate porosity measurements can also be made after the well is drilled without taking samples of the rock by a service company that runs one of three wireline well logs (neutron porosity, formation density, and sonic-velocity logs).

Typical porosity values for an oil reservoir are shown in table 9–1. Natural gas compresses and needs less porosity than an oil reservoir. Very deep gas reservoirs need very little porosity because of the very high pressure.

Table 9–1. Porosity values for an oil reservoir

0–5%	insignificant
5–10%	poor
10–15%	fair
15–20%	good
20–25%	excellent

(Modified from Levorsen, 1967.)

Two types of pores are primary and secondary pores. *Primary pores* are formed on the surface when the sediments are deposited, for example, between the sediment grains. *Secondary pores* are formed in the subsurface by processes such as solution and fracturing.

Porosity is a relatively easy and accurate measurement to make. Because of this, a *porosity cutoff* (a minimum porosity value) is often used to help decide whether to complete an oil well. For sandstones, a typical porosity cutoff is 8 to 10%. Limestones often have less porosity than sandstones but typically have fractures that drain larger areas. For limestones, a typical porosity cutoff of 3 to 5% is used. These values vary depending on the depth and economics of the well.

Permeability is a measure of the ease with which a fluid can flow through a rock. It is measured in units of *darcys* (D) or *millidarcys* (md). A millidarcy is $\frac{1}{1,000}$ of a darcy. A darcy is the permeability that will allow a flow of 1 cubic centimeter per second of a fluid with 1 centipoise viscosity (resistance to flow) through a distance of 1 centimeter through an area of 1 square centimeter under a differential pressure of 1 atmosphere. The greater the permeability of a rock, the easier it is for the fluids to flow through the rock.

The only way to make a quantitative permeability measurement is to drill a core of the reservoir rock and cut a plug. The plug is dried to remove any liquids. An instrument called a *permeameter* is used to measure the permeability of the dried plug by measuring the flow of air or nitrogen through it.

Typical permeability values of an oil reservoir rock are given in table 9–2. Gas is about 50 times more fluid than oil and needs less permeability than an oil reservoir.

Table 9–2. Permeability values for an oil reservoir

1–10 md	poor
10–100 md	good
100–1,000 md	excellent

(Modified from Levorsen, 1967.)

Porosity and permeability in a single sedimentary rock layer are related. In general, the higher the porosity, the greater the permeability (fig. 9–7). Permeability, however, is also controlled by the grain size. The oil or gas flow in rock is most difficult through the narrow connections (*pore throats*) between the pores (fig. 9–8). The smaller the pore throats, the harder it is for the oil or gas to flow. Smaller grain sizes have smaller pore throats. Because of this, porous, coarse-grained rocks such as sandstones that have large pore throats are usually very permeable. A porous, fine-grained rock such as shale or chalk has small pore throats and little or no permeability.

The two most common petroleum reservoir rocks are sandstones and limestones. Most sandstones and limestones, however, are not reservoir rocks. A very low or no permeability sandstone or limestone is called *tight* or *tight sands*. The Spraberry trend in the Midland basin of Texas (fig. 9–9) is a series of oil fields in tight sand reservoir rocks (5 to 18% porosity and 0.01 to 3 md permeability) of the Permian age Spraberry Formation at a depth of 6,800 ft (2,072 m). The field covers an area of 2,500 square miles (6,500 km^2) and has an estimated 10 billion bbl (1.6 billion m^3) of oil in place. The Spraberry Formation is about 1,000 ft (305 m) thick with a total of 300 ft (90 m) of thin layers of fine-grained sandstone in shale and siltstone. It was originally deposited as a deep-sea submarine fan. Because the reservoir rock is fine-grained, it has an average permeability of only 0.5 md.

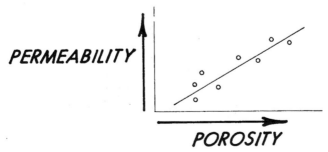

Fig. 9–7. Porosity-permeability relationship in a single reservoir rock

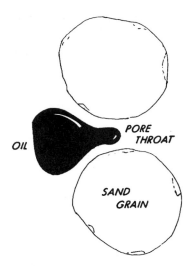

Fig. 9–8. Magnified view of sandstone showing two sand grains and the pore throat between them

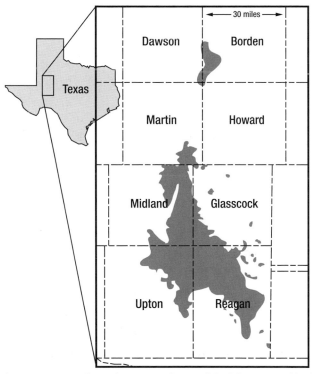

Fig. 9–9. Map of Spraberry trend. (Modified from Handford, 1981.)

Saturation

In an oil or gas reservoir, the oil or gas always shares pore spaces with water (fig. 9–10). The relative amount of the water and oil or gas sharing the pores of the reservoir will vary from reservoir to reservoir and is called *saturation*. It is expressed as a percent and always adds up to 100%. Saturation is why most oil wells pump not only oil but also water, called oil field brine.

Oil field brine is very salty water that shared the pores with the oil. The fluid that occupies the outside of the pore and is in contact with the rock surface is called the *wetting fluid*. Sandstones usually have oil in the center of the pore, and water is on the outside of the pore in contact with the sand grains. Because of this, most sandstones are *water wet* (water coats the sand grains). In contrast, limestones are usually *oil wet* (oil coats the rock surfaces). The percentage oil recovery tends to be greater in sandstone reservoirs than in limestone reservoirs. This is because the fluid in the center of the pore will flow more easily than the fluid on the outside of the pore, which is being held to the rock surface by surface tension. Below the oil-water or gas-water contact, the reservoir is saturated with 100% water (fig. 9–10).

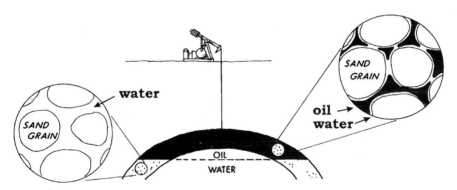

Fig. 9–10. Magnified view of oil and water saturation in the pores of a sandstone oil reservoir. Note the 100% water saturation in the pores below the oil-water contact.

Reservoir Rocks

Sandstones and limestones show a wide variety of textures and were deposited in a variety of environments. *Depositional environment*, where the rock was originally deposited, is an important indicator of how good the rock is as a reservoir rock. Most sandstones and limestones, however, are not reservoir rocks.

Some reservoir rocks, such as limestone reefs and river channel sandstones, were deposited completely encased in shale. Because the shale below is a source rock for gas and oil and the shale above is a caprock, the reef or channel is a gas and oil trap. This is called a *primary stratigraphic trap*.

Sandstone Reservoir Rocks

Sorting is the range of particle sizes in a sedimentary rock (fig. 10–1). A *well-sorted* rock is composed of particles of approximately the same size (fig. 10–1a). A *poorly sorted* rock is composed of particles with a wide range of sizes (fig. 10–1b). Sorting is the most important factor in determining the amount of original pore space in a clastic sedimentary rock. Finer-sized particles in a poorly sorted rock occupy the spaces between the larger-sized particles and reduce the pore volumes. Poorly sorted rocks are lower-quality reservoir rocks than well-sorted rocks. Well-sorted sandstones are called *clean sands.* Because sand grains are light in color, clean sandstones are usually light in color. Poorly sorted sandstones have significant amounts of silt- and clay-sized grains and are called *dirty sands.* Because silt- and clay-sized particles are usually dark in color, dirty sandstones are dark colored.

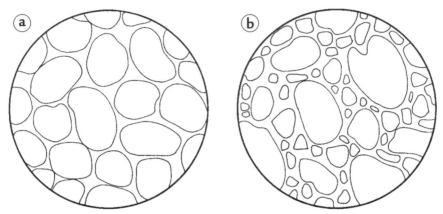

Fig. 10–1. Magnified view of particle sorting: (a) well sorted and (b) poorly sorted

Dune Sandstones

Sand dunes are formed by wind in both desert and coastal environments. The wind picks up and suspends only clay- and silt-sized particles in the air. The only other size particle moved by the wind is fine sand, which is bounced along the ground and deposited locally. Coarser grained sediments cannot be moved by the wind. Because of this, sand dunes are composed of very-well-sorted fine sand and can be excellent reservoir rock.

Sand dunes have a distinctive shape and internal structure. During periods of low wind velocity, wind and bouncing fine sand flow smoothly over any surface irregularity such as a hill (fig. 10–2a). As the wind velocity increases, an eddy current forms downwind from the crest of the hill (fig. 10–2b). The eddy current flows opposite the wind direction, causing the fine sand to pile up on the crest of the hill. Eventually, the sand becomes unstable and avalanches down the backside of the hill, depositing a layer of sand on the avalanche or slip face of the dune (fig. 10–2c). This process continues as long as the wind is blowing fast enough to maintain the eddy current.

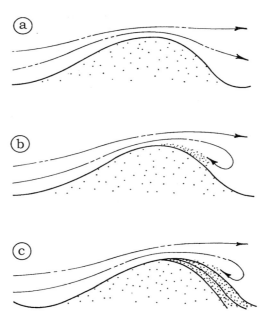

Fig. 10–2. Formation of a sand dune: (a) wind blowing at relatively low speed across a sand dune; (b) wind blowing at a faster speed, forming an eddy current on the backside of the dune; and (c) sand avalanching down the backside of the dune

The sand layers formed by the avalanching sands have a characteristic shape and are called *crossbeds* (fig. 10–3). They are steep at the top (up to 36°) and are almost horizontal at the bottom. As the wind blows, sand is eroded off the windward side of the dune and is deposited on the opposite side of the dune. This causes the dune to migrate in the direction of the slip face and the wind. Dune sand is characterized by internal crossbeds.

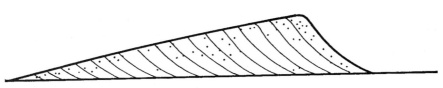

Fig. 10–3. Cross section of sand dune showing crossbeds

Ancient dune sandstones form extensive subsurface sandstone reservoirs that cover large areas. One example is the Nugget Sandstone, the primary reservoir in the Rocky Mountain overthrust belt oil and gas fields of the United States. It was deposited as sand dunes in a desert that covered much of the western part of the United States during the Jurassic time.

During the Permian time, sea level fell to expose the bottom of the North Sea, and a desert climate occurred. A large salt lake surrounded by sand dunes formed in the center of the present-day North Sea. These sand dunes are now buried by other sediments on the bottom of the North Sea and form the Rotliegend Sandstone. It underlies the southern portion of the North Sea and parts of England and Europe, where it is the reservoir for several North Sea gas and oil fields and the giant Groningen gas field in the Netherlands. In the Groningen field, porosities range from 15 to 20% and permeabilities up to 3,000 millidarcys (md).

Shoreline Sandstones

Beaches are long, narrow deposits of well-sorted sand. Waves wash any finer-grained silt and clay particles out of the beach sands. The muddy water is carried offshore where the mud settles out of the water and is deposited in deep water. Waves shape the beach into a long strip of sand.

In the south Texas coastal plain (fig. 10–4), there is a series of buried shoreline sandstones that are oil and gas reservoirs. The Yegua-Jackson beach sands of Eocene age are located inland, and the Frio-Vicksburg beach sands of Oligocene age are located on the Gulf of Mexico side.

The Clinton Sandstone oil and gas fields run north to south through the state of Ohio (fig. 10–5). The reservoir rock is a Silurian age shoreline sandstone completely encased in shale. It was first drilled in the 1890s.

River Sandstones

Most rivers flow through loops and bends (fig. 10–6) called *meanders*. These are caused by friction of water flowing along the bottom of the channel. In a meander, water must flow further along the outside of the meander bend to keep up with the water on the inside of the meander. Water on the outside of the meander bend flows faster and erodes the outer bank. Water on the inside of the meander slows down, depositing the coarsest sediment (sand) that was suspended in the water. The pattern of erosion and deposition on a meander causes the meander form to grow. The crescent-shaped, well-sorted sand bars deposited on the inside of the river meanders (fig. 10–7) are called *point bars*. After the river abandons a meander, a clay plug that will become shale is deposited in the channel.

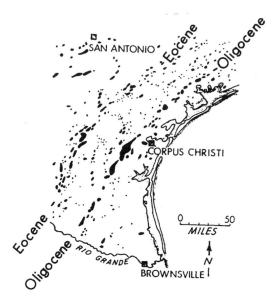

Fig. 10–4. South Texas coastal plain oil fields. (Modified from Landes, 1970.)

Fig. 10–5. Map of Clinton Sandstone gas fields, Ohio. (Modified from State of Ohio, 1966.)

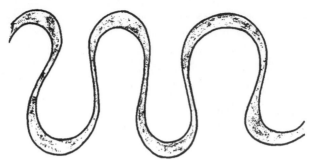

Fig. 10–6. Meandering river

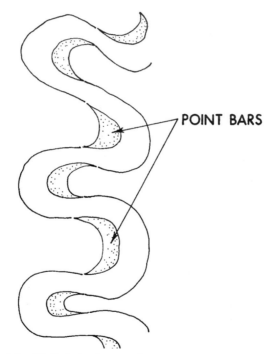

Fig. 10–7. Point bar sands

Buried point bar sandstones are often good oil and gas reservoirs. The Miller Creek field in the Powder River basin of Wyoming (fig. 10-8) is an example. It produces out of the Cretaceous age Fall River Sandstone. Expected production is 5 million bbl (0.8 million m³) of 33 °API gravity oil. The sandstone is tilted 2° to the southwest. The oil is trapped updip in the sandstone against the shale channel fill to the east. There is 35 ft (11 m) of pay with good porosity and permeability.

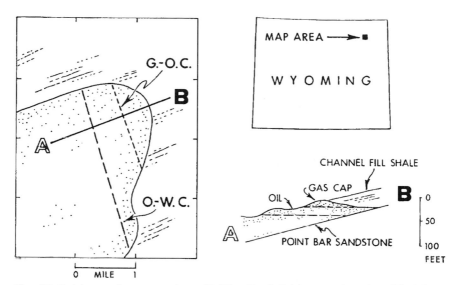

Fig. 10-8. Map and cross section of Miller Creek field, Wyoming. (Modified from Berg, 1968.)

The Bush City oil field of eastern Kansas (fig. 10-9) was formed in a river channel segment. The Pennsylvanian age sandstone reservoir rock is 13 miles (21 km) long, about ¼ mile (0.4 km) wide, and up to 50 ft (15 m) thick. The entire channel segment, which is only about 600 ft (180 m) deep, is productive.

A *braided stream* forms with interconnecting channels separated by sand and gravel bars (fig. 10-10). Braided rivers are caused by an overload of sediments that the river cannot transport. The Triassic age Sadlerochit Sandstone, the primary reservoir rock in the Prudhoe Bay field, Alaska, was deposited partially by a braided river.

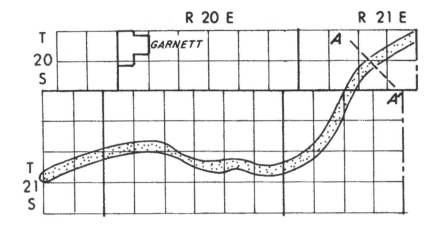

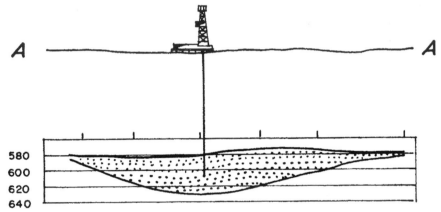

Fig. 10–9. Map and cross section of Bush City oil field, Kansas. (Modified from Charles, 1941.)

Fig. 10–10. Braided stream

Delta Sandstones

A *delta* is a mass of sediments deposited by a river flowing into a body of water, such as a lake or ocean (fig. 10–11). The river often bifurcates or divides into numerous channels, called *distributaries*, on the delta. The distributaries are located on low-lying swamps and marshes that are covered with river water during floods. Two important processes occur on a delta. The river deposits sediments, a constructive force. Waves erode the sediments, a destructive force. The geometry of the delta that forms is a result of the relative importance of river deposition and wave erosion.

A *constructive delta* (fig. 10–12a) is shaped by river deposition. Wave erosion is relatively minor. A constructive delta has lobes of sediments that protrude into the ocean. The Mississippi River Delta is an example. A *destructive delta* (fig. 10–12b) is shaped by wave erosion. It hardly protrudes from the shoreline. Wave erosion forms well-developed beaches in front of a destructive delta. The Niger River Delta and the Nile River Delta are examples.

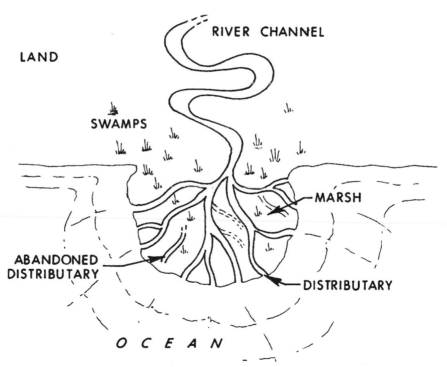

Fig. 10–11. Delta environments

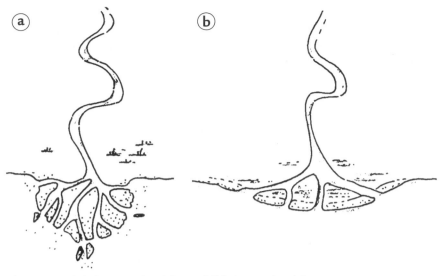

Fig. 10–12. (a) Constructive delta and (b) destructive delta

Ancient deltas are good environments for the formation and accumulation of gas and oil (fig. 10–13). Nutrient-rich river water flowing into the ocean causes large offshore algal blooms. The organic matter eventually falls to the sea bottom, forming an organic mud that is preserved as black shale in front of the delta. Sediments cover the black shale source rock as the delta is deposited out into the ocean. The overlying delta sediments contain beach and river channel sandstone reservoir rocks. As the loose shales compact, the delta surface subsides and is covered with marsh, swamp, and river deposits. The gas and oil form in the underlying black shale source rock and migrate up into the sandstone reservoir rocks.

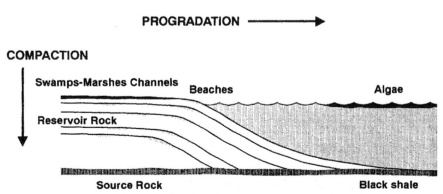

Fig. 10–13. Cross section of delta showing source and reservoir rocks

The Pennsylvanian age Booch Sandstone in Oklahoma (fig. 10–14) is a good reservoir rock. It was deposited as a southward-flowing river channel and distributary channel sandstone on a constructive delta in the Arkoma basin. The Booch Sandstone is productive where the sandstone crosses anticlines.

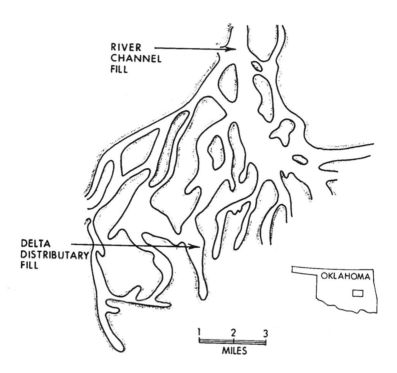

RIVER CHANNEL FILL

DELTA DISTRIBUTARY FILL

OKLAHOMA

1 2 3
MILES

Fig. 10–14. Map of Booch Sandstone, Oklahoma. (Modified from Busch, 1974.)

The Bell Creek oil field reservoir in Montana is a destructive delta. During the Cretaceous, the area bordered by Montana, Wyoming, and South Dakota was occupied by the water-filled Powder River basin. The Black Hills uplift was located to the east. A meandering river flowed out of the Black Hills and emptied into the Powder River basin, forming a destructive delta (fig. 10–15a). The beach and channel sands, called the Muddy Sandstone, are encased in shale and form the reservoir rock for the Bell Creek oil field (fig. 10–15b). The producing sandstone at a depth of 4,500 ft (1,400 m) is 20 to 40 ft (6 to 12 m) thick. The field will ultimately produce more than 200 million bbl (32 million m^3) of oil.

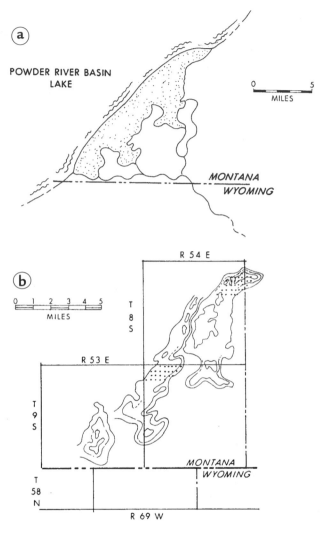

Fig. 10–15. (a) map showing shoreline of Powder River basin, Montana, during Cretaceous time and (b) map of the Bell Creek oil field, Montana. (Modified from McGreger and Biggs, 1970.)

A process that occurs on constructive deltas is *delta switching* (fig. 10–16a). After a delta progrades out into a basin, the river becomes more inefficient as the river must flow over the flat delta to empty into the ocean. During a flood, the river can break through to a shorter, more efficient route to the sea. It will then abandon its older, less efficient route. A new delta will form to the side of the old delta, and the old delta will be eroded back by waves and covered with shallow water. The abandoned delta will also subside due to sediment compaction. The process of delta switching by a river can deposit a thick mass of deltaic sediments along the margin of a basin. On a destructive delta, switching is confined to the major river channel that jumps about the shoreline (fig. 10–16b).

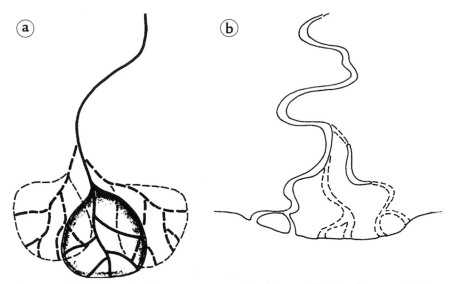

Fig. 10–16. (a) Delta switching on a constructive delta and (b) distributary switching on a destructive delta

The Mississippi River Delta region of Louisiana will produce more than 22 billion bbl (3.5 billion m^3) of oil. Most of the offshore gas and oil production in the Gulf of Mexico is located in shallow water to the southwest of the modern Mississippi River delta. Several million years ago, sea level was lower by about 300 ft (91 m). At that time, the Mississippi River Delta switched back and forth off southwestern Louisiana (fig. 10–17). Rising sea level covered these Miocene, Pliocene, and Pleistocene age deltas. Much of the offshore production is from sandstone reservoirs in the ancient deltas.

By using carbon-14 (^{14}C) age dating, it has been determined there have been six previous deltas (fig. 10–18) of the Mississippi River during the past 5,000 years. Each delta prograded out into the Gulf of Mexico and was abandoned by delta switching. A new route, the Atchafalaya River route by Morgan City, is forming today. This route is only 140 miles (225 km) to the Gulf. The present route by Baton Rouge and New Orleans is inefficient and 300 miles (483 km) long. Some time in the near future, the Mississippi River will abandon its present route and switch down the Atchafalaya River route.

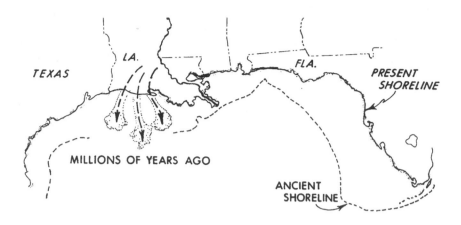

Fig. 10–17. Map of ancient offshore Mississippi River deltas in Louisiana

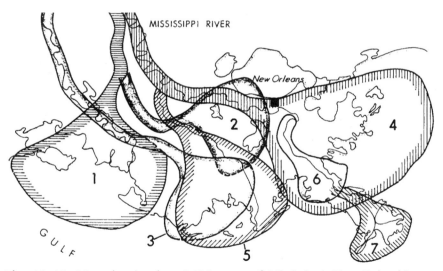

Fig. 10–18. Map showing last 5,000 years of Mississippi River Delta history. (Modified from Kolb and Van Lopik, 1966.)

A map of the Niger River Delta, a destructive delta in central western Africa, shows the present and ancient shorelines (fig. 10–19). The deposition of the delta by the Niger River into the Atlantic Ocean is shown by the progression of shorelines with time. Ancient shorelines (5 and 2 million years ago) located offshore from the present shoreline occurred when sea level was lower. The Niger River Delta is one of the world's greatest oil-producing areas, with more than 41 billion bbl (6.5 billion m³) of recoverable oil in ancient beach and river channel sandstones.

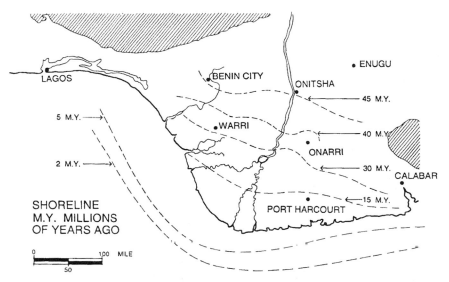

Fig. 10–19. Map of present-day and ancient shorelines and their ages on the Niger River Delta, Nigeria. (Modified from Burke, 1972.)

Granite Wash

Granite wash is a potential reservoir rock formed by the weathering of granite. The rock granite is composed of large, well-sorted, sand-sized mineral grains and weathers to form well-sorted sandstone that can be very thick. After the granite and granite wash have been buried in the subsurface, oil and gas can form in source rocks at a lower elevation. The oil and gas can then migrate up and into the granite wash (fig. 10–20). Granite wash reservoir rocks are common in the subsurface of southern and western Oklahoma and the Texas panhandle. The Elk City field of Oklahoma produces from granite wash where it has been uplifted by an anticline.

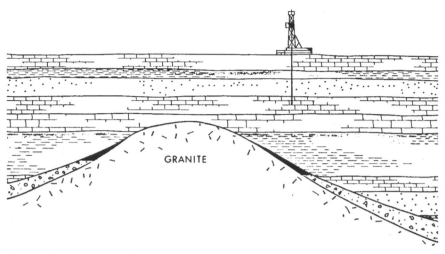

Fig. 10–20. Migration of gas and oil from source rock up into a granite wash reservoir

Carbonate Reservoir Rocks

The sedimentary rock limestone is common in the ancient rock record. This is because lime ($CaCO_3$) is secreted by many animals and plants that live in the ocean and can also precipitate out of seawater. Most limestones were deposited on the bottom of shallow, tropical seas that covered the land many times in the geological past. *Carbonates* are limestones and a related sedimentary rock called dolomite.

Reefs

Reefs are mounds of shells. All reefs have a wave-resistant, calcium carbonate framework of overlapping organic branches formed by a plant or animal. The other plants and animals live in the protection of the framework. Modern reefs often have corals as the framework. During the Paleozoic and Mesozoic eras, corals were not as important as they are today, and sponges, calcareous algae, clams called rudistids, and other organisms formed the reef framework. Modern reefs grow only in clear, shallow, warm waters. There are several types of reefs based on shape (fig. 10–21).

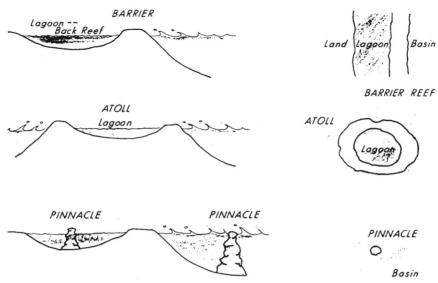

Fig. 10–21. Reef types in cross section and map view

Barrier reefs grow parallel to a shoreline but are separated from the land by a lagoon. Both modern and ancient barrier reefs tend to be large. The Great Barrier reef off northeastern Australia is more than 1,000 miles (1,600 km) long. An *atoll* is a circular or elliptical reef surrounding a central lagoon. Atolls are common in the Pacific Ocean. *Pinnacle reefs* are smaller, steep, cone-shaped reefs.

Ancient reefs are prolific petroleum reservoirs, especially in North America. The reef rock has the most original pore spaces. These spaces are often enhanced in the subsurface when freshwaters percolate through the pores and dissolve the limestone. In the lagoon, limestone mud called *micrite* was deposited. It is not reservoir rock. If the ancient reef is covered with a shale or salt caprock, it forms a primary stratigraphic gas and oil trap.

During the Permian period, the climate of the world was hot and dry, a desert climate. The eastern and central portions of North America were exposed above sea level. Large areas were covered with sand dunes and salt flats. The last area covered with water was three deep tropical-water basins located in west Texas, New Mexico, and Mexico (fig. 10–22). They are the Midland, Marfa, and Delaware basins, known as the *Permian basins*. During the late Pennsylvanian and throughout most of the Permian time, thick limestones were deposited in and between these basins. A barrier reef grew along the margin of the Delaware basin. This Capitan reef, more than 600 ft (180 m) thick, is exposed today in the Guadalupe Mountains of west

Texas (plate 10–1). To the north of the Capitan reef is the subsurface Abo reef (fig. 10–23). Along the top of the buried Abo reef are numerous oil fields. One of the largest of these fields is the Empire Abo field (fig. 10–24), which will produce more than 250 million bbl (40 million m^3) of oil. The caprock is shale. It is bounded on the north by impermeable micrite limestone deposited in the former lagoon.

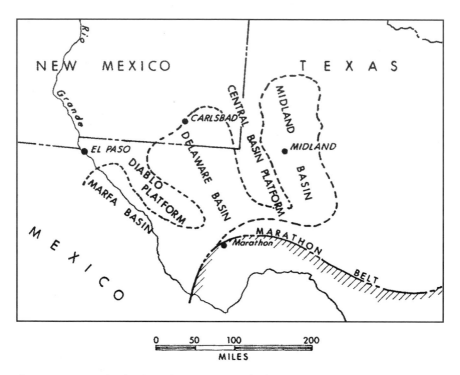

Fig. 10–22. Permian basins of west Texas and New Mexico

Plate 10-1. Capitan reef on Guadalupe Mountains, Midland basin, Texas

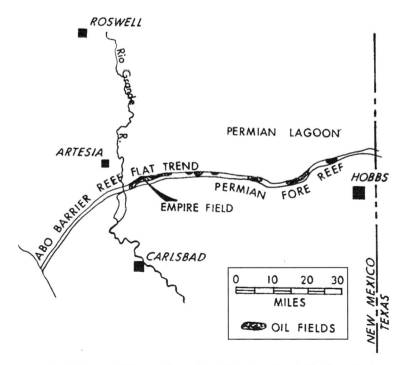

Fig. 10–23. Map of Abo reef trend and Empire Abo field, New Mexico. (Modified from LeMay, 1972.)

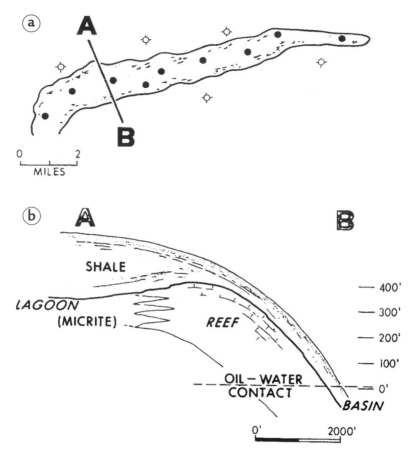

Fig. 10–24. Empire Abo field, New Mexico (a) map and (b) cross section

During the Silurian period, two large basins, the Illinois and Michigan basins, occurred in the north central United States (fig. 10–25). A barrier reef grew along the northern edge of the Michigan basin. This barrier reef, now exposed on the surface of the ground, has been unproductive. However, the hundreds of smaller, pinnacle reefs that grew basinward of the barrier reef are still buried in the subsurface and are productive (fig. 10–26). The pinnacle reefs are hundreds of feet to several miles across and hundreds of feet thick. Each is overlain by a salt layer caprock. More than 1,200 of the pinnacle reefs have been drilled in Michigan. Of these, more than 900 are productive for gas or oil. The Belle River Mills gas field near Detroit (fig. 10–27a) is an example of a Silurian pinnacle reef field. It is 3 miles (5 km) long, 1 mile (1.6 km) wide, and 400 ft (120 m) high (fig. 10–27b). This field will eventually produce 50 Bcf (1.4 billion m^3) of natural gas.

Fig. 10–25. Map of Illinois and Michigan basins

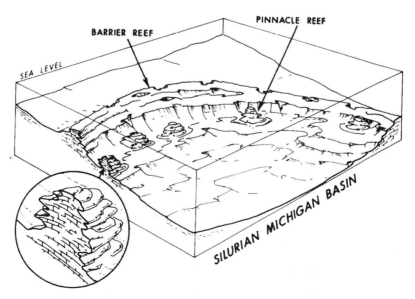

Fig. 10–26. Pinnacle reefs growing basinward from the barrier reef of the Michigan basin

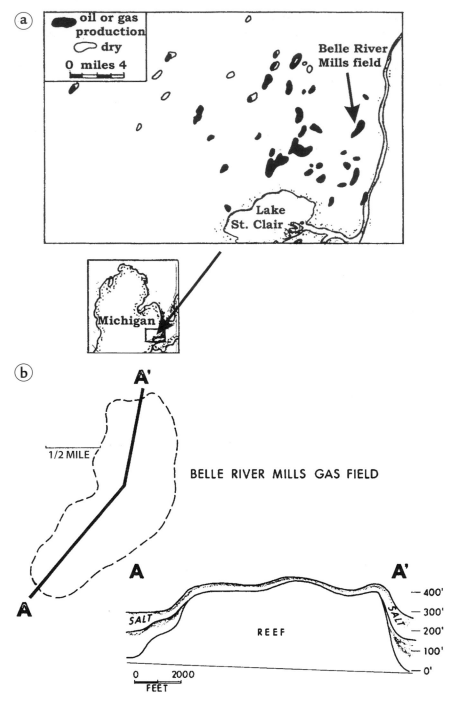

Fig. 10–27. Map showing (a) location of pinnacle reefs near Detroit, Michigan, and (b) cross section of Belle River Mills gas field. (Modified from Gill, 1985.)

A large portion of conventional oil production in Canada comes from buried Devonian age atolls and barrier reefs, located primarily in Alberta (fig. 10–28). Leduc was the first of these to be discovered, in 1947. It was located by seismic exploration but was originally thought to be an anticline until it was drilled. The Redwater oil field of Alberta, the second largest conventional oil field in Canada, was discovered in 1948. A map (fig. 10–29a) shows the elliptical shape of the large, subsurface atoll that covers 200 square miles (520 sq km) and is 15 miles (24 km) across. The reef rock, the reservoir rock, is located on the outside, whereas the lagoonal micrite limestone that is not reservoir rock is located on the inside of the reef (fig. 10–29b). The reef is tilted down to the west. The oil occurs in the highest part of the reservoir rock located to the east. The field will eventually yield 850 million bbl (135 million m^3) of oil.

Fig. 10–28. Map of Devonian reefs, Alberta. (Modified from Procter, Taylor, and Wade, 1983.)

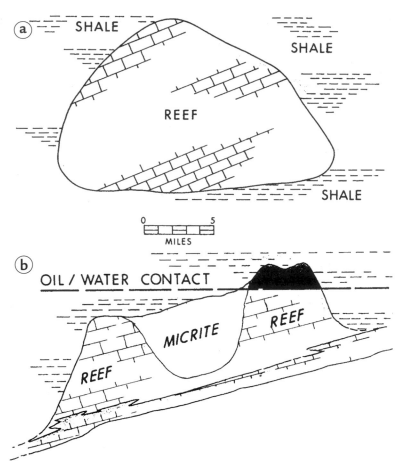

Fig. 10–29. Redwater oil field, Alberta, (a) map and (b) cross section. (Modified from Jardine, Andrews, Wishart, and Young, 1977.)

Limestone Platforms

A *limestone platform* is a large area covered with shallow tropical seas where limestones are being deposited. The Bahaman banks, a modern example of a limestone platform, is located to the southeast of Florida. Sand-, silt- and clay-sized limestone particles are being deposited and reworked by waves and currents. On some limestone platforms, strong tidal currents flow back and forth. The tropical water is saturated with calcium carbonate. As it flows over the limestone platform, calcium carbonate precipitates out of the water forming sand- and silt-sized spheres called *oolites*. Limestones composed of oolites are called *oolitic limestones*. They have excellent original porosity and can be good reservoir rock.

The Magnolia oil field in Arkansas (fig. 10–30) was discovered in 1938 by surface mapping and seismic exploration. The trap is an anticline about 6 miles (10 km) long and 1½ miles (2.5 km) wide. It produces from the Reynolds Oolite Member of the Jurassic age Smackover Formation, which is 300 ft (91 m) thick. Production has totaled 140 million bbl (22 million m³) of oil.

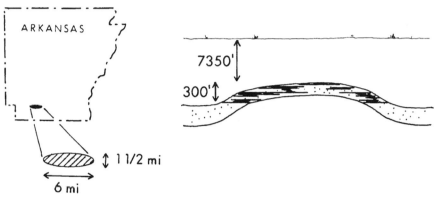

Fig. 10–30. Map and east-west cross section of Magnolia oil field, Arkansas

A *rimmed platform* is formed by a beach of broken seashells and/or oolites deposited by waves along the platform margin. The rim can be good reservoir rock. During the Permian time, the Midland, Marfa, and Delaware basins in west Texas and New Mexico were separated by the Diablo and Central basin platforms. A rim was deposited along the eastern margin of the Central basin platform (fig. 10–31). This rim forms the reservoir rock for many of the major west Texas oil fields such as Yates, McElroy, and Means (fig. 10–32). The McElroy field will produce 601 million bbl (96 million m³) of oil.

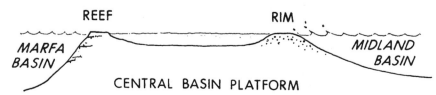

Fig. 10–31. East-west cross section of Central basin platform, west Texas during Permian time

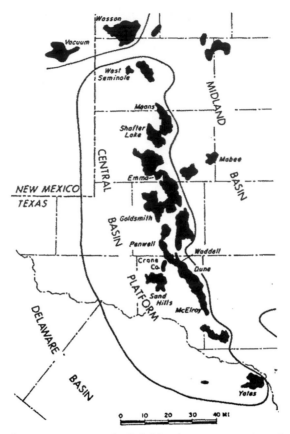

Fig. 10-32. Map of oil fields along eastern rim of Central basin platform, west Texas. (Modified from Harris and Walker, 1990.)

Karst Limestone

Limestone is very soluble in freshwater, especially in warm, humid climates. This is because rain and soil water absorb carbon dioxide, a common gas, that forms carbonic acid that dissolves limestones. A highly dissolved limestone is called *karst limestone*, which can have excellent porosity and permeability. Some karst limestones have caves, but more often they have solution pores up to several inches in diameter called *vugs*. The great Middle East oil fields are primarily in limestone reservoir rocks. Their excellent porosity and permeability come from vugs in the karst limestone.

The prolific Golden Lane oil fields, both onshore and offshore, near Tampico, Mexico (fig. 10–33), produce from karst limestone reservoir rock. The reservoir rock is a large, Cretaceous age atoll. The limestone atoll (El Abra reef) was buried in the subsurface by sediments. Later, the sedimentary rocks covering the reef were eroded, exposing the reef. The top of the reef was dissolved, forming cavernous pores. After that, the reef was again buried in the subsurface by sediments. Finally, the oil migrated up into the karst limestone. The fields were discovered in the early 1900s by drilling on oil seeps. One field on the trend, Cerro Azul, has 1¼ billion bbl (200 million m^3) of recoverable oil. The El Abra reef limestone is extremely porous and permeable. One well (Potero del Llano #4) on the reef produced a total of more than 115,000,000 bbl (18 million m^3) of oil over 40 years.

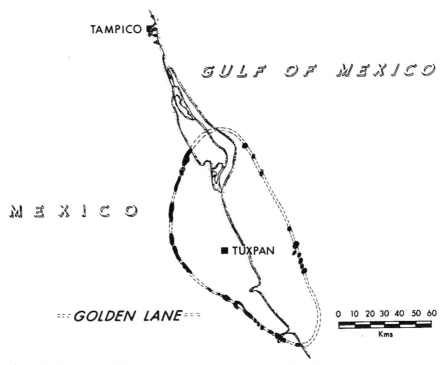

Fig. 10–33. Map of Golden Lane oil fields, Mexico (Modified from Viniegra and Castillo-Tejero, 1970.)

The Dollarhide Pool oil field, located north of Odessa, Texas, is in a limestone cave that ranges from 3 to 16 ft (1 to 5 m) high. The reservoir rock is the Silurian age Fusselman Limestone. Fifteen wells have been drilled into this cave.

Limestones caves cannot be detected from the surface using the seismic method and are usually found by chance when drilling into karst limestone. Karst limestone, however, often occurs directly below a subsurface angular unconformity. An angular unconformity is an ancient erosional surface that was exposed on the surface. Any limestone directly below it is often highly dissolved. In the Sooner trend of oil fields to the southwest of Oklahoma City, Oklahoma, one of the reservoir rocks is the Hunton Limestone, a karst limestone directly under an angular unconformity.

Chalk

Chalks are extremely fine-grained limestones composed of microfossil shells. Single-celled animals called foraminifera (plate 4–1a) and plants called coccolithophores grow floating in tropical seas. Both have shells of calcium carbonate. When they die, their shells fall to the ocean bottom. If they are not dissolved or diluted by other sediments, a chalk is formed. During the Cretaceous time, several chalks were deposited that include the Austin Chalk of Texas and Louisiana; the Niobrara Chalk of South Dakota, Nebraska, and Kansas; the Selma Chalk of Alabama; the chalk cliffs of Dover, England; and the Ekofisk Chalk underlying the North Sea. Because of the very fine-grained sediment size, chalks can have very high porosities but extremely low permeabilities. They can be productive, however, if they are naturally fractured.

Dolomite

Dolomite $[CaMg(CO_3)_2]$ is a sedimentary rock that forms primarily by the alteration of a preexisting limestone. Magnesium-rich waters percolating through subsurface limestone ($CaCO_3$) replace calcium (Ca) atoms with magnesium (Mg) atoms to form dolomite. Often entire beds of permeable limestone have been transformed into dolomite. In impermeable limestones such as micrite or crystalline limestone, dolomite formation can be restricted to areas along fractures through which the magnesium waters were able to flow.

Dolomite is difficult to distinguish from limestone in the field. Dolomite and limestone have a similar crystal shape, color, and hardness. Limestone will bubble in cold, diluted acid. Dolomite, however, will bubble only in hot, concentrated acid. Rhomb-shaped crystals that will not bubble in cold, diluted acid are the best field indication of dolomite. In the laboratory, they are easy to identify.

Dolomite is often a good reservoir rock. Limestones tend to lose porosity when they are buried deeper due to compaction. Dolomites, however, are harder and less soluble than limestones and retain porosity. The dolomite crystals are often larger then the limestones particles that they replace, increasing pore and pore throat sizes and the permeability of the rock. The dolomitized oil field reefs of western Canada have an average of 1% more porosity and 10 times greater permeabilities than the limestone reefs.

The Jay field in northwest Florida is located on an anticline at 15,000 ft (4,500 m) deep. Only about half the structure is productive. The original Jurassic age Smackover Limestone of dense micrite is not reservoir rock on the unproductive side of the anticline (fig. 10–34). On the productive side, dolomite has replaced micrite. There are 730 million bbl (116 million m^3) of 51 °API gravity oil in place. Ultimately, more than 346 million bbl (55 million m^3) of oil will be produced from the Jay field. The average net pay is 95 ft (29 m).

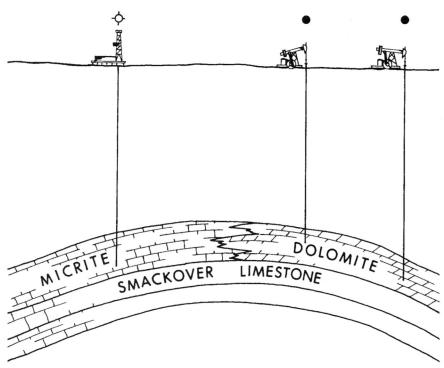

Fig. 10–34. Cross section of Jay field, Florida, showing a dry hole being drilled into the western, nonreservoir, limestone part of the field. (Modified from Ottman, Keyes, and Ziegler, 1976.)

The Scipio-Albian field, discovered in 1957, is the largest oil field in Michigan. It is located in dolomitized Trenton Limestone at a depth of about 3,550 ft (1,080 m). Movement along a large strike-slip fault in the basement (fig. 10–35) folded and fractured the Trenton Limestone above it. The fractures allowed magnesium-rich waters to flow into the limestone and change it into dolomite. Because of this, the field is no wider than a mile (1.6 km) as it follows the subsurface fault (fig. 10–36). The caprock is the Utica Shale that overlies the dolomite. The Scipio-Albian field will ultimately produce more than 150 million bbl (24 million m³) of oil and 200 Bcf (6 billion m³) of gas. The discovery well (#1 Houseknecht) was drilled by the landowner, Ferne Bradford (maiden name of Houseknecht), on the advice of a psychic friend, "Ma" Zulah Larkin. The nearest oil production was more than 50 miles (80 km) away when the well was spudded in 1955. Twenty months later, in 1957, the well encountered the oil reservoir and came in at the rate of 140 bbl of oil per day with considerable gas. The Continental Oil Company had drilled and tested the Scipio oil reservoir in 1943 but decided to plug the well and completed in a more shallow gas zone. The adjacent Stoney Point field, located 5 miles (8 km) to the east and also in dolomite reservoir rock, was not discovered until 1982.

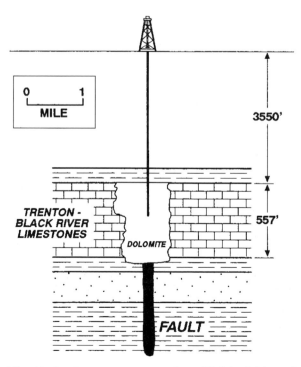

Fig. 10–35. East-west cross section of Scipio field, Michigan

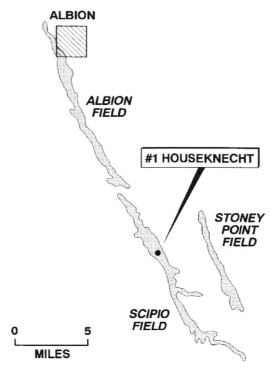

Fig. 10–36. Map of Scipio-Albion trend, Michigan. (Modified from Hurley and Budros, 1990.)

The Yates oil field in west Texas, just to the west of the Pecos River, is a giant oil field that produces from both karst limestone and dolomite. The primary reservoir rock is Permian age Grayburg Dolomite that contains caverns that are up to 21 ft (6 m) high. These were formed when the area was a tropical island, and freshwaters percolated through the limestone. The dolomite formed later. The caprock is salt in the overlying Seven Sisters Formation. There were oil seeps on the Yates Ranch, and a large anticline was mapped on the surface (fig. 10–37). The field was discovered in 1926 by drilling to 997 ft (304 m) where the well blew out. The well was later deepened to 1,032 ft (315 m) where it tested at 72,000 bbl (11,500 m^3) of oil per day. One well on the Yates oil field had an initial production test that yielded more than 8,500 bbl (1,350 m^3) of oil per hour. The field has already produced more than 1.3 billion bbl (207 million m^3) of oil and is expected to produce almost another billion barrels with a recovery of 50% of the oil in the reservoir.

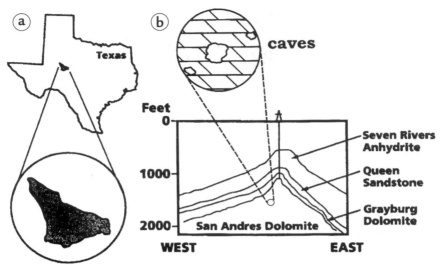

Fig. 10–37. Yates field, west Texas (a) map and (b) cross section. (Modified from Craig, 1988.)

Fractured Reservoirs

Any fractured rock can be a reservoir rock. Even fractured basement rock can be reservoir rock. When an anticline or dome is originally folded, the brittle basement rock is often fractured along the crest of the fold. Oil and gas form in source rock along the flanks of the anticline at a lower elevation than the crest (fig. 10–38) and migrate up into the fractured basement rock.

The Wilmington oil field of California is formed by an anticline and produces out of seven sandstone reservoir rocks and the top of the fractured, metamorphic basement rock. Point Arguello oil field in offshore southern California (fig. 10–39) produces from an anticline in the Miocene age Monterey Formation. The Monterey Formation is composed of alternating layers of fractured chert and shale at a depth of 6,000 to 8,000 ft (1,800 to 2,450 m) below the seafloor. Net pay is about 700 ft (210 m). The average porosity is 15%, and the permeability varies from 0.1 to 3,000 md due to fractures. Without the fractures, the Monterey Formation would not be a reservoir rock. The field will eventually produce 300 million bbl (48 million m³) of oil.

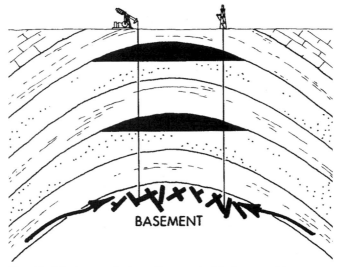

Fig. 10–38. Accumulation of gas and oil in fractured basement rock on an anticline or dome

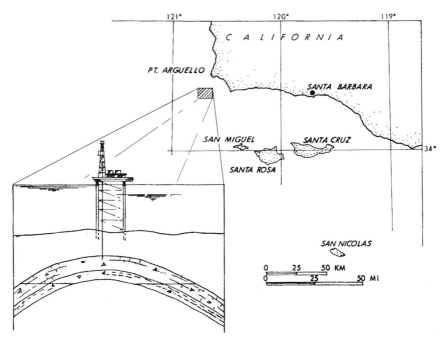

Fig. 10–39. Map and cross section of Point Arguello oil field, offshore California. (Modified from Mero, 1991.)

In the subsurface, to the east of Austin, Texas, there are several ancient volcanoes that produce oil. When the volcanoes erupted during the Cretaceous time, the area was covered with Gulf of Mexico waters. The submarine lava spread out along the ocean bottom, forming mushroom-shaped deposits of basalt. The surface of the basalt was highly fractured and weathered by seawater, giving it porosity and permeability. Sediments then covered the volcanoes that are now buried in the Texas coastal plain. Oil later migrated up into the weathered basalt that is called serpentine by the drillers. Fourteen of these fields, such as the Lyton Springs field (fig. 10–40), have been found.

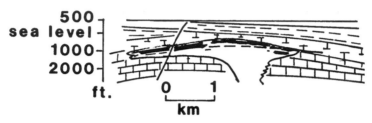

Fig. 10–40. Cross section of Lyton Springs field, Texas. (Modified from Collingwood and Retter, 1926.)

Chalk is a very fine-grained limestone. It has high porosity but low permeability because of its small pores and pore throats. However, where chalk is naturally fractured with joints, it is permeable. The Austin Chalk of Texas and Louisiana, the Niobrara Chalk of Colorado and Kansas, and the Ekofisk Chalk that underlies the North Sea are productive because of the fractures.

Petroleum Traps

A trap concentrates gas and oil in a portion of a reservoir rock. Two types of petroleum traps are structural and stratigraphic. *Structural traps* are formed by deformation of the reservoir rock, such as a fold or fault. *Primary stratigraphic traps* are formed by a reservoir rock, such as a river channel sandstone or limestone reef that is completely encased in shale. The shale below it is the source rock, and the shale above and around it is the caprock. A *secondary stratigraphic trap* is formed by an angular unconformity. A *combination trap* has both structural and stratigraphic elements.

Structural Traps

Anticlines and domes

Many anticlines are asymmetrical, and the crest migrates with depth (fig. 11–1). A well that is drilled on-structure on the surface could be too far off-structure to encounter petroleum at reservoir rock depth. The Salt Creek oil field in the Powder River basin of Wyoming is formed by a large, asymmetrical anticline. The field has five producing zones, all sandstones, of which the Cretaceous age Second Wall Creek Sandstone is the most important. The elliptical-shaped anticline is steepest on the west (where the contours on the structural map are closest together) and gentlest on the east (fig. 11–2). A large oil seep was located over the trap, leading to its discovery in 1907. The field will produce 766 million bbl (122 million m³) of oil and 750 Bcf (21 billion m³) of gas.

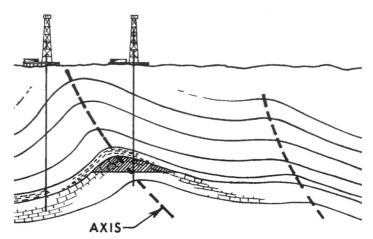

Fig. 11–1. Asymmetric anticline

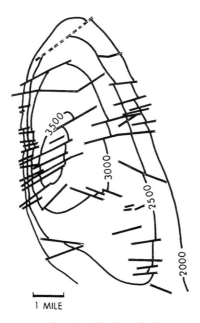

Fig. 11–2. Structural map (depth in ft above sea level) on the reservoir rock, Salt Creek field, Wyoming. The straight lines are faults. (Modified from Barlow and Haun, 1970.)

As reservoir rocks are deformed into an anticline or dome trap, they are often cut by faults (fig. 11–2). The faults can sometimes be barriers to fluid flow and are called *sealing faults*. Sealing faults will divide the structure into individual producing compartments or pools (fig. 11–3). Petroleum production from one side of the sealing fault will not affect production on the other side. Two observations can determine if a fault is a sealing fault and has divided the structure into separate producing compartments. Oil-water and gas-oil contacts in reservoirs are level. If the contacts are at different elevations in the same reservoir rock on opposite sides of the fault, the fault is a sealing fault. If there are different fluid pressures at the same elevation on opposite sides of the fault, the fault is a sealing fault. Many faults in deltas and in coastal and offshore areas are sealing faults because shales are very common and are soft. When the fault moves, the soft shale is smeared along the fault plane to form a clay or shale smear that acts as a seal.

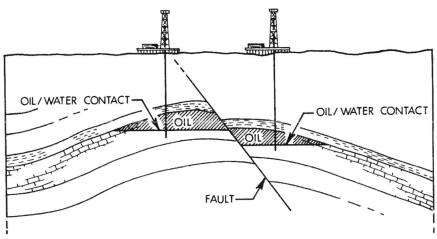

Fig. 11–3. Sealing fault cutting an anticline

The Wilmington field, just to the southeast of Los Angeles, is the largest oil field in California. The trap is a large anticline, 11 miles (18 km) long and 3 miles (5 km) wide (fig. 11–4). The anticline is overlain by an angular unconformity at 2,000 ft (610 m) depth with flat sedimentary rocks above (fig. 11–5). There are seven sandstone-producing zones along with the fractured metamorphic basement that also produces. Miocene to Pliocene age turbidity currents deposited the sandstones.

The anticline is cut by seven sealing faults that separate it into seven producing compartments (figs. 11–4 and 11–5). The three compartments under land were drilled and developed first in the late 1920s. The four southwestern compartments under Long Beach Harbor were never effectively developed until the completion of four artificial islands in the harbor for drilling and production in 1965. The Wilmington field will ultimately produced 2.75 billion bbl (450 million m³) of primarily heavy, sour oil and 1 Tcf (28 million m³) of gas.

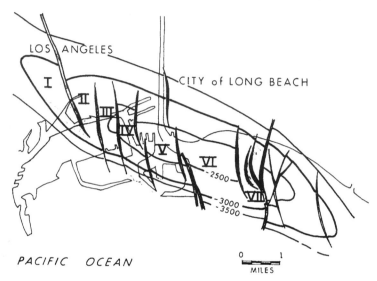

Fig. 11–4. Structural map (depth in feet below sea level) on a sandstone reservoir rock, Wilmington oil field, California. (Modified from Mayuga, 1970.)

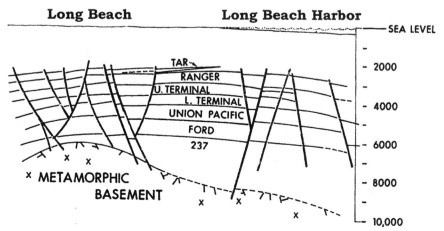

Fig. 11–5. Cross section along the length of the Wilmington oil field, California, showing sandstone reservoir rocks (depth in feet) and faults. (Modified from Mayuga, 1970.)

Growth faults and rollover anticlines

In an area where large volumes of loose sediments are rapidly deposited, such as deltas and coastal plains, a unique type of fault, a *growth* or *down-to-the-basin fault*, forms. The fault is always parallel to and located just inland from the shoreline (fig. 11-6). The weight of the sediments being rapidly deposited along the shoreline pulls the basin side of the fault down. A growth fault is similar to a giant slump. It is called a growth fault because it moves as the sediments are being deposited. This is in contrast to other faults such as normal, reverse, and strike-slip faults that occur in sedimentary rocks millions of years old. It is also called a down-to-the-basin fault, because the basin side is moving down.

Growth faults are unique in that they have a curved fault plane that is concave toward the basin (fig. 11-6). The fault is steep near the surface and becomes less steep with depth. Faults in solid rocks tend to have straight fault planes. The growth fault, however, occurs in loose sediments. As the sediments are buried, weight compacts the sediments, causing the steep, near-surface fault plane to become flatter with depth. Another unique and very important characteristic is a *rollover anticline*, a large, potential gas and oil trap, which commonly occurs on the basin side of the growth fault. This anticline is caused by the curved fault plane that is almost horizontal with depth (fig. 11-7a). As the growth fault moves, a gap forms between the near-surface sediments on either side of the fault (fig. 11-7b). The relatively loose sediments roll over and fall into the gap to form the anticline (fig. 11-7c).

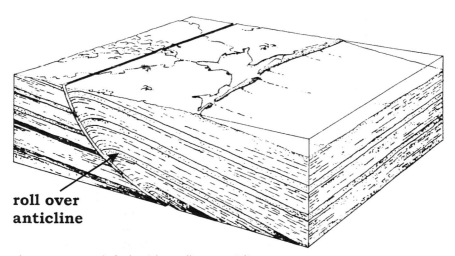

roll over anticline

Fig. 11-6. Growth fault with a rollover anticline

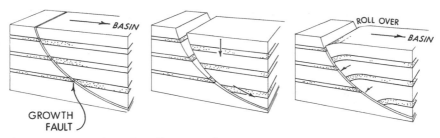

Fig. 11–7. Formation of a rollover anticline

Rollover anticlines are prolific petroleum traps along the Gulf of Mexico coastal plain and in the Mississippi and Niger River deltas. Growth faults will become inactive and buried in the subsurface as the shoreline is deposited out into the basin. Ancient growth faults are found both inland and offshore (fig. 11–8). Offshore growth faults were active during times of lower sea levels.

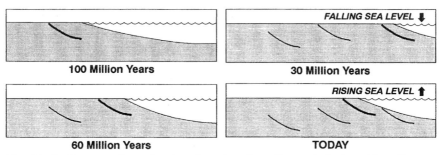

Fig. 11–8. Cross sections showing the migration of growth faults with sediments being deposited in a basin with rising and falling sea levels. The thick line is the active growth fault, and the thin lines are the inactive growth faults.

The Vicksburg oil and gas field trend in the South Texas coastal plain is along a buried, inactive growth fault (fig. 11–9). Rollover anticlines form a series of gas and oil fields that parallel the Vicksburg fault on the Gulf of Mexico side. The trend will produce more than 3 billion bbl (0.5 billion m^3) of oil and 20 Tcf (0.5 trillion m^3) of gas. Exploration and production started in 1934 with the discovery of the Tom O'Conner field. The trap is a rollover anticline, 10 by 3 miles (16 by 5 km) in size (fig. 11–10) on the down-thrown (Gulf of Mexico) side of the Vicksburg fault. The field has several Oligocene to Pliocene age sandstone reservoirs. Because the coastal plain sands are relatively young and have not been buried very deeply, porosities average 31% and permeabilities range up to 6,500 millidarcys (md). The

Tom O'Conner field will ultimately produce more than 500 million bbl (80 million m³) of oil and 1 Tcf (28 million m³) of gas.

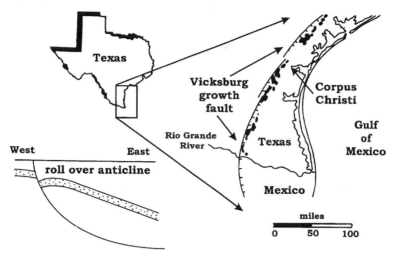

Fig. 11–9. Map and cross section of the Vicksburg fault trend, Texas Gulf Coast. (Modified from Stanley, 1970.)

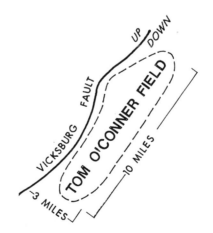

Fig. 11–10. Map of the Tom O'Conner field, Texas. (Modified from Mills, 1970.)

Rollover anticlines on growth faults are often cut by smaller faults called *secondary faults* (fig. 11–11) that displace the reservoir rocks. These secondary faults often are sealing faults that divide the anticline into numerous smaller reservoirs.

A cross section of the Niger River Delta, Nigeria (fig. 11–12) shows the underlying Akata Formation, the black shale source rock. It was deposited in deep waters off the delta but has been covered by younger delta sediments. The reservoir rocks are in the overlying Agbada Formation, which was deposited in shallow water and on the delta. The formation consists primarily of shale, but it contains numerous distributary and river channel and beach sandstone reservoir rocks. The traps are rollover anticlines that are often cut by secondary faults that are sealing faults. The only active growth faults are the ones located just inland from the present shoreline. The growth faults that are buried and located further inland were active when the shoreline was inland. The inactive growth faults located offshore were active when sea level was lower (fig. 11–12).

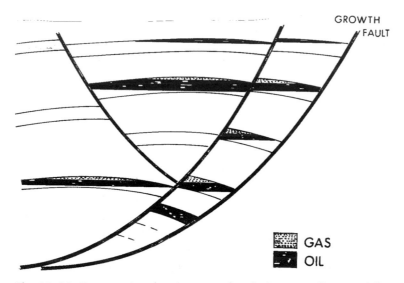

Fig. 11–11. Cross section showing secondary faults on a rollover anticline

Fig. 11–12. North-south cross section of the Niger River Delta

On the top of the delta is the Benin Formation. It consists of river channel sands and marsh and swamp deposits that are still being deposited today. It is estimated that the delta contains 34.5 billion bbl (5.5 billion m^3) of recoverable oil and 94 Tcf (2.7 billion m^3) of recoverable gas. Most fields are relatively small. There are more than100 fields, each with more than 50 million bbl (8 million m^3) of recoverable oil in the delta. However, only one has a billion bbl (160 million m^3) of recoverable oil.

Hibernia, the largest conventional oil field in eastern Canada, is a rollover anticline trap located offshore from Newfoundland (fig. 11–13a). The Murre Fault, a growth fault, was active when sea level was lower, and the area was exposed. The rollover anticline is cut by several secondary faults (fig. 11–13b). The reservoir rocks are the Cretaceous age Avalon (7,200 ft or 2,200 m deep) and Hibernia (12,200 ft or 3,700 m deep) sandstones. The Avalon Sandstone averages 20% porosity and 220 md permeability, whereas the Hibernia Sandstone averages 16% porosity and 700 md permeability. The field will produce 1.2 billion bbl (190 million m^3) of 32 to 35 °API gravity, sweet oil.

Drag folds

Drag folds are formed by friction generated along a fault plane when a fault moves. Friction causes the beds on either side of the fault to be dragged up on one side and down on the other side of the fault (fig. 11–14). Most mountain ranges on land were formed by compressional forces, and the rocks display compressional features such as folds, reverse faults, and thrust faults. The thrust faults occur in zones called *overthrust* or *disturbed belts* that parallel the mountain ranges. The most drilled of these overthrust belts is the Rocky Mountain or Western overthrust or disturbed belt of the United States and Canada (fig. 11–15). Thrust faulting occurred from the Cretaceous through Eocene time during the formation of the mountains.

A west-to-east cross section shows the deformation of the earth's crust (fig. 11–16). Several of the thrust faults moved tens of miles horizontally. In Wyoming and Utah, the 1,000 ft (305 m) thick Jurassic age Nugget Sandstone has been deformed into large subsurface drag folds along the thrust faults. These are the targets for Rocky Mountain overthrust drilling.

The Painter reservoir field, Wyoming, discovered in 1977, is a typical overthrust field. A more detailed, east-west cross section (fig. 11–17) shows the numerous thrust faults and the drag fold trap in the Nugget Sandstone at 10,000 ft (3,050 m) below the surface. A larger-scale, east-west cross section of the field (fig. 11–18) shows more faulting and the complex deformation in more detail. The oil and gas pay zone is more than 770 ft (235 m) thick.

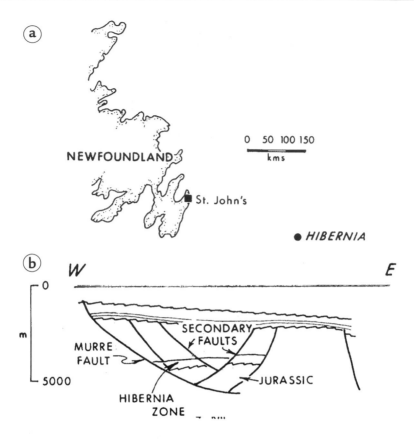

Fig. 11–13. (a) Map and (b) cross section of the Hibernia oil field, offshore eastern Canada. (Modified from MacKay and Tankard, 1990.)

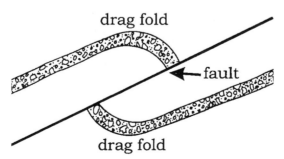

Fig. 11–14. Cross section showing drag folds along a fault

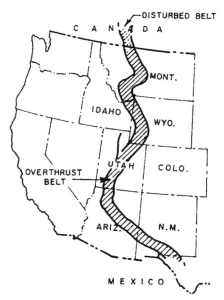

Fig. 11–15. Map of Rocky Mountain or Western overthrust belt

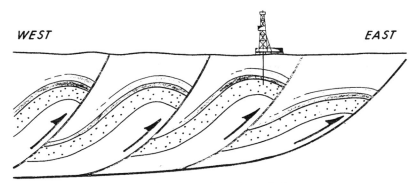

Fig. 11–16. East-west cross section of overthrust belt

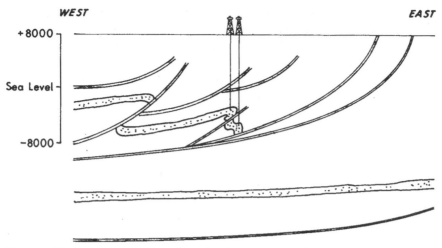

Fig. 11–17. East-west cross section of Painter reservoir field, Wyoming (modified from Lamb, 1980)

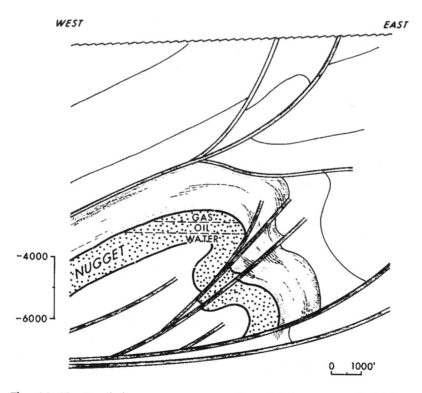

Fig. 11–18. Detailed east-west cross section of Painter reservoir field, Wyoming. (Modified from Lamb, 1980.)

Each drag fold field is located just to the west of a thrust fault (fig. 11-19). The largest oil field is the Anschutz Ranch east field, which will produce 180 million bbl (29 million m^3) of oil and 4 Tcf (110 million m^3) of gas. The largest gas field is Whitney Canyon field, which will produce 5.9 Tcf (167 m^3) of sour gas and 115 million bbl (18 million m^3) of oil. Production on the same disturbed belt in Alberta includes the Turner Valley and Jumping Pond fields. The Canadian production is primarily sour wet gas from Mississippian age limestones.

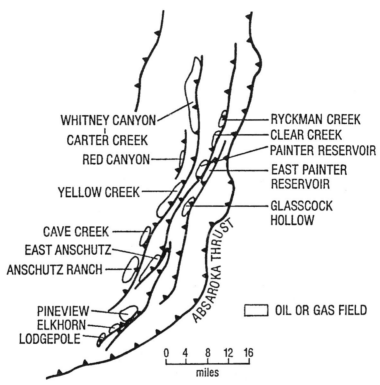

Fig. 11–19. Map of oil and gas fields in the Western overthrust belt of Wyoming and Utah. (Modified from Sieverding and Royse, 1990.)

The location of drag folds in the subsurface cannot be predicted from rock outcrops on the surface because of the intense deformation. It was not until the 1970s that improvements in seismic acquisition and processing opened exploration and drilling in overthrust belts.

Tilted fault blocks

Buried, tilted fault blocks can form large petroleum traps. During the geological past, horizontal sedimentary rocks (fig. 11–20a) were broken by normal faults into large, tilted fault blocks (fig. 11–20b). Some of the blocks can contain reservoir rocks. Later, the seas covered the tilted fault blocks and deposited caprocks of shale or salt on them (fig. 11–20c). The oil and gas then formed and migrated up along the reservoir rock to below the sealing fault or caprock.

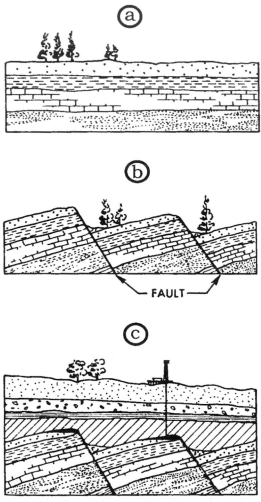

Fig. 11–20. Formation of tilted fault block traps

The Statfjord oil field, the largest oil field in the North Sea, is located in both the United Kingdom and Norway sectors. It was discovered by seismic exploration and drilled in 1974. The trap is a westward-tilted fault block (fig. 11–21). The two very porous and permeable sandstone reservoir rocks (Brent and Statfjord sandstones) are several hundred feet thick. The seal is the overlying shale. The ultimate production will be 3 billion bbl (500 million m^3) of light, sweet crude oil.

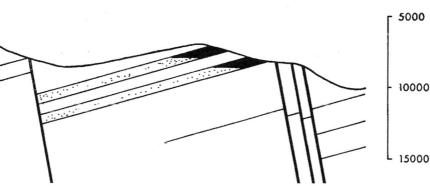

Fig. 11–21. Cross section of Statfjord field, North Sea. (Modified from Kirk, 1980.)

Stratigraphic Traps

Secondary stratigraphic traps—angular unconformities

An angular unconformity can form a giant gas and oil trap when a reservoir rock is terminated under an angular unconformity that is overlain by a seal. The two largest oil fields in the United States, the east Texas field, with more than 5 billion bbl (0.8 billion m^3) of oil already produced, and the Prudhoe Bay field, with 13 billion bbl (2.1 billion m^3) of produced and yet-to-be-produced oil, are in angular unconformity traps.

For centuries, the Native Alaskan peoples had known and used the numerous oil seeps on the tundra along the Arctic coast of Alaska for fuel. Very little exploration for oil occurred on the North Slope of Alaska until the 1960s, because the Arctic Ocean is frozen most of the year, and tankers cannot get into that area to take the oil out. If oil were discovered, the only way to get the oil to market would be to lay an 800-mile (1,290-km) pipeline across the state of Alaska to a southern port (Valdez). That trans-Alaska pipeline would cost billions of dollars. Because of this, only a giant oil field with more than a billion barrels of recoverable oil would be commercially viable.

During the early 1960s, seismic surveys were run over the North Slope, and the angular unconformity trap was revealed (fig. 11–22). The obvious traps, however, were the anticlines in the foothills of the Brooks Range. Six exploratory wells were drilled into the anticlines during the early 1960s. All were dry holes. Apparently the timing was not right, and the oil had already migrated through that area before the anticlines formed. Atlantic Richfield then drilled a wildcat well on the angular unconformity during the winter of 1967–1968 when the ground was frozen and could support a drilling rig. They put a 5% probability of success on a commercial discovery. The nearest well to it was a dry hole located 60 miles (97 km) away. At a depth of 8,200 ft (2,500 m), they discovered the Prudhoe Bay reservoir.

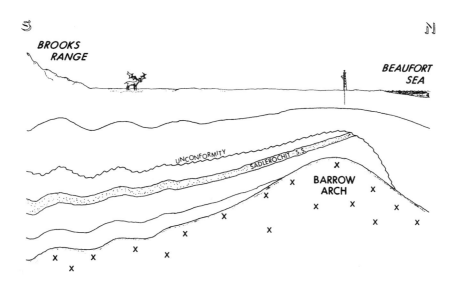

Fig. 11–22. North-south cross section of Prudhoe Bay oil field, Alaska. (Modified from Morgridge and Smith, 1972.)

In the Prudhoe Bay oil field, 13 billion bbl (2.1 billion m³) of oil will be produced from the Permian-Triassic age Sadlerochit Sandstone, the primary reservoir that is about 500 ft (150 m) thick, along with several smaller reservoirs including some limestones. The sandstone was deposited as horizontal layers of river and delta sands when shallow seas covered the area. Deposition of other sediment layers covered the sandstone and buried it in the subsurface. Later, an uplift occurred along the northern coast of Alaska, the Barrow arch, raising the Sadlerochit Sandstone. This exposed the sandstone to erosion that removed it from the top of the arch. Seas later invaded the area, covering the angular unconformity with shale and other sediments, burying it in the subsurface. The oil and gas later formed in a black shale source rock. It migrated up along the unconformity until it met and filled the Sadlerochit Sandstone below the unconformity. Shale on the unconformity is the caprock.

There were 25 billion bbl (4 billion m³) of 27 °API gravity, sour (1.04% S) crude oil in place with an enormous free gas cap (27 Tcf or 760 million m³) above the oil (fig. 11–23). Because of the size of the free gas cap that covers the top of the trap, even leases far to the sides of the trap produce oil.

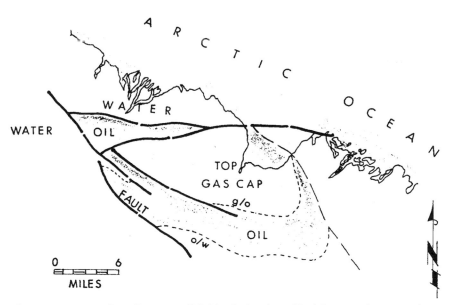

Fig. 11–23. Map of Prudhoe Bay oil field, Alaska. (Modified from Jamison, Brockett, and McIntosh, 1980.)

Primary stratigraphic traps

Reefs, beaches, river channels and incised valley fills, and updip pinch-outs of sandstones form primary stratigraphic traps.

Reefs. Reefs are prolific gas and oil traps in North America. Permian age reefs in west Texas and New Mexico (see chap. 10, figs. 10–22, 10–23, and 10–24), Devonian age reefs of Alberta (see chap. 10, figs. 10–28 and 10–29), and Cretaceous age reefs of Mexico (see chap. 10, fig. 10–33) form giant oil fields. Petroleum production can come not only from the reef but also from a compaction anticline overlying the reef (fig. 11–24). The *compaction anticline* forms in porous sediments, such as sands and shales, deposited on a hard rock mound or ridge, such as a limestone reef or bedrock hill. The sediments are deposited thicker to the sides of the reef than directly over the top. When the sediments are buried deeper, the weight of the overlying sediments compacts the loose sediments. The reef, composed of resistant limestone, compacts less. Because more compaction occurs in the thicker sediments along the flanks of the reef, a broad anticline forms in the sediments over the reef. Any reservoir rocks in the sedimentary rocks overlying the reef can trap petroleum.

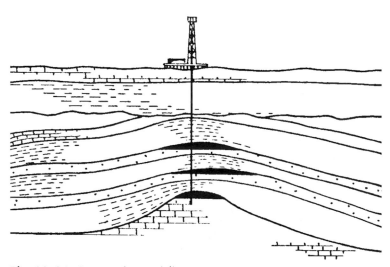

Fig. 11–24. Compaction anticline

Horseshoe atoll, a buried Pennsylvanian-Permian age reef, is located in northeastern Midland basin, part of the Permian basins (see chap. 10, fig. 10–22). The circular reef (fig. 11–25) covers a large area. High points along the reef form oil fields. The largest field on Horseshoe atoll is the

Kelly-Snyder oil field, which will ultimately produce 1.7 billion bbl (270 million m³) of oil. It is located at a depth of 5,000 ft (1,500 m). A cross section through the field (fig. 11–26) shows that the major production comes from the reef that has pores enhanced by solution. Production also comes from the compaction anticline in thin sandstone layers in the overlying shales of the Canyon Formation.

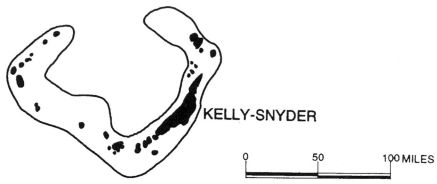

Fig. 11–25. Map of the Horseshoe atoll, Texas, showing oil fields. (Modified from Stafford, 1959.)

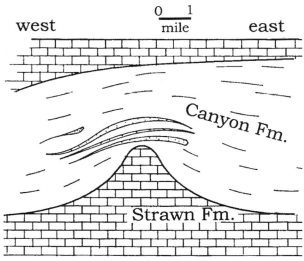

Fig. 11–26. East-west cross section of Snyder field, Texas, showing reef and overlying compaction anticline. (Modified from Stafford, 1957).

The Leduc oil and gas field of Alberta (fig. 11–27) produces from both a dolomitized reef (Leduc Formation, the D-3 zone) and the younger, overlying dolomite layers in the Nisku Formation, the D-2 zone. The Nisku Formation contains shales, sandstones, and dolomites that were compacted over the reef. The field will produce more than 200 million bbl (32 million m³) of oil. The D-2 pay averages 63 ft (19 m), whereas the D-3 pay averages 35 ft (11 m).

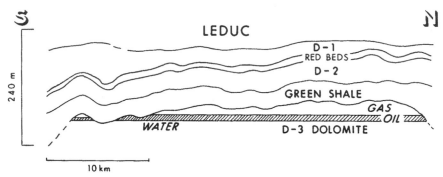

Fig. 11–27. North-south cross section of Leduc oil field, Alberta. (Modified from Baugh, 1951.)

The Cotton Valley lime pinnacle reef trend of east Texas (fig. 11–28) is a gas play. The Jurassic age reefs are 400 to 500 ft (120 to 150 m) high and 15,000 ft (5,000 m) deep. The reservoir quality part of the reef, the sweet spot, covers only 20 to 80 acres (0.1 to 0.3 km²) and is 1,000 to 2,000 ft (305 to 610 m) on a side. Individual reef reserves range from 0.8 to 105 Bcf (0.02 to 3 billion m³) of sour gas with CO_2. The first pinnacle reef discovery was made in 1980, but because of the small drilling target at such great depth, it was not until 3-D seismic exploration became common in the 1990s that the trend was fully exploited.

River channel sands. Most river channel sandstones are deposited and preserved as *incised valley fills* during a fall and rise of sea level. During an ancient sea level fall, the river incises (erodes) a valley. During the following sea level rise, the valley is filled with sands. If the sands are overlain by a caprock, it can form a gas or oil trap.

The Stockholm Southwest field in Kansas, discovered in 1979, is an example of an incised valley fill (fig. 11–29). The Pennsylvanian age Stockholm Sandstone is the reservoir rock at a depth of about 5,000 ft (1,500 m). The caprock is the overlying shale, and the source rock is a deeper shale. The field contains 27 million bbl (4 million m³) of light, sweet oil and will eventually produce 11 million bbl (2 million m³) of oil.

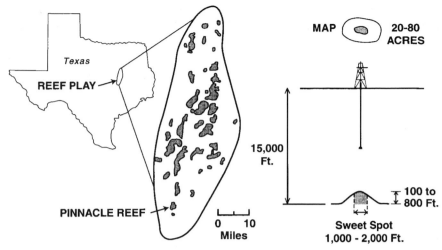

Fig. 11–28. Map of Cotton Valley Lime pinnacle reef trend, east Texas with cross section of pinnacle reef

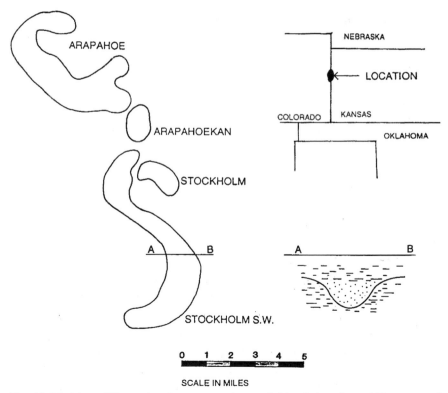

SCALE IN MILES

Fig. 11–29. Map of Pennsylvanian age sandstone fields, Colorado and Kansas, and west-east cross section of Stockholm southwest field. (Modified from Shumard, 1991.)

Beach sands. Beach sands, called *buttress sands*, can be deposited on an angular unconformity during rising seas (fig. 11–30) and form giant oil and gas field reservoirs. The Bolivar coastal fields of Lake Maracaibo, Venezuela (fig. 11–31a), containing more than 30 billion bbl (5 billion m³) of recoverable oil, produce from buttress sands. A cross section (fig. 11–31b) from west to east shows the Oligocene age angular unconformity dipping down to the west under Lake Maracaibo. Miocene age reservoir sandstones (buttress sands) directly overlay the unconformity under Lake Maracaibo. Some of the oil leaks out along the unconformity to form a line of oil seeps along the eastern shore of Lake Maracaibo. The Bolivar coastal fields were discovered in 1917 and produce from wells on platforms in the shallow lake.

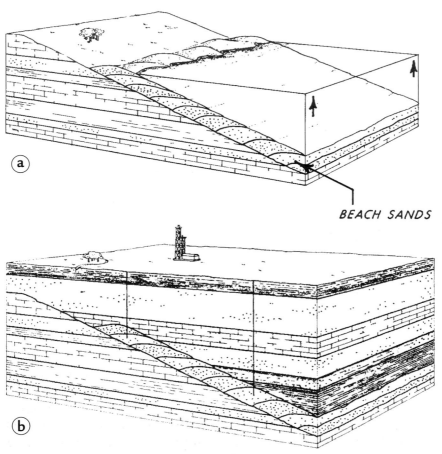

BEACH SANDS

Fig. 11–30. (a) Rising seas depositing beach sands and (b) buttress sands

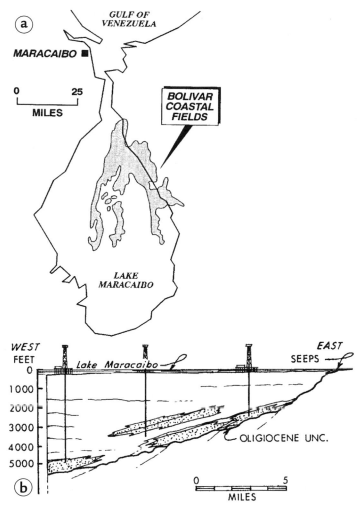

Fig. 11–31. (a) Map and (b) east-west cross section of Bolivar coastal oil fields, Lake Maracaibo, Venezuela. (Modified from Martinez, 1970.)

Pembina, the largest oil field in Canada (fig. 11–32), also produces from a beach sandstone and conglomerate reservoir, the Cretaceous age Cardium Formation that overlies an unconformity. The field was discovered in 1953 by drilling to a deeper seismic anomaly that was thought to be a Devonian age reef. The reef was not there, but the wellsite geologist tested the 20 feet (6 m) of oil sandstone encountered higher in the well and discovered the field. The field covers 900 sq miles (2,330 sq km), making it one of the largest conventional fields in the world by surface area. There are 7.5 billion bbl (1.2 billion m³) of oil in place. Pembina will eventually yield 1.56 billion bbl (250 million m³) of sweet, 37 °API gravity oil. It is an unsaturated oil field with no gas cap, and the average net pay is only about 20 ft (6 m).

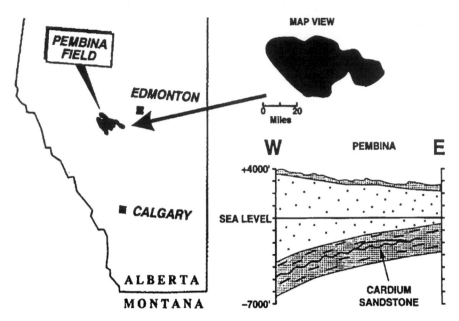

Fig. 11–32. Map and east-west cross section of Pembina oil field, Alberta

Combination Traps

Combination traps have both structural and stratigraphic trapping elements such as the Hugoton-Panhandle field (see chap. 7, fig.7–10).

Bald-headed structures

When an anticline or dome is formed (fig.11–33a), the crest of the structure is exposed to erosion. Most or all potential reservoir rocks can be removed from the top of the structure (fig. 11–33b). Seas later cover the area, and sediments are deposited, burying the eroded structure in the subsurface (fig. 11–33c). When the petroleum migrates up the reservoir rocks, it is trapped below the angular unconformity. Because the crest of the structure is barren, but the flanks are productive, it is called a *bald-headed structure* or *anticline*.

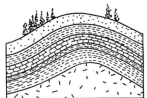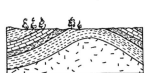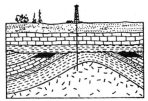

Fig. 11–33. Formation of a bald-headed anticline

The Oklahoma City field, below part of Oklahoma City, is almost a bald-headed structure. A large north-south, strike-slip fault—the Nemaha fault—caused folding of the sedimentary rocks into a steep dome west of the fault (fig. 11–34) in the geological past. Most of the potential reservoir rocks were eroded off the top of the structure, leaving an angular unconformity and only the Arbuckle Limestone (dolomite) as a reservoir rock on-structure over the granite basement rock. Seas then covered the area and deposited more sedimentary rocks. The unconformity and structure were uplifted again, forming a broad dome on the surface just southeast of downtown Oklahoma City. Gas and oil then migrated up the reservoir rocks and were trapped in 29 reservoir rocks under the angular unconformity and in the Arbuckle Limestone on top of the dome.

In 1928, a geologist working for Indian Territory Illuminating Oil Company mapped the subtle dome on the surface (fig. 11–35), and the company drilled the first well on structure. The well came in as a gusher, flowing 6,500 bbl (1,000 m^3) of oil per day from the Ordovician age dolomite in the Arbuckle Limestone at 6,600 ft (2,000 m). Although highest on the structure (figs. 11–34 and 11–36), the Arbuckle Limestone has produced only about 18 million bbl (3 million m^3) of oil. The discovery well went to water by the end of 1928 and had to be abandoned. Drilling out from the center of the structure located the more prolific pay zones.

In 1930, Indian Territory Illuminating Oil Company drilled the No. 1 Mary Sudik well, located far to the south of the Arbuckle discovery well. It was the first well into the Ordovician age Wilcox Sand, a very well-sorted sandstone with 20 to 30% porosity. The well blew out at the rate of 200 MMcf (6 million m^3) of gas and 20,000 bbl (3,200 m^3) of oil per day. Because the natural cement in the Wilcox sand was very weak, large quantities of loose sand blew up the well with the gas and oil, preventing the blowout preventers from being thrown to control the well. Norman, Oklahoma, to the south, and then Oklahoma City to the north, were covered with oil for 11 days. It was named the Wild Mary Sudik Well. The well was finally capped by lowering and clamping a valve to the top of it.

The Wilcox Sand is the most prolific reservoir rock on the structure (figs. 11–34 and 11–37), producing more than 250 million bbl (40 million m^3) of oil. The Oklahoma City field has 755 million bbl (120 million m^3) of recoverable oil. The oil is 37 °API gravity with no sulfur. Both the Oklahoma City and east Texas oil fields were developed at the same time in the early 1930s with unrestricted production. The market was glutted with crude oil, and the price fell to 16 cents a barrel.

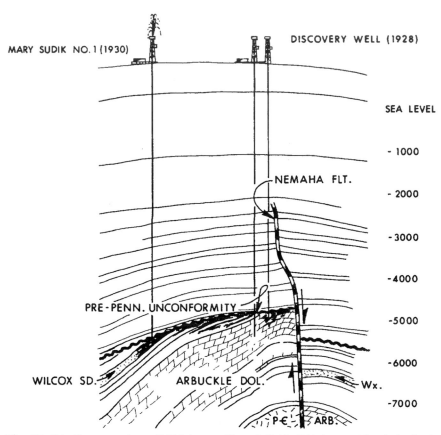

Fig. 11–34. Cross section of Oklahoma City oil field, Oklahoma (depth in feet). (Modified from Gatewood, 1970.)

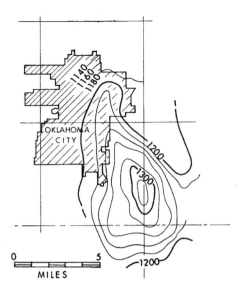

Fig. 11–35. Structural map on Garber Sandstone showing the dome located southeast of Oklahoma City. (Modified from Gatewood, 1970.)

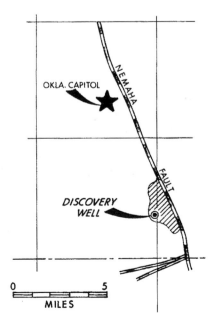

Fig. 11–36. Map of Arbuckle production, Oklahoma City oil field, Oklahoma. (Modified from Gatewood, 1970.)

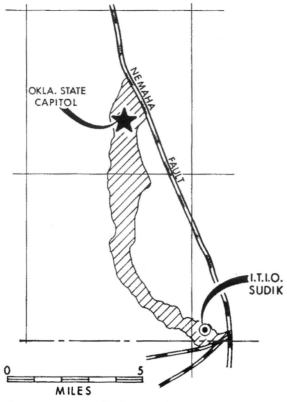

Fig. 11–37. Map of Wilcox production, Oklahoma City oil field, Oklahoma. (Modified from Gatewood, 1970.)

Salt domes

A *salt dome* is a large mass of salt, often miles across, rising from a subsurface salt layer through overlying sedimentary rocks to form a plug-shaped structure. Salt, composed primarily of halite, is a solid that can flow slowly as a very viscous liquid under pressure. A salt layer is formed by the evaporation of water. When sands and shales are later deposited on the salt layer, the weight of the overlying sediments presses down on the salt layer. The salt starts to flow and lifts up a weak area in the overlying sedimentary rocks. As the salt rises, it uplifts and pierces overlying sedimentary rocks to form a *piercement dome*. Because the salt is lighter in density than the surrounding sediments, buoyancy also helps the salt rise.

The salt is composed primarily of halite, which is highly soluble. Large amounts of salt are dissolved as the rising salt dome comes in contact with water in the overlying sediments. However, 1 to 5% of the salt is insoluble anhydrite. As the salt dissolves, an insoluble layer called the *caprock* builds up on the top of the dome (fig. 11–38). This caprock ranges from 100 to 1,000 ft (30 to 300 m) thick. Some of the anhydrite is altered by bacterial and chemical reactions to gypsum, limestone, dolomite, and sulfur. The caprock is often highly fractured and has vugular pores.

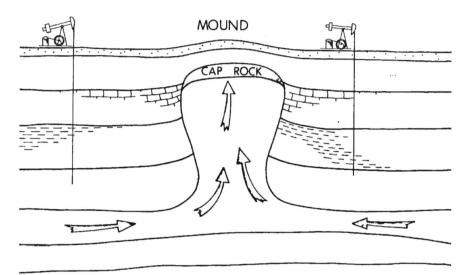

Fig. 11–38. Cross section of a salt dome

Many subsurface salt domes have mounds, 1 to 2 miles (1.6 to 3.2 km) in diameter, on the surface above the rising dome. The highest of these, Avery Island, Louisiana, rises 150 ft (46 m), and the salt is only 16 ft (5 m) below the surface.

There are two areas for drilling on salt domes (fig. 11–39). Above the salt dome, any shallow reservoir rocks such as sandstones are domed. Near the top of the salt dome, uplift has caused the overlying sedimentary rocks to be faulted with normal faults that sometimes form grabens. These form fault traps. Because the caprock is often fractured and porous, it can be productive reservoir rock. Deep along the flanks of the salt dome, reservoir rocks that were uplifted and pierced form traps against the impermeable salt dome.

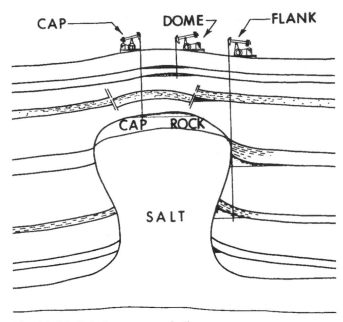

Fig. 11–39. Salt dome gas and oil traps

Individual salt domes often have numerous petroleum traps. Bay Marchand is a salt dome in the shallow waters of the Gulf of Mexico south of New Orleans (fig. 11–40). This salt dome has more than 125 separate producing reservoirs that have been discovered and will eventually yield 615 million bbl (98 million m³) of oil. Bay Marchand is part of a 27-mile (43-km) long salt ridge that also includes the Caillou Island and Timbalier Bay salt domes.

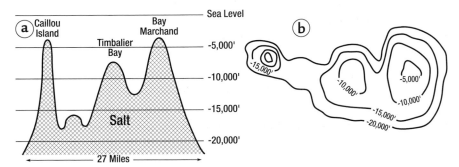

Fig. 11–40. (a) Cross section and (b) map of Bay Marchand, Timbalier Bay, and Caillou Island salt domes, offshore Louisiana. (Modified from Frey and Grimes, 1970.)

Some salt domes form overhangs of salt on the top of the dome. Sandstones that were bent up under the salt overhangs form prolific oil and gas reservoirs. The Barbers Hill salt dome of Texas (fig. 11–41) has salt overhangs on all four sides. It will produce more than 100 million bbl (16 million m^3) of oil from under the overhangs.

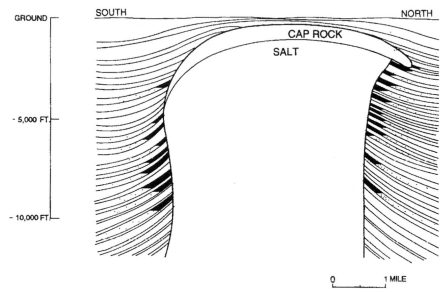

Fig. 11–41. North-south cross section of Barbers Hill salt dome, Texas. (Modified from Halbouty, 1979.)

A large part of the bottom of the northern Gulf of Mexico is underlain by patches of the Jurassic age Louann Salt. The Louann Salt also underlies extensive areas of the Texas, Louisiana, and Mississippi coastal plains that were part of the Gulf of Mexico when the salt was deposited. This salt has formed more than 500 salt domes throughout the Gulf coastal plain and on the bottom of the Gulf of Mexico.

The Spindletop salt dome near Beaumont, Texas (fig. 11–42), was the first salt dome successfully drilled for petroleum (1901). Up to that time, no highly productive wells had been discovered anywhere in the world, and no one had ever drilled a gusher west of the Mississippi River; all the wells had to be pumped.

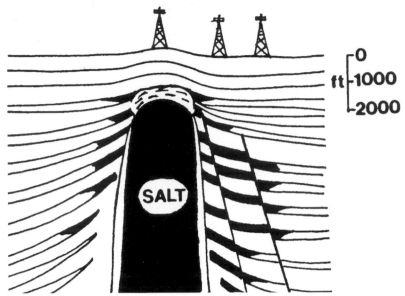

Fig. 11–42. Cross section of Spindletop salt dome, Texas. (Modified from Halbouty, 1979.)

A large mound with gas seeps marked the surface expression of the Spindletop salt dome. A well on that mound was drilled into the Spindletop caprock at just below 1,000 ft (300 m) in 1901. The caprock was very cavernous dolomite with high-pressure oil and gas in it. The well blew out as the world's first great gusher at the estimated rate of 100,000 bbl (16,000 m^3) of green-black oil per day to a height of 100 ft (30 m) above the derrick. It came in at a rate greater than the production rate of all the other oil wells in the United States combined.

Within 9 months, 64 more wells were completed into the Spindletop caprock, each producing a gusher. Caprock production from Spindletop rapidly declined and was depleted by 1903 because of unrestricted production. It was not until 1925 that the deeper oil traps along the flanks of Spindletop were discovered. Spindletop eventually produced more than 49 million bbl (8 million m^3) of oil from the caprock and 82 million bbl (13 million m^3) from the flank traps. It was the world's first great oil field.

Spindletop oil resulted in the creation of more than 100 companies, including Gulf and the Texas Company (Texaco), to drill, produce, transport, refine, and/or market the oil. Crude oil had become cheap and plentiful, and gasoline refined from crude oil for internal combustion engines used in early automobiles had become popular. Salt domes also occur in Kansas, Utah, and Michigan, but they are unproductive.

A thick salt layer, the Permian age Zechstein Salt, underlies the North Sea and forms salt domes. Until the giant Groningen gas field was discovered on land in the Netherlands in 1959, no exploration was done in the North Sea. During the late 1960s, more than 200 exploratory wells were drilled in the North Sea. Some small gas fields were found in the southern United Kingdom sector, but most were dry holes. Many companies were abandoning the North Sea when a well being drilled by Phillips Petroleum in 1969 encountered a 600-ft (183-m) column of oil above the Ekofisk salt dome in the Norwegian sector. It was the first major oil discovery in the North Sea. The reservoir rock, the Ekofisk Chalk, has 25 to 48% porosity but is permeable only because of fractures that were probably caused by the salt dome uplift. Chalk permeability is only 1 to 5 md, but with fractures, it is 1 to 100 md. The caprock is the overlying shale. Ultimate oil production from the Ekofisk field will be 1.7 billion bbl (270 million m^3) and 3.9 Tcf (110 million m^3) of gas.

Petroleum Exploration—
Geological and Geochemical

The API reported that 5,797 exploratory wells were drilled in the United States during 2008. They averaged 7,099 ft (2,164 m) deep and had a 65% success rate.

A *geologist* is a scientist who studies the earth by examining rocks and interpreting their history. A *petroleum geologist* specializes in the exploration for and development of petroleum reservoirs. An *exploration geologist* searches for new gas and oil fields. A *development geologist* plans the drilling of wells to exploit a field. A *petroleum geochemist* uses chemistry to explore for and develop petroleum reservoirs.

Seeps

The first commercial oil well in the United States was drilled in 1859 to a depth of 69½ ft (21 m) with a cable tool drilling rig along the banks of Oil Creek in Pennsylvania (plate 12-1). The driller was William "Uncle Billy" Smith, and the operator was Edwin L. Drake for the Seneca Oil Company. The site was chosen because of a natural oil seep, and the purpose of the drilling was to increase the flow of oil to the surface. The well initially flowed 20 barrels (3 m³) of oil per day from the sandstone reservoir rock. In the previous year (1858), an oil pit was dug to a depth of 60 ft (18 m) on an oil seep at Oilsprings, Ontario, Canada, by James M. Williams. The pit was lined with timber to prevent caving.

Plate 12–1. Drake well (cable tool rig), Titusville, Pennsylvania

How petroleum formed, migrated through subsurface rocks, and accumulated in traps was not understood at that time. For the next 50 years, exploration wells were randomly drilled or located next to seeps—a technique that was relatively successful. Drillers selected the drillsites while envisioning large, flowing underground rivers and subterranean crevasses filled with oil. Geologists were seldom used to select drillsites. Once an oil field was discovered, the "closeology" principle applied. The closer a proposed well was to a producing well, the better the proposed well was.

Every major petroleum-bearing basin of the world has numerous *oil seeps* where oil is naturally leaking on the surface (plate 12–2). This is because not all the oil is trapped as it migrates up from the source rock (fig. 12–1). Even if the petroleum is trapped, the caprock above the reservoir has often been fractured by folding or another process. Some of the gas and oil from the trap leaks through the fractures and onto the surface to form a seep above the *leaky trap* (fig. 12–1). This is why drilling on or near oil seeps was so successful. When oil seeps on the surface, three processes degrade the crude oil to heavy oil, tar, and asphalt. These are water washing, evaporation, and bacteria. Water flowing by the oil will "wash it" by dissolving and removing some of the soluble, lighter fractions. Heating by the sun causes evaporation of the lighter fractions. Bacteria on the surface consume the lighter fractions. The effect of these processes is to degrade the oil into heavy oil, tar, and asphalt.

Plate 12–2. (a) Natural oil seep, Dorset coast, England (b) close-up of seep

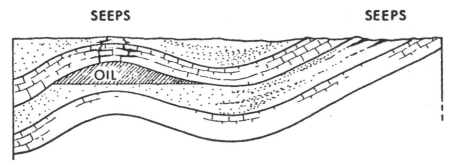

SEEPS **SEEPS**

Fig. 12–1. Oil seeps related to a leaky oil trap

Geological Techniques

In the early 1900s, it was finally accepted that oil accumulated in high areas of reservoir rocks such as anticlines and domes. This was known as the *anticlinal theory*, and it was originally suggested in the late 1800s but not immediately accepted. Oil companies then hired geologists to map sedimentary the rock layers cropping out on the surface. Bull's eye (see chap. 5, fig. 5–11) and lobate patterns (chap. 5, fig. 5–9) were used to locate subsurface traps such as domes and anticlines. With the development of airplanes and aerial photography, surface mapping became more efficient. Vertical black-and-white aerial photographs were made into geologic maps. Spot checks of selected areas on the surface, called *ground truthing*, are used to identify the rocks seen on the aerial photographs. Aerial photography can also be done with radar. The ground image is similar to a black-and-white photograph (plate 12–3). Radar has the advantage that it can see through clouds and can be run at night.

Since the 1970s, satellites have been orbiting the earth at distances of several hundred miles in space. Some of these satellites photograph the surface of the earth in infrared and visible light and transmit these images back to earth. Many of these are spy satellites. However, some satellites operate under the open-sky policy, and the images for almost everywhere on the earth are made available at a uniform price to anyone without restrictions. The images can be digitally enhanced by computer. They are useful in mapping remote areas and for giving a different perspective to a previously mapped area.

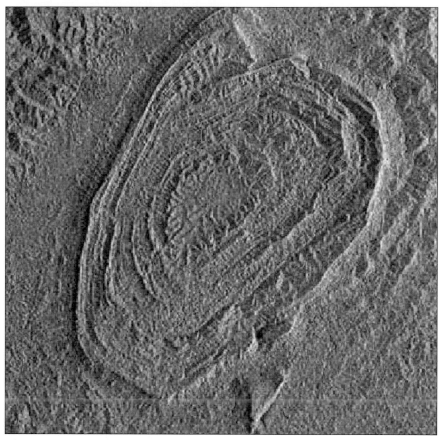

Plate 12–3. Radar image of Kalimantan, Indonesia (note the eroded dome and fault). (Courtesy of RADARSAT International.)

Geological reasoning can be used to find gas and oil. During the mid-1950s, several wells were drilled in eastern New Mexico. It was recognized from well samples that all wells drilled to the north of a line had drilled through lagoonal facies limestones that were deposited during a specific time in the Permian (fig. 12–2). All the wells to the south of that line had drilled through relatively deep-water reef facies limestone deposits called *fore reef* that were deposited at exactly the same time. Thus, there had to be a reef located between these two facies. In 1957, the Pan American Petroleum Company drilled a well between the two facies and discovered the Empire Abo field (see chap. 10, figs. 10–23 and 10–24). The first few wells missed the top of the reef and had a relatively thin pay from 20 to 60 ft (6 to 18 m) thick. A later well, however, was drilled into the reef crest and had 725 ft (221 m) of pay, proving it to be a major discovery with more than 250 million bbl (40 million m^3) of recoverable oil.

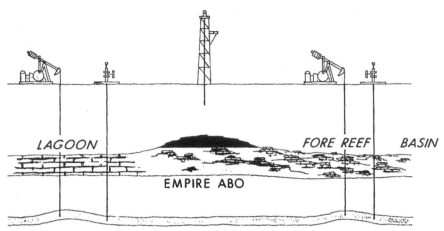

Fig. 12–2. The recognition of reef facies that led to the discovery of the Empire Abo oil field, New Mexico. (Modified from LeMay, 1972.)

In a frontier basin where relatively few wells have been drilled, a geologist starts by looking for large structural traps. First, the size and shape of the basin is determined. Rock outcrops along the margins of the basin are examined and described. Most rocks that occur in the basin subsurface crop out along the edge of the basin. The *stratigraphy* (i.e., sequence of rock layers) in the basin is established to identify potential source rocks, reservoir rocks, and seals. Structures that can be identified by field mapping and reconnaissance seismic surveys are located.

In mature areas that have been relatively well drilled, geologists spend most of their effort in subsurface mapping and constructing cross sections. Most of the large structural traps have been found, leaving the more subtle stratigraphic traps yet to be discovered.

Subsurface maps

Structural and isopach maps are made of potential reservoir rocks with scales ranging from the entire basin to a single field. Every time a new well is drilled in that area, more information is obtained. These new data are then plotted on the maps, and the maps are recontoured and reinterpreted.

Two criteria must be met on a structural map to locate a drillsite. It must be a high area on the reservoir rock, and it must have four-sided closure. If the reservoir rock is filled with water, then any gas or oil, being lighter than water, will migrate to the high area in the reservoir rock. *Four-sided closure* means that the trap (high area) goes down on all four sides. A water glass holds water because it has four-sided closure; that is, the glass goes down to the bottom on all four sides. A gas and oil trap is like an upside down water glass, because gas and oil rise. A high area on a structural map has four-sided closure if the contour lines come all the way around on all sides (fig. 12–3). Without four-sided closure, the gas and oil would leak out of the side where the contour lines do not come around.

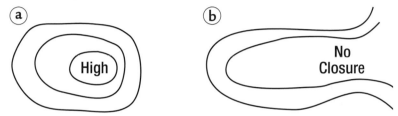

Fig. 12–3. Structural map contours showing (a) closure and (b) no closure

Petroleum traps were originally filled with water. The gas and oil flow into the trap later to replace the water. Because gas and oil are lighter than water, the trap is filled from the top downward. The trap can be filled down to a level, called the *spill point*, at which it cannot hold any more (fig. 12–4). It is the highest point on the rim of an anticline or dome. The vertical distance from the crest of the reservoir rock down to the spill point is the *closure* of the trap. Closure is the maximum vertical amount of gas and oil that the trap can theoretically hold. It is a measure of the potential size of the field. On a structural map (fig. 12–4a), the closure is the vertical distance defined by the contours that come all the way around the high area. The spill point is located where the contours do not come all the way around.

The thickness of the reservoir rock in the trap can be estimated from an isopach map. Thickness along with closure allows you to estimate the maximum amount of gas or oil the trap can hold before you drill a well.

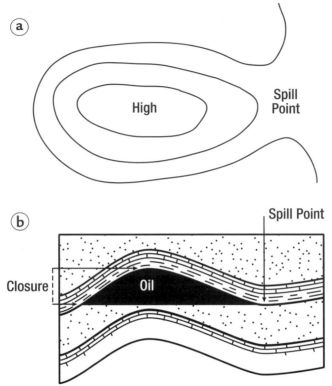

Fig. 12–4. Description of trap (closure and spill point) (a) structural map (b) cross section

Correlation

Constructing a cross section, a vertical slice or panel of the subsurface rocks, by correlation is used to find gas and oil traps. *Correlation* is the matching of rock layers from one well to another. When a well is drilled, a record of the rock layers in the well is made on a well log. The rock layers between well logs are correlated to make the cross section (fig. 12–5). The correlation is started with a marker bed or key horizon. A *marker bed* is a distinctive rock layer that is easy to identify. Volcanic ash layers; thin beds of coal, limestone, or sandstone; and fossil zones are good marker beds. A *key horizon* is the top or bottom of a thick, distinctive rock layer. After correlating the marker bed or key horizon, the rock layers above and below the maker bed or key horizon can then be correlated on physical similarity and their position in the sequence of layers. Lines are drawn along the *contacts*, the boundaries between rock layers. In areas complicated by faulting, facies changes, or unconformities, the most accurate correlation is done by matching microfossils between wells (fig. 12–6).

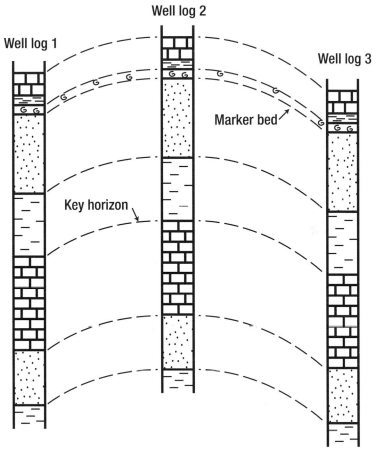

Fig. 12–5. Correlation of well logs

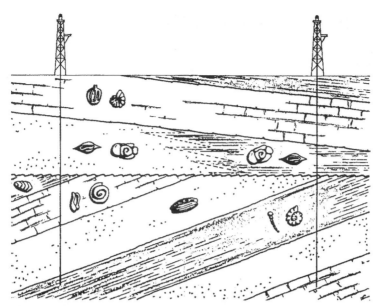

Fig. 12–6. The use of microfossils for correlation between wells

There are two types of cross sections, depending on how the well logs are arranged to make the correlation. Each well log must be arranged or *hung* along a common horizontal surface going through the well logs. A *structural cross section* is made by hanging the well logs by modern sea level in each well (fig. 12–7). The well logs are then correlated. Structures such as folds and faults are illustrated on structural cross sections. Structural cross sections are used to find structural petroleum traps. A *stratigraphic cross section* is made by hanging the well logs from the same marker bed or key horizon in each well (fig. 12–8). Because the marker bed or key horizon was originally deposited horizontally, a stratigraphic cross section restores the rocks to their original horizontal position before they were deformed. Stratigraphic cross sections are used to illustrate the relationship between rock layers such as facies changes and to locate stratigraphic petroleum traps such as reefs.

A *fence diagram* is used to show how wells correlate in three dimensions (fig. 12–9). The diagram is arranged like a map. North is at the top, south at the bottom, east to the right, and west to the left. Each well is located (*spotted*) on the map. The well log for each well is drawn vertically under the well's position. The rock layers are then correlated from one well log to another. Each set of correlations forms a panel. The entire diagram is called a fence diagram.

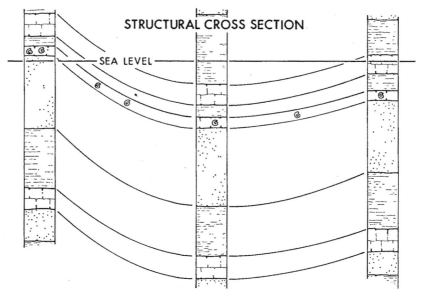

Fig. 12–7. A structural cross section with well logs hung from modern sea level

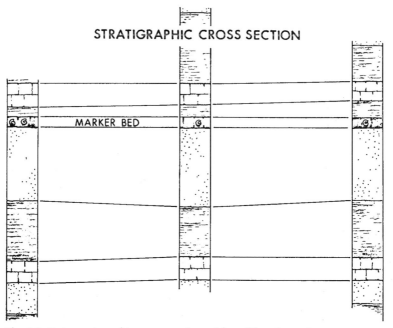

Fig. 12–8. A stratigraphic cross section with well logs hung from a common marker bed

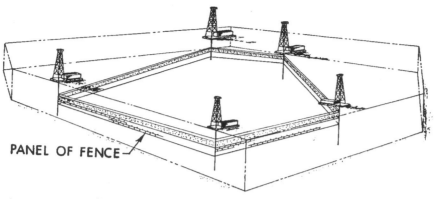

PANEL OF FENCE

Fig. 12–9. Fence diagram

With *sequence stratigraphy*, correlation between wells log is made using unconformities rather than rock layers. It recognizes that unconformities, ancient erosional surfaces, represent an instant of geological time. These unconformities were formed when sea level was relatively low and the land was exposed to erosion during the cyclic rise and fall of sea level (see chap. 2, fig. 2-7). The rocks between two unconformities were deposited during a specific interval of time called a *parasequence set*. This corresponds to the fourth-order cycle of sea level rise and fall. Parasequence sets can be subdivided into smaller time interval units bounded by smaller unconformities called parasequences that correspond to the fifth-order cycles of sea level rise and fall. Geologists can better predict where source rocks and reservoir rocks are located in parasequence sets and parasequences (fig. 12-10) than they can using traditional rock formations.

A lot of subsurface information from wells, no matter who drilled them, eventually becomes public. A company drilling an exploratory well in a new area might want to keep as much information as secret as possible by running a *tight hole*. The information can then be used by the tight hole operator to locate other wellsites and leasing any land that is still open (*loose acreage*). However, state, provincial, and federal laws require that a specific suite of well logs be released to the government regulatory agency within a certain time. This time limit varies; it is one year in the states of Oklahoma and Texas. For US offshore federal land, it is 5 or 10 years. After being released to the government, these logs immediately become public information. In many countries, there is a national oil company that partners with any foreign oil company drilling in that country. Because the foreign oil companies have a common partner in the national oil company, well information can be obtained through that source.

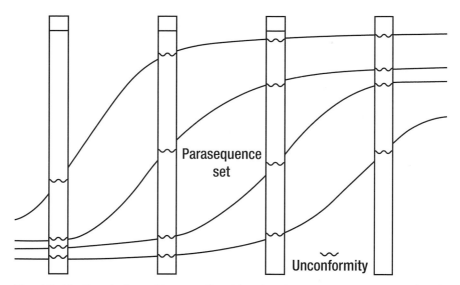

Fig. 12-10. Correlation with unconformities showing a parasequence set that is composed of smaller parasequences bounded by smaller unconformities

An important source of subsurface well information consists of wireline well logs. *Well log libraries* collect copies of wireline well logs for regional areas. There is a well log library in almost every oil patch city in the United States and Canada. Well log library members can review, copy, and even check out well logs and other well information from the library. The major oil companies have digitized the well logs for their areas of interest on computers. A company geologist can access those logs on a workstation.

Large oil companies use *scouts* who gather information on any petroleum-related activity in their assigned region. A scout uses all kinds of "ethical" methods to find out where competitors are exploring, leasing, and drilling. Every time a well is drilled, the scout obtains as much information as possible about the well. The scout fills out a form, called a *scout ticket*. It usually includes the well name, location, operator, spud and completion dates, casing and cement data, production test data, completion information, the tops of certain zones or formations, and a chronology of the well. Because a single company scout cannot keep an eye on all the activity going on in his or her assigned region, scouts from different oil companies meet periodically to coordinate their efforts and exchange information in *scout checks or meetings*. A *czar* or *bull scout* is elected to conduct the meetings. An organization known today as the International Oil Scouts Association annually publishes the statistical information gathered by the scouts.

Commercial scouting firms publish daily, weekly, and monthly reports on regional drilling activity. Some publish *completion cards* for each well drilled in the United States and Canada. The information on the completion card includes name, location, spud date, total depth drilled, depths to the tops of formations in the wells, intervals completed, completion techniques and initial petroleum production. The source of this information is scout tickets and government regulatory agencies. For a fee, the firm will provide completion cards for a regional area and update the information periodically. IHS, a commercial firm, has information on more than 3.6 million wells in the United States and more than 700 thousands wells in Canada that were drilled for gas and oil.

Well cutting libraries collect well cutting for a regional area. Well cuttings can be sold to the library, and well cuttings in the library can be examined for a fee. *Well core libraries* collect cores drilled in that state, province, or country. In Canada, all cores that are cut must be eventually submitted to a central facility in each province for storage and examination.

Each state in the United States and each province in Canada has a *geological survey* that publishes reports and maps on the petroleum geology of that area. The federal government in almost every country has a geological survey that uses professional geologists to make geological reports on that country.

The API assigns a 10- or 12-digit number called the *API number* to every well drilled in the United States. Digits 1 and 2 are state codes; digits 3 through 5 are for county, parish, or offshore; digits 6 through 10 identify the well; and digits 11 and 12 record a well property such as sidetracking.

Geochemical Techniques

Geochemistry is the application of chemistry to the study of the earth. Traces of hydrocarbons in soil and water are often good indications of the proximity of a petroleum trap (fig. 12–11). In an exploratory area, surface samples of waters and soils are taken. These samples are analyzed in the laboratory with instruments such as gas chromatographs for minute traces of hydrocarbons. Many subsurface petroleum reservoirs are leaky and have obvious seeps on the surface. Some traps, however, are not as leaky and have only *microseeps* on the surface that cannot be detected visually. The microseeps often occur in a pattern called a *hydrocarbon halo* that outlines the subsurface trap. Seeps are also common in the ocean. Ships towing water-sampling equipment and shipboard hydrocarbon-sensing devices called *sniffers* are used to detect their locations.

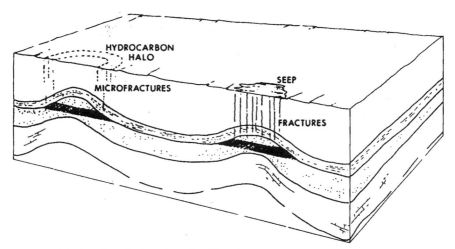

Fig. 12–11. Geochemical exploration for microseeps

Water samples can be taken from subsurface rocks for chemical analysis. The subsurface water salinity must be known before some wireline well log calculations, such as oil saturation, can be accurately made. Traces of hydrocarbons in the subsurface waters of a dry hole could indicate the presence of a petroleum reservoir in the area. No traces of hydrocarbons were found in subsurface waters from wells drilled in the 1960s on the Destin anticline located offshore from the Florida panhandle. This discouraged further exploration in that area.

Vitrinite reflectance is a method used to determine the maturity of a source rock. Vitrinite is a type of plant organic matter often found in black shale. The source rock sample is polished and then examined under a reflectance microscope. The percentage of light reflected from the vitrinite is dependent on the maturity of the source rock. Vitrinite reflectance can determine if oil and gas have been generated. If all the source rock samples from an unexplored basin show that hydrocarbons were never generated in that basin, further exploration would be discouraged.

Geochemistry can also be used to identify the source rock for a specific crude oil. The crude oils in traps can then be correlated with source rocks to determine the migration path for the petroleum. The migration path would be an excellent area for future exploration for undiscovered oil and gas fields.

Plays and Trends

A *play* is a combination of trap, reservoir rock, and seal that has been shown by previously discovered fields to contain commercial petroleum deposits in an area. An example is the Tuscaloosa trend play in Louisiana. The Cretaceous age Tuscaloosa Sandstone generally ranges from 35 to 200 ft (11 to 61 m) thick in Louisiana and is reservoir rock quality. A shale seal overlies it. In the Louisiana coastal plain, large growth (down to the basin) faults cut the Tuscaloosa Sandstone. On the basin (Gulf of Mexico) side of the growth fault, the Tuscaloosa Sandstone can form a rollover anticline trap. By drilling 16,000 to 22,000 ft (4,900 to 6,700 m) to the Tuscaloosa Sandstone rollovers, gas and condensate fields can be discovered. This is known because several fields of this type have already been discovered.

A *trend* or *fairway* is the area along which the play has been proven and more fields could be found. The Tuscaloosa trend extends from Texas, through Louisiana, and into Mississippi (fig. 12–12). The trend was opened up by the discovery of the False River field in 1974.

Fig. 12–12. Tuscaloosa trend for deep, wet gas in Louisiana

A *prospect* is the exact location where the geological and economic conditions are favorable for drilling an exploratory well. A prospect can be presented by using prospect maps that illustrate the reasoning for selecting that drilling location. The maps include at least a structure and isopach map of the drilling target and a map of test results and fluid recoveries from wells in the area. An economic analysis of the prospect should include reserves and risk calculations.

There are four major geological factors (essential elements) in the success of a particular prospect. First, there must have been a source rock that generated petroleum. Second, there must be a reservoir rock to hold the petroleum. Third, there must be a trap. This includes a reservoir rock configuration that has four-sided closure, a seal on the reservoir rock, and no breach of the trap. Fourth, the timing must be right. The trap had to be in position before the petroleum migrated through the area.

Geologists have put together petroleum systems for all petroleum-producing areas of the world. A *petroleum system* is a volume of sedimentary rocks that includes a source rock that is or has generated oil and gas and all the seeps and accumulations of that oil and gas. It can be an entire basin or part of a basin. The essential elements that include source rock, reservoir rock, seal, and overlying sediments in that petroleum system are identified. The timing of processes that include trap formation and petroleum generation, migration, and accumulation for that petroleum system are determined. The petroleum system is named after the source rock and reservoir rock such as the Mandal-Ekofisk petroleum system of the North Sea and the Akata-Agbada petroleum system of the Niger River Delta of Nigeria.

Petroleum Exploration—
Geophysical

Geophysics is the application of physics and mathematics to the study of the earth. *Geophysicists*, who are trained in mathematics and physics, commonly use three surface methods—gravity, magnetic, and seismic—to explore the subsurface. At the present, seismic exploration is where most of the exploration money is spent and most of the technological advances are being made.

Gravity and Magnetic Exploration

Gravity meters and magnetometers are relatively inexpensive, portable, and easy-to-use instruments. A *gravity meter* or *gravimeter* measures the acceleration of the earth's gravity at that location. A *magnetometer* measures the strength of the earth's magnetic field at that location. Both are small enough to be transported in the back of a pickup truck. A magnetometer can be mounted in a stinger on the back of an airplane to conduct an *aeromagnetic survey* (plate 13-1) that is fast and efficient and does not need permission from the land owners. The magnetometer can also operate while being towed behind a boat. The gravity meter does not work well in either an airplane or the ocean because of vibrations.

Plate 13–1. Aeromagnetics—the magnetometer is in the stinger behind the airplane. (Courtesy of Fugro.)

The gravity meter is very sensitive to the density of the rocks in the subsurface. It measures gravity in units of acceleration called *milligals*. Over a typical area of earth's crust with 5,000 ft (1,525 m) of sedimentary rocks underlain by basement rock that is very dense, the gravity measurement is predictable (fig. 13–1). A mass of relatively light rocks such as a salt dome or porous reef can be detected by the gravity meter because of values over it that are lower than normal gravity. A mass of relatively heavy rocks near the surface such as basement rock in the core of a dome or anticline can be detected by higher-than-normal gravity values.

The magnetometer measures the earth's magnetic field in units called *gauss* or *nanoteslas*. It is very sensitive to rocks containing a very magnetic mineral called magnetite. If a large mass of magnetite-bearing rock (e.g., basement rock) occurs near the surface, it is detected by a larger magnetic force than the normal, regional value (fig. 13–2). The magnetometer is primarily used to detect variations of basement rock depth and composition. It can be used to estimate the thickness of sedimentary rocks filling a basin and to locate faults that displace basement rock.

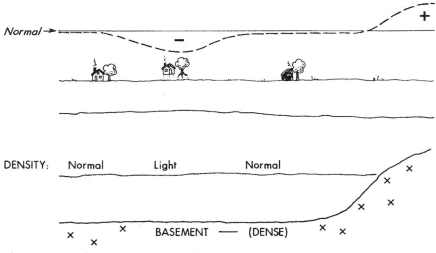

Fig. 13-1. Gravity meter measurements over an area

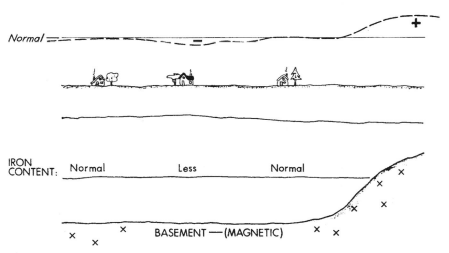

Fig. 13-2. Magnetometer measurements over an area

In order to explore the subsurface of an area using a gravity meter, a grid pattern of points is located on the surface. A gravity reading is made at each point. The gravity values are then plotted on a base map and contoured similar to a topographic map. With an aeromagnetic survey, the plane flies in two sets of parallel lines that intersect at right angles. The magnetic data are also contoured. Most of the area will have "normal" gravity and magnetic measurements. Anomalies of abnormally high (maximum) or low (minimum) gravity and magnetics are noted.

A subsurface salt dome is seen as a surface anomaly of relatively low gravity and magnetics because the salt is light in density and has no magnetite mineral grains compared to the surrounding sedimentary rocks (fig. 13–3). Many salt domes in the coastal areas of Texas and Louisiana were discovered in the 1920s by gravity meter surveys. A subsurface reef can have a gravity anomaly that ranges from abnormally high to abnormally low. The abnormally high anomaly is caused by a dense (nonreservoir) limestone reef, and the abnormally low anomaly is caused by a porous limestone reef. Magnetics are generally not useful in locating reefs.

A dome or anticline can be identified by both a high gravity and high magnetic anomaly. This is caused by dense, magnetite-bearing basement rock that is close to the surface in the center of the structure (Fig. 13–4).

A subsurface dip-slip fault can produce a sharp change in both gravity and magnetic values along a line because the basement rock is higher on one side of the fault than on the other side (fig. 13–5). Small features at shallow depths and large features at deep depth have similarly sized gravity and magnetic anomalies. Because of this, it is difficult to determine the size and depth of the feature causing the anomaly.

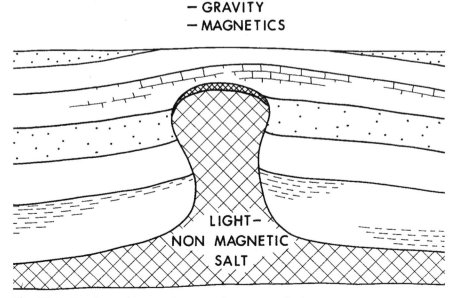

Fig. 13–3. Gravity and magnetic anomalies over a salt dome

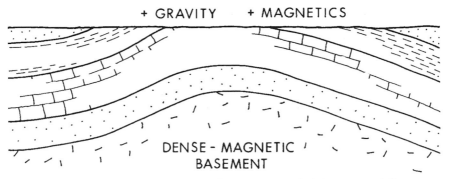

+ GRAVITY   + MAGNETICS

DENSE - MAGNETIC
BASEMENT

Fig. 13–4. Gravity and magnetic anomalies over an eroded dome or anticline

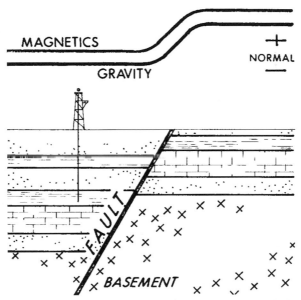

MAGNETICS

GRAVITY

NORMAL

FAULT

BASEMENT

Fig. 13–5. Gravity and magnetic anomalies over a fault

Seismic Exploration

The first oil field found by seismic exploration alone was the Seminole field of Oklahoma in 1928. The seismic data at that time were recorded by analog in the field on a sheet of paper. The printout was noisy and not very accurate. The greatest improvements in petroleum exploration in the last several decades have involved new seismic acquisition techniques and computer processing of digital seismic data.

Acquisition

The seismic method uses impulses of sound energy that are put into the earth. The energy travels down through the subsurface rocks, is reflected off subsurface rock layers, and returns to the surface to be recorded. Seismic exploration uses subsurface echoes to image the shape of subsurface sedimentary rocks and locates petroleum traps. A source and a detector are used. The source emits an impulse of sound energy either at or near the surface of the ground or at the surface of the ocean. The sound energy is reflected off subsurface rock layers. The maximum reflection energy occurs when the angle of incidence between the seismic source and reflector is equal to the angle of reflection between the reflector and seismic detector (fig. 13-6). Only about 2 to 4% of each sound impulse is reflected off each layer, and the remaining sound impulse goes further into the rock, to be reflected off deeper and deeper layers. The reflected sound energy from each layer returns to the surface, where the detector records it.

The detector on the surface records both the *signal*, wanted direct (primary) reflections from the subsurface rock layers, and *noise*, unwanted energy. Noise can be caused by surface traffic, wind, surface and air waves, and subsurface reflections that are not direct reflections from subsurface rock layers. A high signal/noise ratio is desired. A *noise survey*, a small seismic survey, can be run first to determine the nature of noise in that area and plan the optimum seismic program to reduce noise.

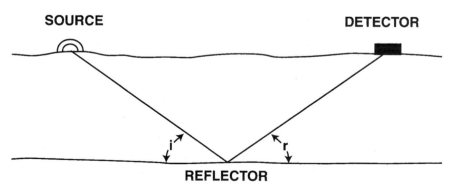

Fig. 13-6. A seismic reflection with the angle of incidence (i) equal to the angle of reflection (r)

The location of the seismic source is called the *shot point*. On land, the most common seismic sources are explosives and vibroseis. Dynamite was the first seismic source used and is still the most common explosive source used today. Explosives are used today where the surface is covered with loose sediments, swamps, or marshes. When using explosives, a small, truck-mounted drilling rig often accompanies the seismic crew to drill a *shot hole*, usually 60 to 100 ft (18 to 30 m) deep, to a point below the soil. The explosives are planted in solid rock on the bottom of the hole. *Primacord*, a length of explosive cord, can be planted in a trench about 1 ft (0.3 m) deep or suspended in air as a seismic source. Explosives as a seismic source are expensive.

About 70% of the seismic exploration run on land today is done by *vibroseis*, a technique developed by Continental Oil (now ConocoPhillips). In vibroseis, a *vibrator truck* (plate 13–2) with hydraulic motors mounted on the back of a truck and a plate called a *pad* or *baseplate* located on the bottom of the truck bed are used. The vibrator truck drives to the shot point and lowers the pad onto the ground until the back wheels are above the ground and most of the weight of the truck is on the pad. The hydraulic motors use the weight of the truck to shake the ground for a time (*sweep length*), often 7 to 20 seconds. A range of frequencies, called the *sweep*, is imparted into the subsurface. Vibroseis is very portable and can be run in populated areas. Other less common land seismic sources include weight drop, gas gun, land air gun, and guns such as a shotgun.

Plate 13–2. Vibrator truck

To reduce the noise generated from the source, several linked explosive shots or vibrator trucks can be used simultaneously in a *shot point array*. The seismic energy commonly has a usable frequency of 8 to 120 Hertz (Hz) or cycles per second. The human ear can hear 20 to 20,000 Hz.

At sea, a common seismic source is an air gun. The *air gun* is a metal cylinder that is several feet long (plate 13–3). It is towed in the water at a depth of 20 to 30 ft (6 to 9 m) behind the ship. On the ship are air compressors. High-pressure air at 2,000 psi (140 kg/cm^2) is pumped through a flexible, hollow tube into the air gun in the water. On electronic command, ports are opened on the air gun. An expanding, high-pressure air bubble in the water provides a seismic source that is not harmful to marine life.

Air guns are described by the capacity of their air chamber, such as 200 in.^{3} (3,000 cm^3). Air guns of different sizes (*tuned air gun arrays*) are often fired at the same time to cancel any noise from the source such as the air bubble expanding and contracting after the first impulse. The air gun is also used in some applications in swamps and marshes. Other seismic sources used at sea include water gun, sleeve gun, and sparker.

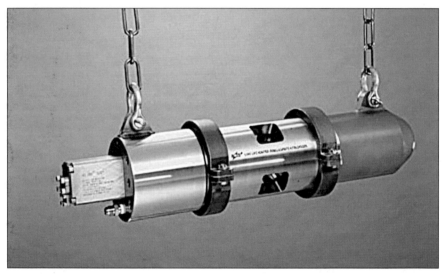

Plate 13–3. Air gun. The air gun is towed in the water behind the seismic ship. The electrical connections and air hose are attached to the back (left) side. (Courtesy of Bolt Associates.)

A *seismic contractor* is a company that owns and operates the seismic equipment and runs the seismic survey. The seismic contractor can run the seismic survey under contract with an exploration company. A *spec survey* can also be run by a seismic contractor. A limited number of exploration companies then pay for and view the nonexclusive seismic records. In another method, several exploration companies share the cost and results of a seismic survey run by a seismic contractor in a *group shoot*.

Before seismic exploration is run on private land in countries such as the United States and Canada, a *permit person* must obtain permission from the surface rights owners of the land. A fee per shot hole or seismic line mile is paid, and damage fees are negotiated. A *survey crew* then cuts a path through the trees and brush (if necessary), accurately locates and flags the shot points and detector locations, and records them in the survey log book. Members of the seismic crew called *jug hustlers* lay the cable and arrange and plant the geophones.

At sea, permitting is not necessary. The ship's crew does the navigation while the seismic crew runs the seismic equipment. Surveying on land and at sea today is done by global positioning using navigational satellites.

The impulse of seismic energy travels down through the subsurface rocks (fig. 13–7), strikes the top of the subsurface layers, and is reflected back to the surface as echoes. The returning echoes are recorded on land by vibration detectors called *geophones* or *jugs*. They detect vertical ground motion and translate it into electrical voltage.

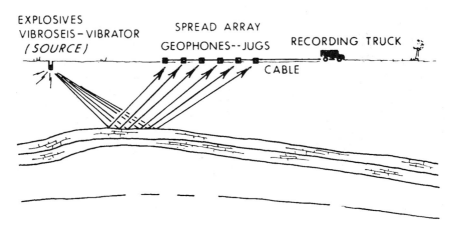

Fig. 13–7. The seismic method on land

The geophone often has a spike on the bottom so it can be planted in the ground (plate 13–4). One to dozens of geophones are connected to form a *group* that records as a single unit called a *channel*. By using several geophones in a group, noise is reduced. The geophones in a group are arranged in a line, several parallel lines, a star, a rectangle, or another geometric pattern. Groups of geophones are deployed in a larger geometric pattern called the *spread*. A common spread called a *linear spread* (fig. 13–8) consists of a long main cable stretched out in a line several miles long. Shorter cables at specific intervals connect the individual geophone groups, which are equidistant, with the main cable. A *split spread*, with the source in the middle of the linear spread, is commonly used on land (fig. 13–8b). Using these methods, a large number of geophone groups can be used to cover a large area of the subsurface with each seismic shot (fig. 13-7). A reading of 96 channel or trace data means 96 geophone groups were used for each shot point. The geophones are all connected to a lead cable that goes to the recording truck or doghouse. The data can also be transmitted digitally by a radio telemetry system that uses radio signals to make the connection. The *recording truck* has an enclosure on the back called a *doghouse*, which contains equipment used to digitally record and store the seismic data.

Plate 13–4. Geophone. (Courtesy of American Petroleum Institute.)

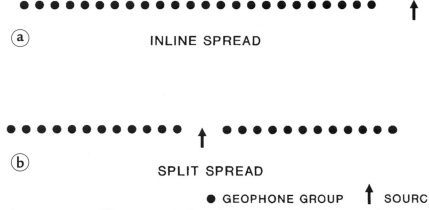

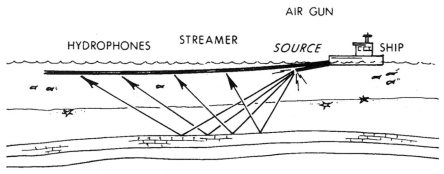

Fig. 13–8. Map of linear spread of geophone groups: (a) inline or end on spread and (b) split spread

The *roll along technique* is often used to move the geophones on land. After each seismic shot, a portion of the geophone cable is detached from one end of the linear spread and moved to the other end. The shot point is then moved an equal distance in the same direction.

At sea, the source is towed in the water behind a boat (fig. 13–9) that travels at about 5 knots (5 nautical miles per hour). The seismic energy is powerful enough for much of the sound to penetrate the ocean bottom. The returning reflections are recorded on vibration detectors, called *hydrophones*, contained in a long plastic tube, the *streamer*, that is towed behind the boat. Wires are run from the hydrophones through the streamer to the doghouse on the ship where the recording equipment and computers are located. The streamer is filled with a clear liquid such as kerosene to be neutrally buoyant. It is strung out for many miles in a straight line behind the boat.

Fig. 13–9. The seismic method at sea

Devices called *birds* or *depth controllers* are used to keep a streamer at a depth of 20 to 50 ft (6 to 16 m). A buoy with a light and positioning equipment is located on the end of the streamer. Offshore seismic data are acquired by an inline spread with the source at the end of a linear spread. To cover a large area, the seismic ship can often tow 12 parallel streamers and four source arrays (plate 13–5). Some ships can tow more than 30 streamers.

Plate 13–5. Seismic ship towing air gun arrays and streamers. (Courtesy of Western Geophysical Division of Baker Hughes.)

In a variation called *seafloor seismic*, the hydrophone streamer is located on the ocean bottom. The seismic source ship travels parallel to the ocean bottom cable or cables. This is used where there are obstructions such as production platforms, in shallow water, or areas of limited access.

Common-depth-point (CDP) or *common-mid-point* (CMP) stacking is a process used to improve the signal/noise ratio by reinforcing actual reflections and minimizing random noise. It involves recording reflections for each subsurface point from different source and detector distances (*offsets*) and combining (*stacking*) the reflections (*traces*). The number of times that each subsurface point is recorded is called the *fold*. It is the number of reflections (traces) that are combined in stacking to produce one stacked reflection. A 48-fold or 4,800% stack uses 48 reflections (traces) off the same subsurface point at different offset distances to form one stacked reflection (trace).

Cableless, cable-free, or no cable nodal seismic acquisition is a trend in modern seismic acquisition to eliminate the long cables that connect geophones and hydrophones with the recording equipment. A *node* is a self-contained battery-powered seismic detector containing a very accurate clock and seismic data recording instruments used on the land surface or seabed. It is completely enclosed in a hard-impact plastic case. The node can usually be remotely started and stopped. Land nodes can contain a global positioning system and a spike on the bottom to plant it. A land node is typically 6 in. (15 cm) high, 5 in. (13 cm) in diameter, and weighs less than 5 lbs (2.3 kg). After recording, the node is retrieved and the sesmic data downloaded.

Seismic exploration is most expensive on land, especially in rugged terrain. It is less expensive and of better quality at sea.

Seismic record

A *seismic* or *record section*, on which the seismic data are displayed, is similar to a vertical cross section of the earth (fig. 13–10). The original vertical scale is in seconds. Zero seconds is always at or near the surface of the ground or exactly at the surface of the ocean. *Time lines*, usually in $^1/_{100}$ seconds, run horizontally across the section. The $^1/_{10}$ second lines are heavier, and the full second lines are heaviest. Across the top of the record are the shot point locations. On the side of the record is the header. The *header* displays information such as the seismic line number and how the data were acquired and processed.

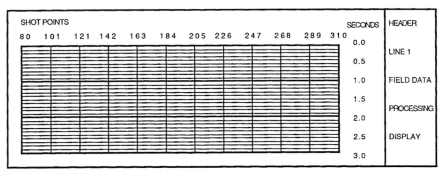

Fig. 13–10. Seismic record format

A *shot point base map* (fig. 13–11) accurately shows the location of the seismic lines and individual shot points.

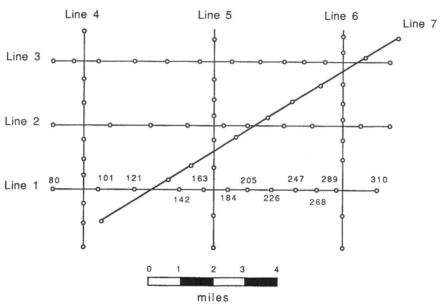

Fig. 13–11. Shot point base map

Depth to a seismic reflector on a seismic record is measured in *milliseconds* ($\frac{1}{1,000}$ of a second). The seismic energy travels down, is reflected off a subsurface rock layer, and returns to the surface. Because it travels twice the depth (down and up), time on a seismic record is *two-way travel time*. The deeper the reflecting layer, the longer it takes for the echo to return to the surface.

Seismic data can be recorded three ways. A *variable area wiggle trace* (fig. 13–12a) uses vertical lines with wiggles to the left or right to record seismic energy. Wiggles to the right are reflections from the subsurface (the geophone detected an upward motion) and are usually shaded black. Wiggles to the left (the geophone detected a downward motion) are left blank. A *variable density display* (fig. 13–12b) uses shades of gray to represent seismic energy amplitude. The darker the shade, the stronger the reflection.

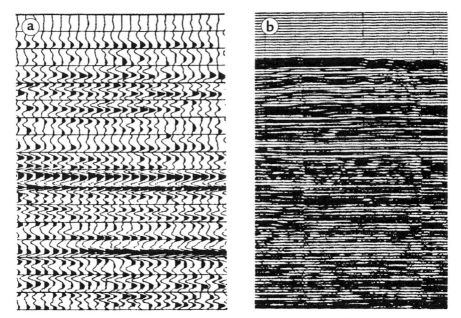

Fig. 13–12. (a) Variable area wiggle trace and (b) variable density display

Colored seismic displays have become common (plate 13–6). The human eye can distinguish many different colors and see more information on a colored display. Seismic interpreters can identify far more subtle trends on a colored display as well. In one method, peaks and troughs of reflections are colored blue and red on a white background (plate 13–6a). In another method, a larger spectrum of colors is used. An example is cyan-blue-white-red-yellow, with cyan the maximum peak amplitude and yellow the maximum trough amplitude.

Interpretation

Each reflection that can be traced across a seismic section is called a *seismic horizon*. Layered rocks on a seismic record are sedimentary rocks (fig. 13–13). Salt domes and reefs, however, do not show layered reflectors. The area with no good, continuous reflections below the layered sedimentary rocks is basement rock; it has short, discontinuous, and disordered reflections.

Any deformation in the sedimentary rocks such as tilting, faulting, or folding is apparent on a seismic record (fig. 13–14). The primary purpose of seismic exploration is to determine the structure of the subsurface rocks. Reefs are identified on a seismic record as mounds without internal layering (fig. 13–15). Salt domes are seen as unlayered plugs (fig. 13–16). Salt dome edges are defined by uplifted and terminated sedimentary rocks.

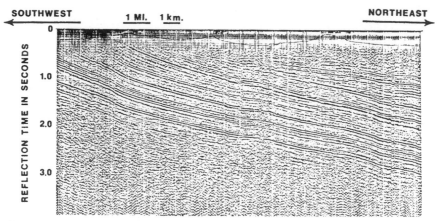

Fig. 13–13. A seismic record in the Wind River basin of Wyoming. Well-layered sedimentary rocks are seen dipping down to the northeast. Unlayered basement rock is located below the sedimentary rocks. (Courtesy of Conoco.)

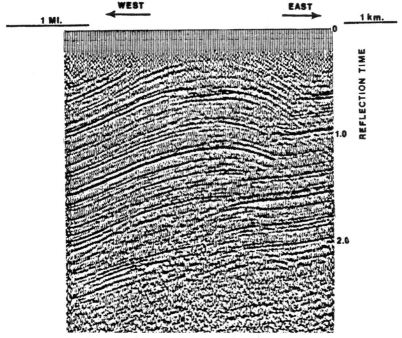

Fig. 13–14. A seismic record in the Big Horn basin of Wyoming. A drag fold occurs on a curving thrust fault. This is the Elk basin oil field. (Courtesy of Conoco.)

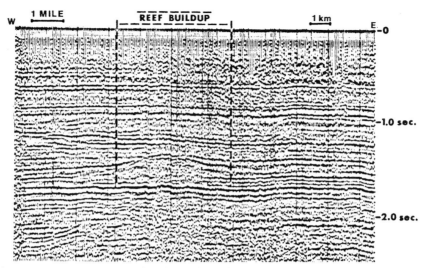

Fig. 13-15. A seismic record in the Midland basin, Texas. A reef, part of Horseshoe atoll, is shown. (Courtesy of Conoco.)

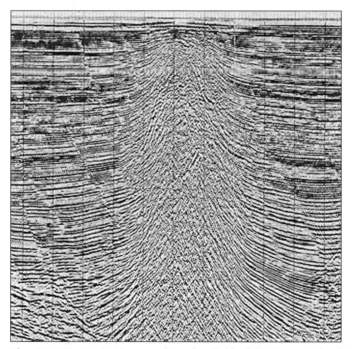

Fig. 13-16. A seismic record in the Gulf of Mexico south of Galveston, Texas. Note the salt dome. (Courtesy of TGS Calibre Geophysical Company and GECO/Prakla.)

Contoured maps of the subsurface can be made using seismic sections. A map of depth in milliseconds to a seismic horizon is called a *time structure map*. It is very similar to a structural map made from well data. An *isotime, isochron*, or *time interval map* uses contours to show the time interval in milliseconds between two seismic horizons (fig. 13–17). It is similar to an isopach map made from well data. If the seismic velocities through the rocks are known, the time structure and isotime maps can be converted into structural and isopach maps with values in feet or meters instead of time in milliseconds.

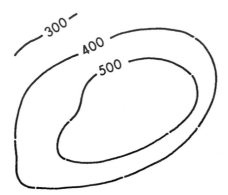

Fig. 13–17. An isochron map contoured in milliseconds

The amplitude of the seismic echo off the top of a surface depends primarily on the contrast in *acoustic impedance* (sound velocity times density) between the upper and lower rock layers that form the surface. The greater the contrast, the larger the reflection. The percent of seismic energy reflected is called the *reflection coefficient*. Typical sedimentary layers have a reflection coefficient of 2 to 4%.

Because gas has a very slow velocity, the slowest velocity sedimentary rock is an unconsolidated gas sand. If overlain by a seal, the acoustic impedance contrast will produce an echo of about 16% of the seismic energy, called a *bright spot*. It is seen as an intense reflector on the seismic profile (fig. 13–18). Bright spots have been used very successfully to locate gas reservoirs and free gas caps on saturated oil fields. Not all bright spots, however, are commercial deposits of natural gas. A *dim spot*, where the reflection amplitude becomes less, occurs over some reefs.

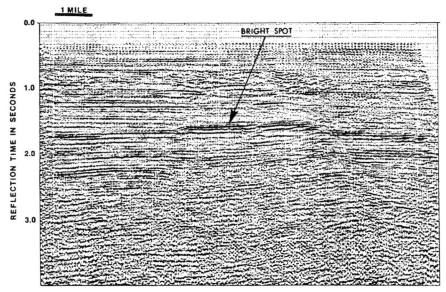

Fig. 13–18. A seismic record in the Gulf Coast over a gas field. Note the bright spot on the gas sand. (Courtesy of Conoco)

If a sonic log is used to compute the velocity, and a density log is used to compute the density of each rock layer in a well that is located on a seismic line or its vicinity, the acoustic impedance of each surface layer can be computed. A *synthetic seismogram*, an artificial, computer-generated seismic record, can be made of computed reflections caused by differences in acoustic impedances. It is compared to the actual seismic record, and the composition of the various rock layers on the seismic record can be identified. The reflection response of seismic energy to various rock layer configurations can be modeled.

Two types of *direct hydrocarbon indicators* on a seismic record are bright spots (fig. 13–18) and flat spots. A *flat spot* is a level, flat seismic reflector in a petroleum trap formed by rock layers that are not flat, such as an anticline. The flat spot is a reflection off a gas-oil contact (plate 13–6a).

The seismic method does not give the fine detail that is seen on well logs and rock outcrops. The large unconformites on seismic records correspond to third-order cycles of sea level rise and fall (fig. 2–7). The rocks deposited between two unconformities on seismic records are called *sequences*. They were deposited during a specific time interval similar to the parasequence sets of sequence stratigraphy (see chap. 12, fig. 12–10). Sequences are composed of parasequence sets that are to small to be seen on seismic records.

Processing

Digital recording of seismic data in the field and computer processing of the seismic data both in the field and in a data center (begun in the 1960s) have greatly improved the accuracy and usefulness of seismic exploration.

A correction (*statics*) is made on the seismic data for elevation changes and the thickness and velocity of the near-surface, loose sediments in the *weathering layer* or *low-velocity zone*.

As the seismic energy travels through the subsurface rocks, the relatively sharp impulse of seismic energy tends to become spread out, and some portions of the energy are lost. *Deconvolution* is a computer process that compresses and restores the recorded subsurface reflections so that they are similar to the original seismic energy impulse. This makes the reflections sharper and reduces some of the noise.

A seismic section is accurate only over flat, horizontal rock layers. Dipping rock layers have a different path for the seismic energy from source to detector than horizontal rock layers in the same position. Because of this, dipping rock layers do not appear on the seismic record in their actual positions. They are shifted to a downdip position and appear flatter than they are. This effect causes anticlines to look larger, and synclines look smaller than they actually are. It causes the rock layers in a deep, steeply dipping syncline to cross, forming a bow tie (fig. 13–19a). Rock layers that sharply terminated against a fault appear to cross with rock layers on the other side of the fault (fig. 13–19b). A computer process called *migration* moves the dipping rock layers into a more accurate position on the seismic record.

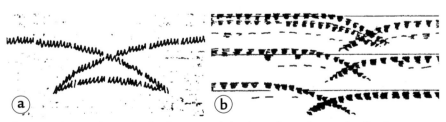

Fig. 13–19. Unmigrated seismic events: (a) a bow tie as the result of a deep, steep syncline and (b) crossing events due to a fault

Many basins such as the Gulf of Mexico and the North Sea have extensive salt layers. Passage of seismic energy through the salt blurs the seismic image of any potential petroleum structures below the salt. A computer processing technique called *prestack migration* of subsalt seismic data results in a clearer seismic image of the deeper structures but involves significantly more computer time. The first subsalt discovery in the Gulf of Mexico was Mahogany in 1993. It is located 80 miles (129 km) offshore from Louisiana in a water depth of 375 ft (114 m). The trap is a faulted anticline with a sandstone reservoir located at a depth of 18,500 ft (4,500 m). The high-pressure sandstone reservoir has up to 33% porosity and 2.5 darcys (D) permeability. The Louann Salt is located above it at a depth of 15,000 ft (4,500 m), and wells have to be drilled through more than 3,000 ft (914 m) of salt to reach the reservoir.

Each seismic line is run to intersect another seismic line (*tie in*) so that the reflections can be correlated from one record to another. If the reflections from two intersecting seismic records do not correlate, it is called a *mis-tie*.

A typical seismic record shows the structure of the subsurface rocks and identifies sedimentary rocks by their characteristic layering. It does not, however, identify the individual sedimentary rocks layers, such as San Andreas Limestone or even the rock type. A seismic record is more valuable when the individual sedimentary rock layers have been identified, and potential reservoir rocks and seals can be traced. To do this, the seismic line is often run (*tied in*) through a well that has been already drilled. The well logs from that well then provide the basis for identifying subsurface rock layers on the seismic record. If no well is available, a *stratigraphic test well*, or *strat test*, is drilled on the seismic line. The primary purpose of the well is to collect subsurface samples and run wireline well logs. This identifies the ages and composition of reflections on the seismic profile.

In *reprocessing*, new methods of computer processing are applied to old digital seismic data. Because new fields can be found by reprocessing old data, the seismic data is never released from the company that owns it. Any information kept secret is called *proprietary*. *Seismic brokers* are used to sell and buy proprietary seismic data.

Time-to-depth conversion

Seismic data are recorded in seconds (*time domain*), and a well log is recorded in feet or meters (*depth domain*). Because of this, the vertical scales on each are different and cannot be directly compared. If the seismic velocity through each rock layer is known, a *time-to-depth conversion* can be made on the seismic data to make it compatible with well-log data.

Two ways to measure seismic velocities are by checkshot survey and vertical seismic profiling. In a *checkshot survey*, a geophone is lowered down the well. The seismic source (e.g., dynamite, air gun, or vibroseis) is then detonated on the surface. The geophone is then raised up the well a distance of 200, 500, or 1,000 ft (60, 150 or 300 m), and another measurement is made. This is repeated until the geophone is on the surface. *Vertical seismic profiling* (*VSP*) is the same as a checkshot survey except that the geophone interval is shorter (50 to 100 ft or 15 to 30 m).

Amplitude versus offset

Amplitude versus offset (*AVO*) is an analysis of seismic data to locate gas reservoirs and help identify the composition of the rock layers. *Offset* is the distance between the seismic source and the receiver. The amplitude of a reflection usually decreases with increasing offset distance. Gas reservoirs and different sedimentary rocks such as sandstones, limestones, and shales have different reflection amplitudes versus offsets. Some increase, and others decrease with offset.

3-D seismic exploration

In the 1980s and 1990s, three-dimensional (3-D) seismic exploration (fig. 13–20) was developed. This method produces a three-dimensional seismic image of the subsurface. On land, 3-D seismic data are often acquired by *swath shooting* with receiver cables laid out in parallel lines, and the shot points run in a perpendicular direction. In the ocean, 3-D seismic exploration is often run with *line shooting* in closely spaced, parallel lines from a single ship towing several arrays of air guns and streamers.

There are many different patterns of sources and geophones that can be used for 3-D seismic acquisition. In *undershooting,* the sources and geophones are not even located on the land being surveyed. The source is located on one side, and the geophones are located on the other side of the land.

The 3-D seismic survey is divided into horizontal squares called *bins.* All reflections whose midpoints fall within a particular bin are combined for common-mid-point (CMP) stacking. The CMP fold is the number of midpoints in each bin. Bins are commonly 55 by 55 ft, 110 by 110 ft, 20 by 20 m, or 30 by 30 m. This technique is similar to how CMP stacking is used in two-dimensional (2-D) seismic imaging to improve the signal/noise ratio by reinforcing actual reflections and minimizing random noise.

After computer processing, a 3-D view of the subsurface is produced. Rock layers are migrated more accurately, and more details are shown than on a 2-D seismic image. A *cube display* is very common (plate 13–6b). The cube can be made transparent so that only the highest amplitude reflectors are shown. The 3-D seismic image on a computer monitor can be rotated and viewed from different directions. A *time* or *horizontal slice* (fig. 13–20 and plate 13–6c) of the subsurface is a flat seismic picture made at a specific depth in time (milliseconds). Various reflectors that intersect the slice are shown. A single seismic reflection can be displayed as a *horizon slice,* and a fault surface can be shown as a *fault slice.*

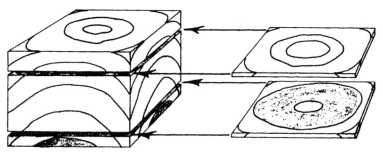

Fig. 13–20. 3-D seismic cube diagram with two time slices

Special rooms called *visualization centers* (and several other names) are used to display the 3-D seismic images (plate 13–7). In one type of room, there are screens on the walls, and the viewers sit in chairs. A computer operator projects the seismic image on the screens and can move the image (plate 13–7a). In another type of room called *immersive*, there are screens on three walls and the floor. The viewer is immersed in the 3-D seismic image and can walk through the subsurface (plate 13–7b). As the viewer turns his or her head and moves, the subsurface perspective moves with the viewer.

Three-dimensional seismic exploration is expensive because of acquisition costs and computer processing. A 3-D seismic survey can have hundreds of gigabytes (10^9) and even terabytes (10^{12}) of information. However, more 3-D seismic exploration, both on land and in the ocean, is being run today than 2-D seismic exploration. It saves money by decreasing the percentage of exploration dry holes. It also saves money during developmental drilling by accurately imaging and defining the subsurface reservoir. The optimum number of developmental wells can then be drilled into the best locations to drain the reservoir efficiently.

4-D and 4-C seismic exploration

Four-dimensional (4-D) or *time-lapse seismic exploration* uses several 3-D seismic surveys over exactly the same producing reservoir at various time intervals (e.g., two years) to trace the flow of fluids though the reservoir. As a reservoir is drained, the temperature, pressure, and composition of the fluids change. Gas bubbles out of the oil, and water replaces gas and oil as they are being produced. Time slices of the reservoir are compared, and changes in the seismic response such as amplitude can document the drainage. Undrained pockets of oil can be located and wells drilled to drain them.

Four-component (4-C) or *multicomponent seismic exploration* records both compressional and shear waves that are given off by a seismic source.

A *compressional wave* (*P-wave*) is how sound travels through the air. Particles through which the compressional wave is traveling move closer together and then farther apart (fig. 13–21a). A *shear wave* (*S-wave*) is like a wave on the surface of the ocean. The particles move up and down (fig. 13–21b). Shear waves are slower than compressional waves and cannot pass through a liquid or gas. The conventional seismic method records only the compressional waves with a one-component geophone. The 4-C seismic method records the compressional wave (one component) and also uses three geophones that are perpendicular to each other to record the shear wave (three components).

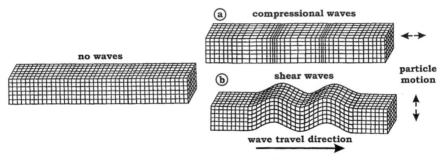

Fig. 13–21. Seismic waves: (a) compressional and (b) shear

Four-component seismic exploration is used to locate and determine the orientation of subsurface fractures. Because P- and S-waves have different velocities through different rocks, 4-C seismic exploration can determine the composition of subsuface rock layers. P-waves are distorted by gas in sedimentary rocks, but S-waves are not. Because of this, recording S-waves produces a more accurate picture of sedimentary rocks that contain gas.

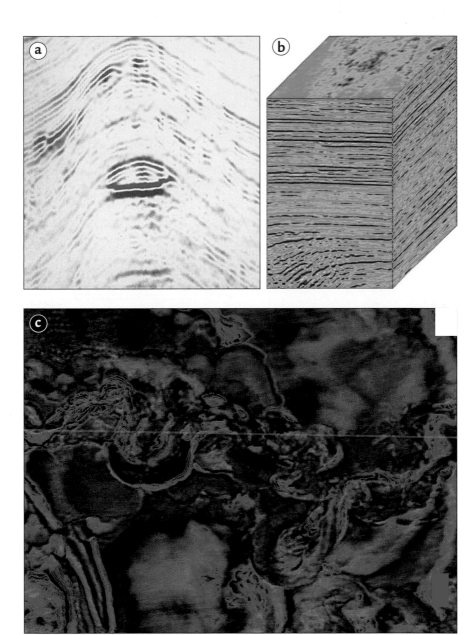

Plate 13–6. Colored seismic records: (a) anticline showing a flat spot on the oil-water contact (courtesy of Geophysical Service Inc.); (b) cube display of 3-D seismic data (courtesy of Bureau of Economic Geology, The University of Texas at Austin); and (c) horizontal slice from a 3-D seismic showing a meandering submarine channel on a submarine fan, offshore Niger River Delta (courtesy of CGGVeritas)

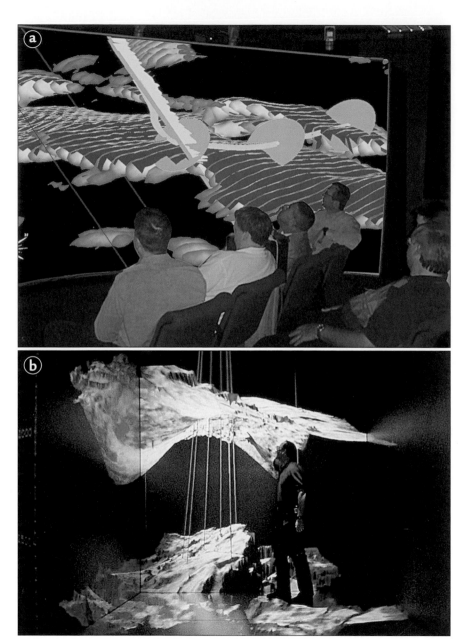

Plate 13–7. (a) 3-D seismic visualization station (courtesy of ARCO and Silicon Graphics) and (b) 3-D seismic visualization station (courtesy of Landmark Graphics, a Halliburton Company)

Drilling Preliminaries

The API reported that 110,721 wells were drilled in the world during 2008. Of those, 54,195 (49%) were drilled in the United States, 18,661 (17%) in Canada, and 1,790 (1.6%) in the Middle East. The average depth of a US onshore well was 6,225 ft (1,897 m) at an average cost per well of $4,351,000, or $699 per foot. The deepest well drilled for petroleum in the world was the Bertha Rogers No. 1 to a depth of 31,441 ft (9,583 m) in Oklahoma during 1974. It was a dry hole.

A geologist will locate the *wellsite* or *drillsite*, a location to drill and possibly find commercial oil and gas. A drilling target or targets are identified. A *drilling target* is a potential reservoir rock such as the Bartlesville Sandstone. Depth to the drilling target is estimated. Geologists can determine whether the drilling target is good reservoir rock based on wells that have already been drilled though the drilling target in that area.

Land and Leasing

In the United States and parts of Canada, private land is called *fee land* and has two separate ownerships: a surface rights owner and a mineral rights owner. The *surface rights owner* can build a house, ranch, or farm on the land. The *mineral rights owner* can explore and drill for gas and oil and owns and can produce the gas and oil. The surface owner and mineral rights owner are not necessarily the same person. The mineral rights could have been separated from the surface rights and sold. In every other country of the world besides the United States and Canada, the federal government owns the mineral rights to all the land. About one-third of the mineral rights on land in the United States is owned by the federal or state government in public land. In Canada, land owned by the federal or provincial government is called *crown land*.

It is the job of a *landman* to identify and locate the mineral rights owner of fee land. This is done by searching through the county or parish courthouse records or by using a commercial title company. Commercial land ownership maps that are frequently updated can be used to determine the status of leases, the names of lessees, and the identity of surface rights and mineral rights owners. A *title opinion* can be obtained from an attorney who determines the mineral rights ownership history of the land. The landman then approaches the mineral rights owner and attempts to persuade the owner to sign a lease.

A *lease* is a legal document that grants the right to explore and drill for gas and oil during the life of that lease. The mineral rights owner, the *lessor*, receives a bonus and a royalty, both negotiated, for signing the lease. A *bonus* is money for signing the lease. Bonuses can be $25 up to $25,000 per acre depending on how promising the land is. A *royalty* is a promised fraction of the gross revenue from any oil and gas that is produced from that land, free and clear of production costs. A standard royalty used to be one-eighth but now is commonly three-sixteenths, one-fifth, or sometimes one-fourth. The exploration company, the *lessee*, is granted the right to explore and drill on that land during the *primary term* of the lease. Primary terms between one and five years are common. If commercial petroleum in paying quantities from that land is not started by the end of the primary term, the lease expires. If petroleum is found and commercial petroleum in paying quantities is established before the end of the primary term, the lessee legally own the gas and oil and has the right to produce it for the life of the field during the *secondary term* of the lease that automatically takes effect.

Leases are printed on standard forms. Two types of leases are paid up and delay rental. In the *delay rental lease*, a specific sum (*delay rental*) must be paid to the lessor if drilling has not commenced by the end of each year during the primary term to maintain the lease. If not, the lease expires. The *paid-up lease* does not require delay rentals.

Leases are real property. They can be bought, sold, and traded. One company can transfer (*farm out*) a lease for a royalty (such as one-fourth) or some other consideration to another company that will drill the lease. Leases accepted from another company are *farmed in*.

Producing properties have both royalty and working interest owners. The *royalty interest* receives a fraction of the gross oil and gas revenue, free and clear of the cost of production. A *landowner's royalty* is granted to the mineral rights owner when the lease is signed. Other royalties might go to the landman, geologist, or promoter who put the deal together. This is called an *overriding royalty* that was created from the working interest. The *net revenue interest* is the percentage, such as 87.5%, that is left after all royalties have been deducted. The *working interest owners* receive the remaining portion of oil and gas revenue after the royalty interest owners have taken their share and production expenses have been paid. Working interest owners are responsible for all drilling and production costs.

After a well is completed as a producer on fee land, the operator of the well fills out a *division order* listing the name of each partial owner of the well (net revenue interest owner), address, percent interest, and how they are to be paid their share of the production revenue. Each interest in the well receives a copy of division order. *Division order analysts* in the land department of the operator's company administer the distribution of production revenue and maintain those records.

Foreign Contracts

An oil company that operates in many countries is called a *multinational* or *international company*. A company that is owned by a federal government and usually operates only in that country is called a *national* or *host company*. There are three phases of oil and gas operations. The *exploration phase* includes geological, geochemical, and geophysical exploration and drilling of exploration wells. The *exploitation phase* involves the development of newly discovered fields. The *production phase* occurs during oil and gas production.

Several types of contracts can be negotiated between a multinational company and a foreign government, such as a concession agreement, production-sharing contract, service contract, and production contract. A contract can involve a specific *concession*, an area of land and/or ocean bottom to be explored during a specific time called the *contract time*.

The oldest contract is a *concession agreement* (also known as a license/ concession agreement or a tax and royalty agreement). A multinational company is granted an exclusive concession and bears the entire cost and risk of exploration, exploitation, and production. The host country is paid bonuses, taxes, and royalties on production. In a variation of this contract, the multinational company still bears the costs and risks of exploration, but the host country will share the cost and risk of exploitation.

A *production-sharing contract* is common today. The multinational company is granted a concession to explore during a specific contract time. During the contract time, the company bears the entire cost of exploration and drilling. If commercial amounts of oil or gas are not found by the end of the contract time, the contract becomes invalid and the company loses all the costs of exploration and drilling. If commercial amounts of gas or oil are found, an agreed-upon share of the gross oil and gas production, called *cost oil*, goes to the company to sell and recover the costs of exploration, drilling, and production. After costs have been recovered, the remaining oil, called *profit oil*, is split by an agreed formula between the multinational company and the host government or company.

A *service contract* provides a contractor with a fee for specific services such as exploration or production. A *production contract* involves a contractor taking over an existing or underdeveloped field and improving production. The contractor is paid a portion of the increased production.

Authority for Expenditure

Before a well is drilled, an *authority* or *authorization for expenditure* (*AFE*) is completed (fig. 14–1). This form estimates the cost of drilling and completing the well, both as a dry hole and a producer. Costs such as drilling intangibles, completion intangibles, and equipment are listed. *Intangibles* are salaries, services, and equipment that cannot be salvaged after the well is drilled. The AFE includes the cost of the drilling rig, mud, logging, testing, cementing, casing, well stimulation, prime movers, pumps, tubing, separator, and other services and expenses. It is used to economically evaluate the well before it is drilled. The operator and any other financial contributors to the well approve the AFE. The operator then uses the AFE as a guideline for expenditures.

The original AFE for the Macondo well that blew out in the Gulf of Mexico was about $96 million. However, due to cost overruns during drilling, the final AFE was more than $154 million.

AFE
(Authority For Expenditure)

	Dry Hole	Completed Well
Drilling Intangibles		
Site Preparation	_____	_____
Drilling Footage	_____	_____
Cement and Surface Casing	_____	_____
Logging	_____	_____
Mud	_____	_____
.	_____	_____
.	_____	_____
Completion Intangibles		
Cement	_____	_____
Perforating	_____	_____
Frac or Acid	_____	_____
.	_____	_____
.	_____	_____
Equipment		
Casing	_____	_____
Tubing	_____	_____
Pump Jack	_____	_____
Separator	_____	_____
Tanks	_____	_____
.	_____	_____
.	_____	_____
Total Well Cost	$_____	$_____

Fig. 14–1. Authority for expenditure (AFE)

Economic Analysis

Two important aspects of evaluating whether a well should be drilled are risk and reserves. *Chance of success, success rate,* or *risk* is an estimate of the decimal or percent chance the well has of finding commercial amounts of gas or oil and being completed as a producer. For example, the chance of success of drilling an oil or gas producer could be 0.65, or 65%. Chance of success can be estimated from historical drilling data from that area and type of well, or a geologist can evaluate a particular drilling prospect by making a detailed risk analysis of all the geological factors that are necessary for a successful well, such as the presence of reservoir rock, seal, and trap. In addition, the geologist needs to calculate the oil or gas reserves. *Reserves* are the estimated amounts of gas or oil that can be produced from that well. These amounts can be calculated from a formula (volumetric or engineering formula, chapter 25).

Two common methods of evaluating a drilling prospect are return on investment (ROI) and payout (PO). *Return on investment* is the net revenue from oil and gas production sales divided by the maximum cash outlay for drilling and completion. Because money is spent immediately for drilling and completion and the revenue from oil and gas production sales is spread out over many years after that, the time value of money must be considered by *discounting*. Tables with an annual discount rate are used to determine what the present value of the money is when received at a later date. For example, $100 of net oil and gas production revenue 10 years from now at a discount rate of 15% is worth only $24.72 in present value today. Present value is used in the ROI calculation that is then multiplied by the chance of success expressed as a decimal to calculate the *risk-adjusted* ROI.

Payout is the time in months or years that it takes net revenue from oil and gas production sales to equal the money expended on drilling and completing the well. It is also discounted for the time value of money (*discounted* PO). Discounting causes the return of investment to be smaller and the payout to be longer. The best drilling prospects have higher risk-adjusted ROIs and shorter discounted payouts.

The economic value of drilling a prospect can also be evaluated by the *internal rate of return* (*IRR*). It is the investment rate, such as 20%, that applied to the cost of drilling and completing a well will equal the net revenue interest from oil and gas production sales over the life of the well. Some companies set a minimum IRR before drilling is approved. If the drilling project does not meet that minimum IRR, the money can be better spend on another investment.

Drilling Contracts

Drilling rigs are usually owned, maintained, and operated by *drilling contractors*. The exploration company signs a drilling contract with the drilling contractor to drill the well to a specific depth or horizon (drilling target) at a specific location (drillsite). The drilling contract includes the *spud date* (when the well is to be started), hole diameters, how much the well can deviate from vertical, drilling mud to be used, logging and testing to be done, casing sizes and depth of each casing string, cementing, how the well is to be completed, the drill collars to be used, and the subsurface formations to be drilled. The drilling contract also contains rates for when the well is not being drilled, such as during standby or logging operations.

There are three common types of drilling contracts. A *footage drilling contract* is very common on land. It is based on a cost per foot to drill down to the contract depth. A *daywork contract* is common offshore and is based on a cost per day to drill down to contract depth. A *turnkey contract* has an exact cost to drill down to the contract depth. It can also have the obligation to complete and equip the well based on well tests. A *combination contract* has a footage rate to a certain depth and a daywork rate below that. Standard drilling contracts by the API and the American Association of Oilwell Drilling Contractors (AAODC) are commonly used.

Service and supply companies are contracted. A *service company* performs a specific service such as wireline logging or mud engineering. A *supply company* furnishes equipment such as casing.

Joint Operating Agreements and Support Agreements

There are several ways to develop an area with a limited budget, to encourage a well to be drilled, or to reduce the financial impact of possibly drilling a dry hole or holes. A company can enter into a joint operating agreement with one or more other companies. A *joint operating agreement (JOA)* can be for drilling a single well or for the development of a larger *working interest area*. The JOA defines the rights and duties of each party including each party's share of expenses. An *operator* who is in charge of the day-to-day operations is identified. After the well or wells are drilled, the JOA defines how the production is to be shared by the parties.

A *support* or *contribution agreement* can be used to encourage and support drilling a well. There are three types. In a *dry-hole agreement*, a party agrees to make a cash contribution if the well being drilled by another party is a dry hole. In return, that party receives the geological and drilling information from that well whether or not the well is a dry hole. In a *bottom-hole contribution agreement*, a party agrees to make a cash contribution to the party drilling a well to a certain depth in return for geological and drilling information on that well. In an *acreage contribution agreement*, a party contributes leases or interests to another party who is drilling a well in that area in return for geological and drilling information on that well.

Site Preparation

To *stake a well*, a surveyor accurately determines the well location and elevation. A *plat* (map) of the site is prepared and registered with the appropriate government agency. A bulldozer can be used to grade an access road to the site and make a turnaround. The bulldozer then clears and levels the site (*drilling pad*). A drilling pad is commonly 4 to 6 acres (0.016 to 0.024 km²) in area and is often covered with gravel. Boards might be laid if the ground is wet. A matting, often made of 3 × 12 in. (7.5 × 30 cm) timbers, can be spread on the surface to support the rig and improve drainage. A large pit, the *reserve pit*, is dug and lined with plastic next to the drilling rig. It will hold unneeded drilling mud, cuttings, and other materials from the well. Provisions are made for a water supply at the drilling site by drilling a water well or laying a water pipeline.

The deeper the well, the larger and stronger the rig has to be to support the drillpipe on the drill floor as the pipe is being pulled out of the well. Each drilling rig is rated for maximum depth. If the well is shallow (< 3,000 ft or 1,000 m), the entire drilling rig comes on the back of a truck or trailer. This is a *truck-mounted* or *portable rig*. If the well is deeper, the rig comes in modules on the back of several tractor-trailers. The modules are specifically designed to be trucked and be quickly fitted together at the drillsite with large pins secured with cotter pins (plate 14–1).

Plate 14–1. Pin with cotter pin used to secure drilling rig modules

For a deep well, a rectangular pit (*cellar*) can be dug (fig. 14–2) and lined with boards or cement. The cellar provides space below the drill floor for the blowout preventers. In very remote areas, a *helirig* is used. The rig is made of specially designed modules that are transported by helicopter.

The rig is assembled during *rig up* and disassembled during *rig down*. The start of drilling a well is called *spudding in*. Spudding in a medium or deep well usually begins with a small truck-mounted rig that drills a large-diameter but shallow hole (20 to 100 ft or 7 to 30 m) called the *conductor hole*. Large-diameter pipe (20 in. or 50 cm), called *conductor casing* or *pipe*, is then run and cemented into the conductor hole (fig. 14-2). In soft ground, the conductor casing can be pile-driven without drilling. The conductor casing stabilizes the top of the well and provides an attachment for the blowout preventers in areas where shallow gas could be encountered.

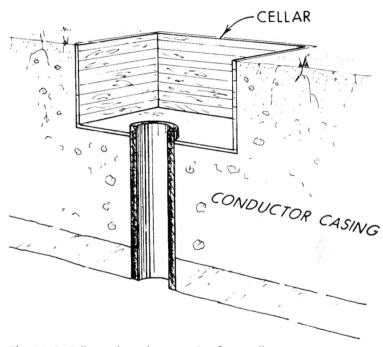

Fig. 14–2. Cellar and conductor casing for a well

Types of Wells

A well drilled to discover a new oil or gas reservoir (plate 14–2) is called a *wildcat* or *exploratory well*. It can be drilled in an area that has no production (*new-field exploratory well*) or to test a new reservoir rock that has no current production in a producing area (*new-pool exploratory well*) that is either shallower (*shallower pool test*) or deeper (*deeper pool test*) than current production. An exploratory well can also be drilled to significantly extend the limits of a discovered field or to significantly extend the limits of a discovered reservoir (*outpost* or *extension test*, or *step-out well*). A *rank wildcat* is drilled at least 2 miles (3 km) away from any known production. If the well does discover a new field, it is called the *discovery well* for that field. As soon as possible after a discovery, the size of the field must be determined. If this is private fee land, it must be determined which leases need to be drilled to maintain the leases and which can be abandoned. If this is an offshore field or in a remote area or foreign country, the size of the field needs to be determined to compute the amount of oil and gas that can be produced (reserves). This will determine if the field is large enough to economically justify further development. Field size is determined by *step-out*, *delineation*, or *appraisal wells* that are drilled to the sides of the discovery well. If the oil-water or gas-water contact can be located on all four sides of the discovery well, the area of the field can be determined. Wells drilled in the known extent of the field are called *developmental wells*. Wells drilled between producing wells in an established field to increase the production rate are called *infill wells*.

Government Regulations

Before the early 1930s in the United States, there were no regulations about how close the wells could be located or how fast the oil could be produced. Today, governments prevent the exploitation of a field with excessive drilling and production rates. Each well in a field is given a *drilling and spacing unit* (*DSU*), a square (sometimes a rectangle) of a certain surface area on which only one well can be drilled and completed (fig. 14–3). An area of 10, 20, or 40 acres is typical for an oil well. Oil viscosity and reservoir permeability are two important factors in determining well spacing. Higher viscosity oils and lower permeability reservoirs need smaller spacing for efficient drainage. Gas wells drain a larger area, and DSUs of 640 acres are common. Usually the well does not have to be located in the center of the DSU but cannot be located on the edge. In some countries production is limited by an allowable to prevent excessive production rates. An *allowable* is the maximum amount of oil and gas production that is permitted from a single well or field during a specific unit of time such as a month.

Plate 14–2. Wildcat well being drilled by a modern rotary drilling rig. (Courtesy of Parker Drilling.)

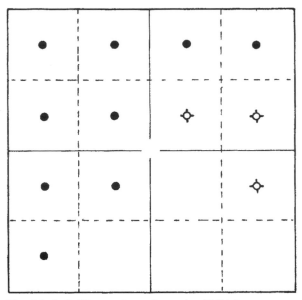

Fig. 14–3. Drilling and spacing units (DSUs)

Cable Tool Rigs

When the first commercial oil well in the United States was drilled at Titusville, Pennsylvania, in 1859, a cable tool rig was used (see chap. 12, plate 12–1). Cable tool rigs had been in use for hundreds of years before that to drill for freshwater or for brines that were evaporated for salt.

A *cable tool rig* is relatively simple (fig. 14–4). The hoisting system consists of a tower with four legs called a derrick (which was originally wooden), 72 to 87 ft (21.9 to 26.5 m) high. An engine, originally a steam engine, causes a wooden walking beam to pivot up and down on the Sampson post. The bit, a solid steel rod about 4 ft (1.2 m) long with a chisel point on it, is suspended down the well from the opposite end of the Sampson post by a rope or cable. As the walking beam pivots up and down, it causes the rope and bit to rise and fall. The bit pounds the well down by pulverizing the rock. The rope or cable is wound around a reel called a bull wheel. The rope or cable goes from the bull wheel, up over a single wheel (crown block) at the top of the derrick, then down to a temper screw on end of the walking beam, and finally down the well to the bit. As the well is pounded deeper, more rope or cable is let out by turning the temper screw.

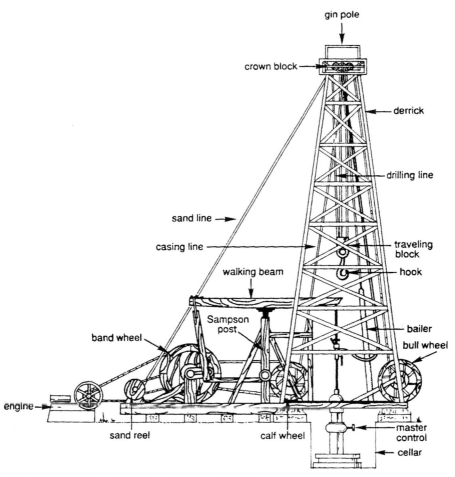

Fig. 14–4. Cable tool drilling rig (Hyne, 1991)

After drilling 3 to 8 ft. (0.9 to 2.4 m), the bottom of the well becomes clogged with rock chips. The bit is then raised, and a bailer is lowered into the well on a sand line to remove the rock chips and water. After the bailer is raised and emptied, the bit is run back into the well to pound deeper. Heavy casing is run down the well from wire rope wound around the calf wheel. The wire rope runs through a multiple sheave crown block at the top of the derrick. The casing (large-diameter pipe) is used in the well to keep water from filling the well and to prevent the sides from caving in. Lighter equipment, such as the bailer, is run in the well on a sand line from the sand reel.

Cable tool drilling is very slow with 25 ft. (7.6 m) per day being an average and 60 ft. (18.3 m) being very good. It does not effectively control subsurface pressures, and blowouts were common during cable tool operations. However, all fields discovered during the 1800s were drilled by cable tool rigs. A cable tool rig in New York drilled a well to a depth of 11,145 ft. (3,397 m) in 1953.

The rotary drilling rig that replaced the cable tool drilling rig was introduced in various areas throughout the world from 1895 to 1930. The greatest advantage of the rotary drilling rig is that it could drill the well considerably faster (several hundred to several thousand feet per day). There were, however, some major problems to be worked out with the early rotary drilling rigs, and they were not immediately accepted. In 1950, there was an equal number of active cable tool and rotary drilling rigs in the United States. Today, almost all wells are drilled with rotary drilling rigs.

Drilling a Well—The Mechanics

Baker Hughes (http://investor.shareholder.com/bhi/rig_counts/rc_index.cfm) reported that there was a total of 3,121 active drilling rigs in the world. Of those, 2,783 (89%) were land rigs, and 1,887 (60%) were in the United States and 415 (13%) were in Canada.

Today, almost all wells are drilled with rotary drilling rigs (plate 15–1). A *rotary drilling rig* rotates a long length of steel pipe with a bit on the end of it to cut the hole called the *wellbore*. The rotary rig consists of four major systems: power, hoisting, rotating, and circulating systems (fig. 15–1).

Power

The *prime movers* are diesel engines that supply power to the rig and are usually located on the ground in back of the rig. Diesel fuel is stored in tanks near the engines. Most of the power is used by the hoisting and circulating systems. Some also goes to the rotating system, rig lights, and other motors.

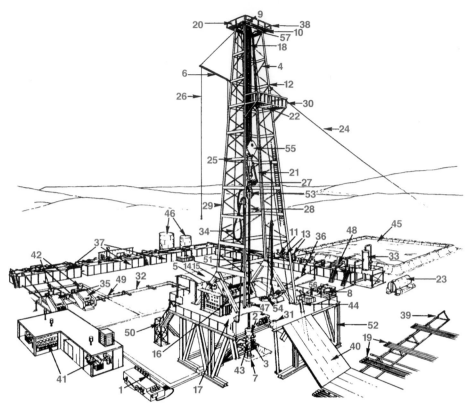

Fig. 15–1. Rotary drilling rig: (1) accumulators, (2) annular blowout preventer, (3) blowout-preventer stack, (4) brace, (5) cathead, (6) cateline boom, (7) cellar, (8) choke manifold, (9) crown block, (10) crown platform, (crow's nest), (11) mud gas separator, (12) derrick or mast, (13) desanders and desilters, (14) dog house, (15) drawworks, (16) driller's console, (17) drill (derrick) floor, (18) drilling (hoisting) line, (19) drillpipe on pipe rack, (20) duck's nest, (21) elevators, (22) finger board, (23) fuel tank, (24) Geronimo line, (25) grit, (26) hoisting (drilling) line, (27) hook, (28) kelly, (29) leg, (30) monkeyboard, (31) mouse hole, (32) mud discharge line, (33) mud/gas separator, (34) mud (rotary) hose, (35) mud pumps (hogs), (36) mud return line, (37) mud tanks, (38) pigpen, (39) pipe rack, (40) pipe ramp, (41) prime movers, (42) pulsation dampeners, (43) ram blowout preventers, (44) rat hole, (45) reserve pit, (46) reserve tanks, (47) rotary table, (48) shale shaker, (49) shock hose, (50) stairways, (51) standpipe, (52) substructure, (53) swivel, (54) tongs, (55) traveling block, (56) trip tank, (57) water table (Hyne, 1991)

Depending on the size of the rig and the drilling depth, there are one, two, or four engines. Each engine is rated by horsepower and fuel consumption. They commonly develop 1,000 to 3,000 horsepower (hp). Power from the diesel engines is transmitted to the rig mechanically by a system of pulleys, belts, shafts, gears, and chains called a *compounder*. A drilling rig powered by only diesel engines with the power transmitted by a compounder is called a *mechanical rig*.

Newer rigs are *diesel-electrical rigs* with the diesel engines coupled to an alternating current (AC) or direct current (DC) generator that supplies electrical power though electrical cable to the rig. The rig floor equipment is driven with more efficient AC or DC electric motors.

Hoisting System

The hoisting system is used to raise and lower and to suspend equipment in the well (fig. 15-2). The *derrick* or *mast* is the steel tower directly above the well that supports the crown block at the top and provides support for the drillpipe to be stacked vertically as it is pulled from the well. If the tower comes on a tractor-trailer and is jacked up as a unit, it is a *mast*. All land rigs use masts. On a *cantilevered mast rig*, the mast is transported in sections, assembled horizontally (fig. 15-3) and then pivoted up to a vertical position using the traveling block and drawworks on the rig. Masts are stabilized by guywires that radiate out from the top of the mast to anchors in the ground. If the tower is erected vertically on the site, it is a *derrick*. All offshore rigs use derricks. Derricks and masts are commonly 80 to 187 ft (24.4 to 57 m) tall to accommodate two, three, or four joints of vertical drillpipe in a stand. They have a square cross section with four vertical *legs* made of structural steel. The horizontal structural members between the legs are called *girts* (fig. 15-4). The diagonal members are *braces*. An inverted, V-shaped opening in the front of the derrick or mast called the *V-door* allows drillpipe and casing to be pulled up the pipe ramp onto the drill floor.

Derricks and masts are rated for maximum drillpipe load. They are also rated for wind load and can commonly withstand winds of 100 to 130 miles per hour (160 to 208 km/hr). The base of the mast or derrick is a flat, steel surface called the *drill*, *derrick*, or *rig floor*, where most of the drilling activity occurs. Two *substructures* made of a steel framework 10 to 30 ft (3 to 10 m) high can be used to raise the drill floor above the ground (plate 15-1). This is done to provide space for wellhead equipment below the drill floor such as the blowout preventers (BOPs) when drilling a deep well.

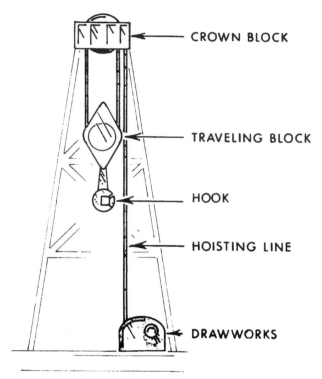

CROWN BLOCK

TRAVELING BLOCK

HOOK

HOISTING LINE

DRAWWORKS

Fig. 15–2. Hoisting system

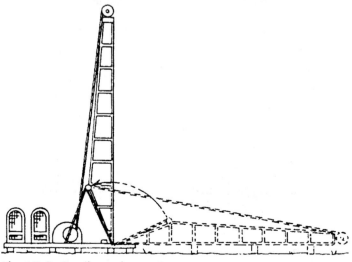

Fig. 15–3. Cantilevered mast rig

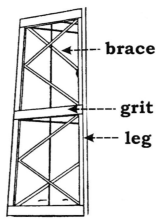

Fig. 15–4. Part of derrick showing vertical legs, horizontal girts, and diagonal braces

Plate 15–1. Rotary drilling rig. (Courtesy of Parker Drilling.)

The *drilling* or *hoisting line* is made of braided steel wire about 1⅛ in. (3 cm) in diameter. The line consists of several strands of braided steel wire wound around a fiber or steel core. The hoisting line is described by the type of core, number of strands around the core, and the individual wires per strand. There are several ways to wrap the strands around the core.

The hoisting line is spooled around a reel on a horizontal shaft in a steel frame called the *drawworks* on the drill floor. The prime movers drive the drawworks to wind and unwind the drilling line. There are several speeds and both forward and reverse on the reel. The driller controls the drawworks from a *brake*, a hand lever, on the drill floor. Drawworks are often rated by input horsepower that commonly range from 500 to 3,000 hp. Small spools called *catheads* are attached to a *catshaft* that runs horizontally through the drawworks. They are used to pull lines such as the jerk or spinning line.

On the drilling rig, there are two sets of wheels (*sheaves*) on horizontal shafts in steel frames called *blocks*. The drilling line from the drawworks goes over a sheave in the *crown block* that is fixed at the top of the derrick or mast. It then goes down to and around a sheave in the *traveling block* that is suspended in the derrick or mast. The drilling line goes back and forth through sheaves in the crown and traveling block 4 to 12 times. The end of the drilling line is fixed to a *deadline anchor* located under the drill floor. After a certain amount of usage, the drilling line is moved 30 ft. (9 m) through the anchor to prevent wear on any particular spot along the line. Below the traveling block is a *hook* for attaching equipment. As the drilling line is reeled in or out of the drawworks, the traveling block and hook rises and falls in the derrick or mast to raise and lower equipment in the well.

Rotating System

The rotating system is used to cut the hole (fig. 15–5). The turning drillpipe, bit, and related equipment are called the *drillstring*. Suspended from the hook directly below the traveling block is the *swivel*. The swivel allows the drillstring that is attached below it to rotate on bearings in the swivel while the weight of the pipe is suspended from the derrick or mast.

Below the swivel is located a very strong, four- or six-sided, high-grade molybdenum steel pipe called the *kelly* that is 40 or 54 ft (12.2 or 16.5 m) long (plate 15–2). The kelly has sides to enable it to be gripped and turned by the rotary table. The kelly turns all the pipe below it to drill the hole. The *rotary table* is a circular table in the drill floor that is turned clockwise (*to the right*) by the prime movers. If it were turned in the opposite direction, the drillpipe would unscrew. The kelly goes through a fitting called the *kelly bushing* (plate 15-2 and fig. 15-6a), which fits onto the *master bushing* (fig. 15-6b) in the rotary table. Rollers in the kelly bushing allow the kelly to slide down through the kelly bushing as the well is drilled deeper. The rotary table, master bushing, kelly bushing, and kelly rotate as a unit.

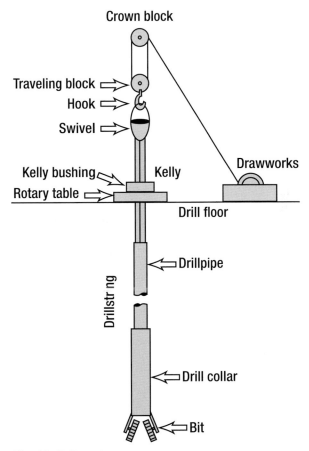

Fig. 15-5. Rotating system

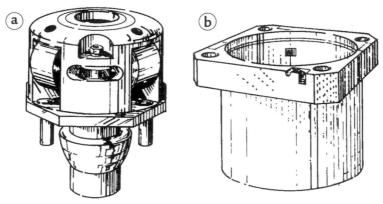

Fig. 15–6. (a) Kelly bushing and (b) master bushing

Plate 15–2. Kelly, kelly bushing, and rotary table on drill floor

Below the kelly is the *drillpipe* (fig. 15–7). The round, heat-treated alloy steel drillpipe ranges from 18 to 45 ft (5.5 to 13.7 m) long but is very commonly 30 ft (9.1 m) long. The pipe ranges in outer diameter from 2⅞ to 5½ in. (7.3 to 14 cm) and is threaded with male (*pin*) connections on each end. A larger diameter section on one end is the *tool joint* that has been screwed and welded onto the drillpipe. It has female (*box*) connections and gives each section of drillpipe a male connection on one end and a female connection on the other. The threaded ends are tapered for easy connection and to keep the drillpipe screwed. Each section of drillpipe is called a *joint*. *Pipe dope*, a lubricant such as grease, is applied to the threads of each joint of pipe as it is being added to the drillstring. The drillpipe wall is thicker with an *upset* where it is threaded on each end to strengthen that area. Most upsets are internal in that they decrease the inner diameter of the pipe.

Fig. 15–7. Joint of drillpipe

The API specifications of drillpipe include three length ranges and five grades for strength. Drillpipe is also described by nominal weight per foot, inner diameter, collapse resistance, internal yield strength, and pipe body yield strength. Drillpipe is reused after each well is drilled and is graded for wear. There are five API drillpipe wear grades. After the drillpipe is worn, it is replaced with new drillpipe.

The kelly must always be located on top of the drillstring. After drilling the well down 30 ft (9.1 m), another joint of drillpipe must be added to the drillstring to make it longer in a process called *making a connection*. The next joint of drillpipe, however, must be added to the bottom of the kelly to keep the kelly at the top of the drillstring. The pipe can be raised from the well by *pipe elevators* that are attached to the bottom of the traveling block and are designed to clamp onto the pipe. The *tongs* and *spinning wrench* are large clamp and wrench devices that are suspended from cables above the drill floor (plate 15–3). They are used to screw (*make up*) and unscrew (*break out*) pipe. A steel wedge with handles (fig. 15–8), the *slips*, can be placed in the rotary table bowl to hold the pipe with teeth and prevent the drillpipe from falling down the hole. The next joint of drillpipe used to make a connection (fig. 15–9a) is kept in a hole in the drill floor called the *mouse hole*.

To make a connection, the drillstring is raised until the entire kelly is above the rotary table. The slips are then put into the rotary table bowl, and the kelly is unscrewed from the top of the drillstring (fig. 15–9b). The kelly is then swung over to the mouse hole and screwed into the next joint of pipe. The drillpipe is then raised out of the mouse hole and swung over to the rotary table where it is screwed into the drillstring (fig. 15–9c). The slips are then removed from the bowl in the rotary table. The well is drilled another 30 ft (9.1 m) deeper, and another connection is made again. A *spinning chain* can be used to wrap around a joint of pipe and be pulled to start screwing two joints together. This, however, is dangerous and is no longer common.

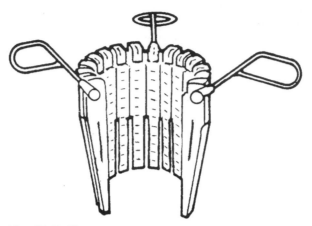

Fig. 15–8. Slips

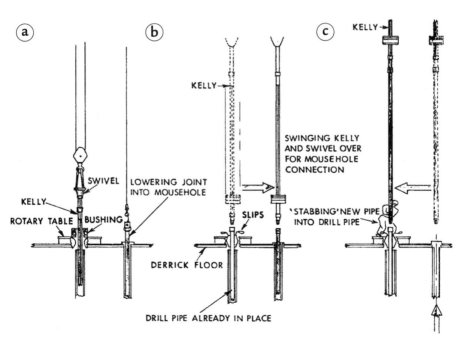

Fig. 15–9. Making a connection. (Modified from Baker, 1979.)

Plate 15-3. Tongs and spinning wrench being used on drillpipe. (Courtesy of API and Chevron.)

Drillpipe is stored horizontally on the *pipe rack* located on the ground next to the front of the rig. Individual joints are dragged up to the drill floor along the pipe ramp and through an opening in the mast or derrick, the *V-door*.

The section of the drillstring below the drillpipe is called the *bottomhole assembly*. It is composed primarily of thicker-walled, heavier, stronger pipes called *drill collars* (fig. 15-10). Drill collars are made of heat-treated alloy steel and are 31 ft. (9.4 m) long. Box and pin connections are cut into each end. A drill collar joint can each weigh 4,000 lbs (1,814 kg) or more. Drill collars are designed to put weight on the bottom of the drillstring to drill straight down and prevent the drillpipe from kinking and breaking; 2 to 10 joints of drill collars are often used.

Fig. 15-10. Drill collar joint

Heavyweight drillpipe is intermediate in strength and weight between drillpipe and drill collars. It has the same outer diameter but a smaller inner diameter than drillpipe and comes in 30½ ft (9.3 m) joints. Heavyweight drillpipe is often run between the drillpipe and drill collars to minimize stress between the two and prevent drillstring failure in that area.

Smaller sections of pipe called *subs* can be run between and below the drill collars in the downhole assembly to do various functions. A *stabilizer* is a common sub (fig. 15–11) that uses blades to contact the well walls. It is designed to keep the drillstring central in the well. A *vibration dampener* or *shock sub* uses rubber, springs, or compressed gas to absorb vibrations from the bit. It is usually run just above the bit. A *bit sub* is used to make the connection between a bit and the drill collar or sub above it. A *crossover sub* is used to make a connection between two different sizes of pipe or two different thread types. A *hole opener* (fig. 15–12) uses roller cones to enlarge the wellbore. A *reamer* has three or six tungsten steel rollers along its sides and is often run above the bit to provide a gauge hole. A *gauge hole* is a hole with a specific minimum diameter. The bit has a pin connection on the top that screws into the *bit sub* on the bottom of the drillstring.

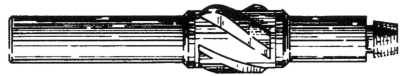

Fig. 15–11. Stabilizer

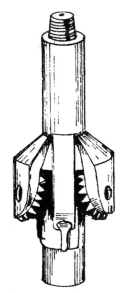

Fig. 15–12. Hole opener

The most common bit that has been used for rotary drilling is a *rotary cone bit* with three rotating cones, called a *tricone bit* (fig. 15–13). The body of the tricone bit is made of three legs of heat-treated steel alloy that have been welded together. Each leg has a jet and channel for drilling mud to flow though it. Each cone is mounted on a bearing pin that protrudes from the leg. The cones rotate on sealed and self-lubricating bearings and rollers in a race. Some surfaces on the bit are hard-faced with tungsten carbide to increase resistance to abrasion wear. As the bit is turned on the bottom of the drillstring, the cones rotate. Teeth or buttons on the cones are designed to either flake or crush the rock on the bottom of the well. The rock chips that are formed are called *well cuttings* or *cuttings*.

There are hundreds of different tricone bits that are classified as either milled teeth or insert bits. The *milled-teeth* or *steel-tooth* type of tricone bit has teeth machined out of the solid cones and are designed to flake the rock (plate 15–4). The teeth of adjacent cones fit between each other to clean out cuttings. This bit is used for relatively soft (long, widely spaced teeth) and medium (short, closely spaced teeth) hardness rocks.

The *insert* or *button* type of tricone bit has holes drilled into the solid cones (plate 15–5). Buttons of tungsten carbide stick out of the holes and are designed to crush the rock. It is used to drill relatively hard rocks.

A *diamond bit* is made of solid steel with no moving parts (fig. 15–14). Hundreds of industrial diamonds are attached in geometric patterns on the bottom and sides of the bit. Grooves in the face of the bit (*watercourses*) allow drilling mud to flow out of the center of the bit, across the diamonds, and to the side of the bit. Diamond bits are used for drilling very hard rocks.

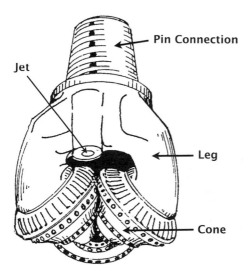

Fig. 15–13. Tricone drill bit

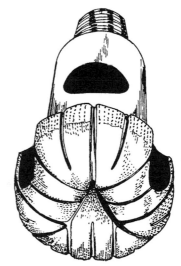

Fig. 15–14. Diamond bit

Plate 15–4. Milled-teeth tricone drill bit. (Courtesy of API.)

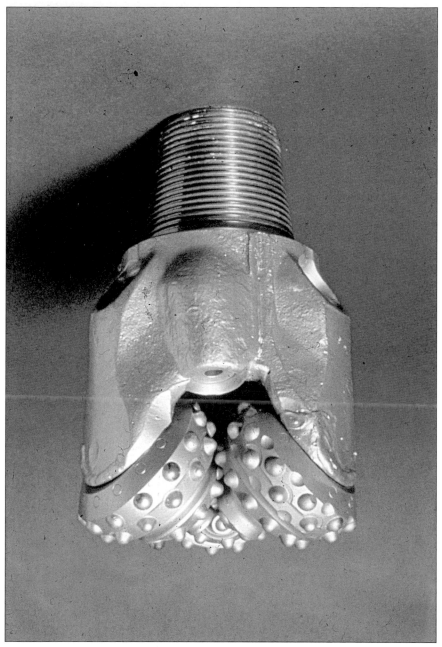

Plate 15–5. Insert tricone drill bit. (Courtesy of API.)

A *polycrystalline diamond compact* (PDC) bit is also a solid metal diamond bit with no moving parts (fig. 15–15). It has protruding metal cutters, typically ½ to 1 in. (1.3 to 2.5 cm) in diameter, on the bottom. The cutters are faced with blanks of synthetic industrial diamonds that have been cemented on. This bit is designed to continuously shear away the rocks to produce cuttings. These are the most expensive bits. A PDC bit, however, can often be used for several hundred hours and drill more footage in a well than other types of bits. PDC bits were first introduced in the 1970s and have become very popular. In the future, most wells will probably be drilled with PDC bits.

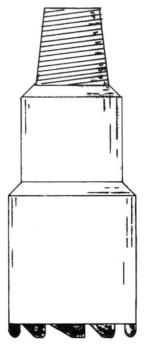

Fig. 15–15. Polycrystalline diamond compact (PDC) bit

Bits commonly come in diameters ranging from 3¾ to 26 in. (9.5 to 66 cm). The International Association of Drilling Contractors identifies bits with three numbers, such as 334. These numbers are based on (1) cutting structure such as milled teeth or inserts, (2) hardness of formation to be drilled, and (3) mechanical design. The bit is turned at a rate of 50 to 100 rpm. It is generally turned slightly faster at shallow depths and slower at deeper depths. Not all the weight of the drillstring is let down on the bit because that would crush it. The larger the bit, the more weight is applied. About 3,000 to 10,000 psi per inch of bit diameter is used.

A tricone bit wears out after 8 to 200 hours of rotation, with 24 to 48 hours being common. A worn bit can be detected from the change in noise that the drillpipe makes from the drill floor and by a decrease in rate of penetration. The bit is changed by *making a trip* during which the drillstring is pulled from the well, the bit changed, and the drillstring is run back in the well (fig. 15–16). To start, the kelly is raised above the rotary table. The slips are put in the rotary table bowl, and the kelly is unscrewed from the top of the drillstring. The kelly is then put in a hole in the drill floor called the *rat hole*. A member of the crew called the *derrick operator* climbs up and onto a small platform called the *monkeyboard* located 90 ft. (27.4 m) above the drill floor near the top of the derrick or mast. The drillpipe is then pulled from the well in a process called *tripping out*, unscrewed, and stacked in the derrick or mast (plate 14–2). The pipe is usually pulled and unscrewed three joints (a *tribble*) at a time, and is called a *stand*. A stand could also be two or four joints on some rigs. The derrick operator guides the stands between the fingers in a *finger board* located under the monkeyboard. The bottom of the stand rests on the drill floor. The bit is then changed, and the pipe run back into the hole in a process called *tripping in*. This process takes rig time, and the deeper the well, the longer making a trip takes.

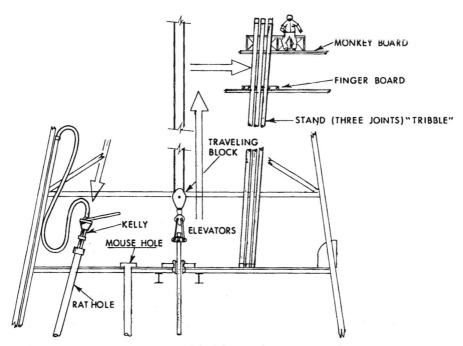

Fig. 15–16. Changing a bit. (Modified from Baker, 1979.)

To screw or unscrew a bit, a *bit breaker* (fig. 15–17) is placed in the rotary table to grip the bit. The rotary table is then turned to screw and unscrew the bit.

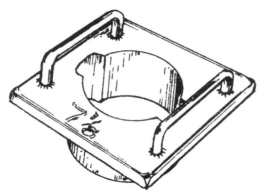

Fig. 15–17. Bit breaker

Circulating System

The circulating system pumps drilling mud in and back out of the well hole. Drilling mud is stored in several steel *mud tanks* on the ground beside the rig (plate 15–6). The drilling mud is kept mixed in the tanks by rotating paddles on a shaft called a *mud agitator* or by a high pressure jet in a *mud gun*. Large pumps driven by the prime movers, called *mud hogs*, use pistons in cylinders to pump the drilling mud from the mud tank. Mud pumps are either duplex or triplex. A *duplex pump* uses two double-acting pistons in cylinders that drive the mud on both the forward and backward strokes. A *triplex pump* uses three single-acting pistons in cylinders that drive the mud only during the forward stroke.

The mud flows from the pumps through a long rubber tube, the *mud hose*, and into the swivel. The drilling mud then flows down through the hollow, rotating drillstring and jets out through the holes in the drilling bit on the bottom of the well. The holes on tricone drill bits, called *nozzles* or *jets*, are located between each pair of cones (fig. 15–13). On a diamond bit, the drilling mud flows through the bit into grooves, called *watercourses*, on the face of the bit, and across the diamonds (fig. 15–14). The drilling mud picks the rock chips (cuttings) off the bottom of the well and flows up the well in the space (*annulus*) between the rotating drillstring and well walls. At the top of the well, the mud flows through the BOPs, along the *mud return line* and on to a series of vibrating screens made of woven screen cloth in a steel frame called the *shale shaker*. The shale shaker is located on the mud tanks and is designed to separate the coarser well cuttings from the drilling mud. It can be either single or double deck. The *double-deck shaker* has a coarser screen located above a finer screen. The screens are tilted 10° from horizontal to cause the cuttings to vibrate down the screen and into the reserve pit. The mud then flows though other solids control devices such as cone-shaped *desanders* and *desilters* (plate 15–7) that centrifuge the mud to remove finer particles. The mud then flows back into the mud tanks to be recirculated down the well.

The mud tanks are 6 ft (1.8 m) high, up to 8 ft (2.4 m) wide, and are usually 26 ft (7.9 m) long. They have two, three, four, or more compartments. A common mud tank configuration has the *shaker tank* receiving the drilling mud from the well after the cuttings have been removed. The drilling mud flows from the shaker tank to the *reserve tank* and then to the *suction tank*. Drilling mud from the suction tank goes to the mud hogs.

On the shaker tank is a *mud gas separator* that removes any subsurface gas that was dissolved in the returning drilling mud. Adjacent to the mud tanks but away from the rig is a large earthen pit called the *reserve pit* (plate 15–6). It holds discarded mud for reuse and the cuttings from the shale shakers.

Plate 15–6. Mud tanks and reserve pit

Plate 15–7. Desanders and desilters

Drilling mud is a mixture of special clay with either water (*water-based drilling mud*), oil (*oil-based drilling mud*), a mixture of oil and water (*emulsion mud*), or a synthetic organic matter and water mixture (*synthetic-based drilling mud*). The water in water-based drilling mud is usually fresh but can be saline. Oil-based drilling mud is made from diesel, mineral, or synthetic oil and brine. It has excellent bit-lubricating properties and does not affect the formations being drilled. It is, however, expensive, hard to dispose of after drilling, and can be flammable. The emulsion mud with 8 to 12% oil in water has advantages of both. The synthetic-based drilling mud has the oil-based drilling mud advantages and is relatively easy to dispose of. Water-based drilling mud made with freshwater is commonly used on land and synthetic-based drilling mud is commonly used offshore.

Water-based drilling mud is usually made of freshwater and bentonite. *Bentonite* is a type of clay that forms a colloid and will stay suspended in the water a very long time after agitation has stopped. Drilling mud viscosity and density can be increased by adding more bentonite (*mud up*) or decreased by adding water (*water back*). Chemicals mixed with the mud for various effects are called *additives*. A mass of additives put into drilling mud to remedy a situation is called a *pill*. Heavier drilling mud used to exert more pressure in the well is made by mixing in high-density substances called *weighting material* such as barite ($BaSO_4$) or galena (PbS).

Other mud additives include the following:
- alkalinity or pH control agents
- bactericides
- defoamers
- emulsifiers
- flocculants
- filtrate reducers
- foaming agents
- shale control agents
- surface active agents
- thinners
- lost-circulation material

The clay and additives are brought onto the drillsite in dry sacks and are stored in the *mud house*. They can be added to the mud in the mud tanks through a *hopper*, a funnel-shaped device. Drilling mud is described by weight. Freshwater, for reference, weighs 8.3 pounds per gallon. Typical water-based bentonite drilling mud weighs 9 to 10 pounds per gallon. A very heavy drilling mud designed to exert a greater pressure on the bottom of the well can weight 15 to 20 pounds per gallon.

The density, viscosity, and other properties of the drilling mud are frequently checked during drilling by a *mud man* or *drilling fluids engineer*. The mud person is usually a service company employee. A *Marsh funnel* (fig. 15–18) is used to determine mud viscosity (the resistance to flow). The time the mud takes to flow through the funnel in seconds is calibrated to mud viscosity. In the laboratory, an instrument called a *viscometer* is used to determine mud viscosity and gel strength. *Gel strength* is the ability of the mud to suspend solids. A *filtration test* is made by passing the mud through a filter in a filter press. It measures the thickness and consistency of the solids (mud cake) on the filter and the amount of liquid (filtrate) that passed through the filter. The pH (alkalinity) and solid content of the mud are also measured.

Circulating drilling mud in a well serves several purposes. The mud removes well cuttings from the bottom of the well to allow drilling to continue. Without removing the cuttings, drilling would have to stop every few feet to remove the cuttings that clog up the bottom of the well as had to be done on a cable tool drilling rig. As the mud flows across the bit, it cleans the cuttings from the teeth. The drilling mud also cools and lubricates the bit. In very soft sediments, the jetting action of the drilling mud squirting out of the bit helps cut the well.

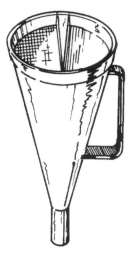

Fig. 15–18. Marsh funnel

The drilling mud also controls pressures in the well and prevents blowouts. At the bottom of the well, there are two fluid pressures on two different fluids. Pressure on fluids in the pores of the rock (reservoir or fluid pressure) tries to force the fluids to flow through the rock and into the well (fig. 15–19). Pressure exerted by the weight of the drilling mud filling the well tries to force the drilling mud into the surrounding rocks. If the pressure on the fluid in the subsurface rock is greater than the pressure of the drilling mud (*underbalance*), water, gas, or oil will flow out of the rock into the well. This can cause the sides of the well to cave in (*sluff in*), trapping the equipment. In extreme cases, it can cause a blowout where fluids flow uncontrolled and often violently onto the surface.

In order to control subsurface fluid pressure, the weight of the drilling mud is adjusted to exert a greater pressure on the bottom of the well than the pressure on the fluid in the rocks (*overbalance*). Some of the drilling mud is then forced into the surrounding rocks during drilling. The rocks act as a filter, and the solid mud particles are plastered to the sides of the well to form a *filter* or *mud cake* as the fluids enter the rock. The filter cake is very hard. It stabilizes the sides of the well and prevents subsurface fluids from flowing into the well.

After a well has been drilled, the drilling mud is disposed of and not reused. If it is freshwater-based drilling mud, it can be spread on the adjacent land to fertilize the crops (*land farming*). Saltwater-based, oil-based, and emulsion drilling muds, however, usually have to be trucked away to a disposal site. On an offshore drilling rig, a barge can be used to bring the mud ashore to a disposal site.

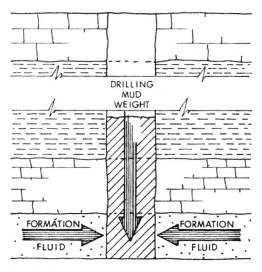

Fig. 15–19. Pressures on the bottom of a well

The *kelly cock* (fig. 15–20), a valve, is run either above or below the kelly. It allows drilling mud to be circulated down the drillstring but can be closed by hand with a hexagonal wrench to prevent fluids from flowing up the drillstring. The kelly cock is closed when a connection is being made to prevent mud from spilling out of the kelly.

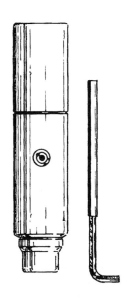

Fig. 15–20. Kelly cock with wrench

Blowout preventers (BOPs) are used to close off the top of the well. They are bolted (*nippled up*) to the top of the well and are located below the drill floor (fig. 15–21). Arranged vertically in the BOP stack are a series of rams and a preventer. *Blind rams* are two large blocks of steel that can close over the well. Because they have flat surfaces, they can be thrown only when the drillstring is not in the well. *Pipe rams* are two large blocks of steel that are designed to close around pipe in the well. They have inserts cut into the surfaces to fit around a specific size pipe. *Variable-bore rams* can be thrown around a range of pipe sizes. Blind, pipe, and variable-bore rams have rubber coatings on their metal closing surfaces to prevent pipe damage and improve the seal. *Shear rams* are designed to cut across any pipe in the well to close the well quickly. They are used primarily on offshore wells.

The *annular preventer* is made of synthetic rubber with steel ribs in a doughnut-shape that fills a steel body. It is compressed by pistons to fit around any size and shaped equipment in the well. It goes in a short cylinder on the top of the BOP stack, and it is the first one closed if there is pipe in the well. If it is not effective, the rams are thrown.

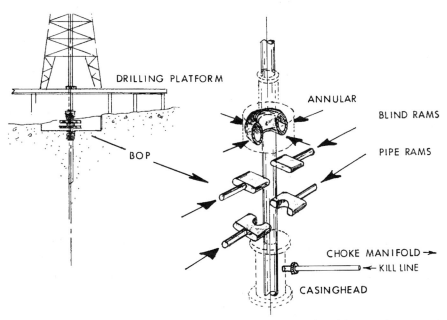

Fig. 15–21. BOP stack showing rams and preventers. (Modified from Baker, 1979.)

A typical BOP stack has an annular preventer at the top with one or more rams in line below it. Between the rams and the annular preventer are drilling spools. A *drilling spool* is a steel, spool-shaped fitting that permits attachment of kill and chokelines to the stack.

The power to activate the BOPs is stored pneumatically in cylinders called the *accumulator* that is mounted on skids next to the rig. The cylinders contain hydraulic fluid and nitrogen gas. A charging pump always keeps pressure in the cylinders so that the BOPs can be thrown even if the rig's prime movers are down. There is a *BOP panel* on the drill floor and another one a safe distance away from the rig. Handles on the panel are used to throw individual rams and the annular preventer.

BOP stacks are designed to API standards. The API describes BOPs by (1) working pressure, (2) inside bore diameter, and (3) type of rams and annual preventer. *Working pressure* is the maximum pressure that the equipment is designed to operate under. As the well is drilled deeper, bigger BOPs with higher working pressures are installed to replace those with lower working pressures. Working pressures range from 2,000 to 15,000 psi (140 to 1,050 kg/cm^2). The *substructure*, a steel framework, raises the elevation of the drill floor to make space for the BOP stack (plate 15–1). A cellar (see chap. 14, fig. 14–2) can also be used to provide space for the BOP stack.

The *choke manifold* is a series of pipes, automatic valves, gauges, and chokes on the ground next to the drilling rig. It is connected to the BOP stack outlet by a *chokeline*. The choke manifold is used to direct flow from the well to the reserve pit, burning pit, mud tank, or mud-conditioning equipment. It can be used to relieve pressure buildup in a well after the BOP stack has been thrown and to circulate heavier drilling mud into the well through a *kill line*. It is operated from a control panel on the drill floor.

Drilling Operations

Operating a drilling rig is very expensive. Except for small, shallow rigs, drilling operations occur 24 hours a day. Three 8-hour shifts of workers usually operate the rig. Each shift is called a *tour* (pronounced *tower*). The *graveyard* or *morning tour* is from midnight to 8 a.m. The *day tour* is from 8 a.m. to 4 p.m. The *evening tour* is from 4 p.m. to midnight. In remote areas and on offshore rigs, two 12-hour tours are used each day.

The *drilling contractor* is the company that owns and operates the rig. The *tool pusher* is a drilling company employee who is similar in authority to the captain of a ship. The tool pusher supervises the drilling operations and usually lives at the drillsite 24 hours a day. If the drilling contract specifies it, the tool pusher will make sure the drilling supplies are ordered and delivered to the rig on time.

The *operator* is the company that organizes and finances the drilling, selects the drillsite, and contracts with the drilling contractor to drill the well. The operator has a *company representative* at the drillsite who works with the tool pusher to make sure the well is being drilled to specifications. Each morning the tool pusher and/or company representative compiles the results of the past 24 hours of drilling into a *daily drilling* or *morning report*. The report is transmitted back to the *drilling superintendent* at the drilling contractor's office and to the operator. The report includes the depth of the well at 6 a.m., footage drilled during the last 24 hours, rig time spent on different activities, supplies used, and other drilling and geological data.

If the drilling contract specifies it, an employee of the operator called the *materials person* will be responsible for calculating the amount and ordering and supervising the timely delivery of supplies to the rig.

A *driller* is in charge of each tour. The driller operates the drilling machinery from a drilling console on the drill floor and gives orders to the crew (plate 15–8). The *drilling console* contains instruments and controls for the rotary table, drawworks, mud pumps, and chain drives. Dials on the console include a weight indicator, mud pump pressure gauge, rotary tachometer, rotary torque indicator, pump stroke indicator, and rate-of-penetration recorder. The *weight indicator* shows the weight of the drillstring suspended from the hook on the bottom of the traveling block. Too much weight could cause the derrick or mast to collapse. The *rotary tachometer* shows the rotary table speed. The *rotary torque indicator* measures stress on the drillpipe and is used to prevent pipe twist-off. The *rate-of-penetration indicator* shows how fast the drill is penetrating the rocks, usually recorded in minutes per foot or meter. The driller stands in front of the drilling console with one hand on the drawworks *brake*, a control lever used to raise and lower equipment and apply weight to the bit. After each tour, the driller fills out a *tour report* describing the activities during that tour. Tour reports are used to make the daily drilling reports.

The *derrick operator* is second in command of each tour. The derrick operator can also monitor the drilling mud and circulating equipment. On the drill floor, there are two to four *roughnecks* or *rotary helpers*, depending on the size of the rig. They handle and maintain the drilling equipment. On a large rig, one person called the *motor operator* is responsible for the prime movers.

Plate 15–8. Driller and drilling console. (Courtesy of API.)

Modern Rotary Drilling Rigs

Modern drilling rigs are designed to drill more efficiently and cut down on nonproductive rig time. The time on modern rigs is distributed with about 10% rig moving, 40% drilling, 15% tripping in and out, 8% running casing and cement, 6% circulating and coring, and 22% on other tasks.

On modern drilling rigs the drillstring is turned by a *top drive* or *power swivel* (plate 15-9). It is a large electrical or hydraulic motor that generates more than 1,000 hp. The top drive or power swivel is hung from the hook on the traveling block or is an integral part of the derrick or mast and turns a shaft into which the drillstring is screwed. It moves up and down the derrick or mast while drilling. Drilling is faster and safer with a top drive than with a rotary table. While making a connection with a top drive, drillpipe is added to the drillstring three joints at a time instead of one to save rig time. Slips are still used in the master bushing on a stationary rotary table to prevent the drillstring from falling down the well when making a connection.

Plate 15-9. Top drive

On very modern diesel-electric rigs, AC instead of DC electric motors are used to drive the floor equipment such as rotary table and drawworks. Variable frequency drives are used to adjust the AC power frequency to control the speed of the AC motors. AC motors maintain horsepower and produce more torque at both high and low speeds compared to DC motors.

AC motors also apply a more consistent weight on bit during drilling. An AC motor can also be used as a brake on the drawworks to replace the mechanical band brakes.

On some modern, land rigs, the driller sits in a glass-enclosed, climate-controlled driller's cabin overlooking the drill floor. The driller can monitor all the rig operations from the driller's cabin and operate the machinery with touch screens and joy sticks on a console (plate 15–10).

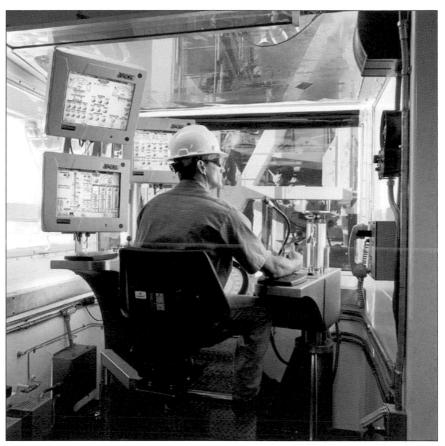

Plate 15–10. A driller's cabin overlooking the drill floor on a modern drilling rig. (Courtesy of Helmerich & Payne.)

Very fine mesh screens on modern shale shakers eliminate the need for desanders. The returning drilling mud and well cuttings go from the shale shaker screens to the mud gas separator or directly to the desilters. The modern mud tanks can be round on the bottom instead of flat to facilitate cleaning. *Closed loop drilling* is used in environmentally sensitive areas. No reserve pit is used and the drilling fluid and well cutting are contained in a closed circulating system with very little or no discharge.

Pipe handling on the drill floor is done mechanically using *iron roughnecks*. A high-torque spinning wrench and rotary table are used to make up or break out pipe. The iron roughneck can be operated manually from a floor-mounted local control or be fully automated from the driller's control console.

The crew on a modern rotary rig is the same with a driller, derrick operator, motor operator and two roughnecks. The rig up and rig down times and the time from spud to total depth are significantly reduced on modern rigs. The more efficient drilling rate is measured in several thousand feet per day.

Drilling Problems

Subsurface Conditions

Both temperature and pressure increase with depth. The rate of temperature increase is called the *geothermal gradient*. It averages 2°F/100 ft (3.6°C/100 m) for the earth but varies between 0.5° to 5°F/100 ft (1° to 9°C/100 m). In a well, the geothermal gradient can be measured by a *temperature bomb* that is run to the bottom of the well to record the temperature or by a *temperature log* that continuously records temperatures as it is being raised in the well. An average geothermal gradient for a sedimentary basin is about 1.4°F/100 ft (2.5°C/100 m). The average surface temperature in Oklahoma and most of Texas is 55°F (13°C). Oil in a reservoir at 10,000 ft (3,000 m) in the Anadarko basin of Oklahoma should have a temperature of about 195°F or 90°C (fig. 16-1). When oil and gas come up a well, they are hot.

There are two separate pressures in the subsurface. The pressure on the rock is called *earth* or *lithostatic pressure*. It increases at an average rate of 100 psi/100 ft (23 kg/cm^2/100 m). The pressure on the fluids in the pores of the rock is *reservoir, formation,* or *fluid pressure*. It depends on the density of the overlying water but averages 45 psi/100 ft (10 kg/cm^2/100 m). The pressure on oil in reservoir at 10,000 ft (3,000 m) should be about 4,500 psi (316 kg/cm^2/100 m) (fig. 16-2). This *normal* or *hydrostatic pressure* on the fluids in the rock is caused by the weight of the overlying water in the pores of the rocks. During drilling, the pressure of the drilling mud is usually slightly higher (overbalance) than hydrostatic pressure to keep the fluids back in the rocks.

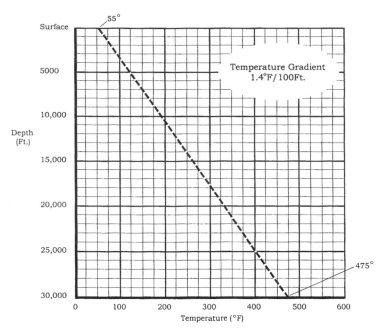

Fig. 16–1. Geothermal gradient

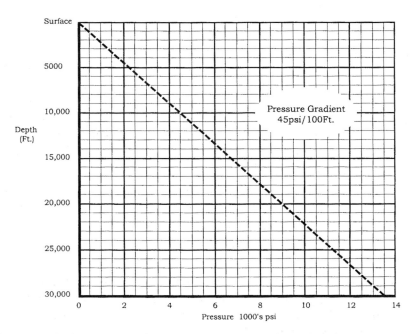

Fig. 16–2. Fluid pressure gradient

Problems While Drilling

Fishing

A common drilling problem is that something breaks in or falls down the well during drilling. For example, the drillstring twists off and falls to the bottom. A cone can break off the tricone bit, or a tool such as a pipe wrench can fall from the rig floor into the well. This is called a *fish* or *junk* and cannot be drilled with a normal drill bit. Drilling is suspended and a special tool called a *fishing tool* is leased from a service company to grapple for the fish in a process called *fishing*.

To retrieve pipe in a well, either a spear or an overshot is screwed into the bottom of a *fishing string* composed of drillpipe and run into the well. The *spear* is designed to fit into and grip the inside of the pipe (fig. 16–3a), and the *overshot* fits around and grips the outside of the pipe (fig. 16–3b). A *washover pipe* or *washpipe* consists of a section of large-diameter pipe (casing) with a lower cutting edge (fig. 16–3c). The cutting edge grinds (*dresses*) the surface of a fish smooth. Drilling mud is pumped through the washover pipe to clear debris from around the fish to prepare it for another fishing tool. A *tapered mill reamer* (fig. 16 –3d) is rotated to open collapsed casing and mill irregular-shaped fish. A *junk mill* (fig. 16–3e) is run and rotated on a fishing string to grind a fish into small pieces. The pieces can then be retrieved by a *junk* or *boot basket* (fig. 16–3f) that is run on a fishing string to just above the bottom of the well. Drilling mud is then pumped down, either down the center or along the outside of the fishing string. Fluid turbulence picks up the metal pieces and they fall into a container or basket. A *wireline spear* (fig. 16–3g) uses barbs to hook broken wireline. Both *permanent* and *electric magnets* (fig. 16–3h) can retrieve magnetic fish. An *impression block* (fig. 16–3i) is used to determine the nature of a fish in the well to select the correct fishing tool. It is a weight with soft lead or wax on the bottom. The tool is run in the well on a wireline or tubing string to give an impression of the fish.

A *jar* is often used in the fishing string above the tool. The jar is a section of pipe that either mechanically or hydraulically imparts a sharp upward or downward jolt to the tool on command (fig. 16–4). Explosives can be used to blow up the junk. The pieces are then retrieved with a magnet or junk basket.

Fishing takes time (often days) while the drilling rig is not drilling. The operator, however, is still paying for the rig during fishing. Many drilling contractors sell *fishing insurance*. For a fee, the operator is not financially responsible for fishing operations.

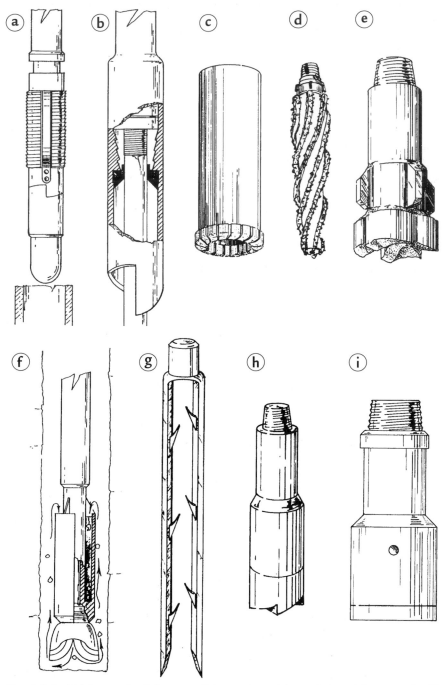

Fig. 16–3. Fishing tools: (a) spear, (b) overshot, (c) washpipe, (d) tapered mill reamer, (e) junk mill, (f) boot basket, (g) wireline spear, (h) fishing magnet, (i) impression block

Fig. 16–4. Jar

Stuck pipe

The drillstring can become stuck in a well due to either differential wall pipe sticking or mechanical problems. This is called *stuck pipe*. During *differential wall pipe sticking*, the drillpipe adheres to the well walls due to suction. The driller first tries to free the pipe by sudden jarring. The impact can be provided by a jar in the drillstring. A lubricant called a *spotting fluid*, often a mixture of diesel or mineral oil and a surfactant, can be applied along the well walls. The drilling mud can also be made lighter to decrease the suction.

Mechanical pipe sticking is often caused by a dogleg in the well. A *dogleg* is any deviation in the well greater than 3° per 100 ft. (30 m). Doglegs are caused by drilling through dipping hard rock layers or a change in the weight on the bit during drilling. A dogleg can result in *keyseating*, the formation of a wellbore cross section in the form of a key hole (fig. 16–5). It is caused by the drillpipe abrading a groove in the side of the well that is smaller that the hole drilled by the bit. Larger diameter drill collars cannot pass through the keyseat. The well has to be enlarged by reamers.

Ledges are hard rock layers that ring the wellbore and can cause pipe sticking. They are formed when drilling through alternating layers of hard and soft rocks. The soft rock washes out above and below the hard rock layers to form ledges.

A *stuck-point indicator tool* or *stuck-pipe log* can be run to determine exactly where (*stuck point*) the pipe is stuck. A *back-off operation* is performed as a last resort. The stuck pipe can be cut with either a string shot or a chemical cutter. A *string shot* uses an explosive cord that is detonated one joint above the stuck point while the pipe is being unscrewed. A *chemical cutter* (fig. 16–6) is run on a wireline and activated by an electrical signal. It uses a chemical propellant, a hot, corrosive fluid that jets out of the cutter under high pressure to slice through the pipe. After the pipe is cut, a washpipe (fig. 16–3c) is then run on a fishing string to wash around the stuck pipe and detach it from the well wall.

Wall sticking can be prevented by using *spiral-grooved drill collars* (fig. 16–7). The three grooves, located 120° apart, decrease the area of pipe in contact with the well walls but have little effect on the weight and strength of the pipe.

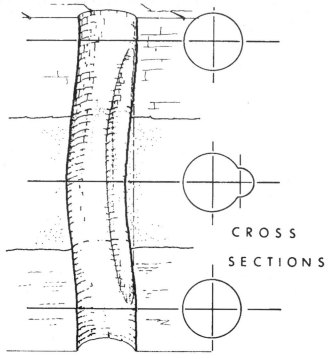

Fig. 16–5. Keyseat

Fig. 16–6. Chemical cutter

Fig. 16–7. Spiral-groove drill collar

Sloughing shale

Sloughing shale is soft shale along the wellbore that adsorbs water from the drilling mud. It expands out into the well and falls to the bottom of the well in large balls that are not easily removed by the circulating drilling mud. Chemicals such as potassium salts added to the drilling mud or oil-based drilling muds are used to inhibit sloughing shales.

Lost circulation

If a very porous, cavernous, or highly fractured zone is encountered while drilling, an excessive amount of drilling mud is lost to that zone during *lost circulation*. The zone is called a *thief* or *lost-circulation zone*. A pill of *lost-circulation additive* or *control agent* can be mixed with the drilling mud and pumped down the well to clog up the lost-circulation zone.

Lost-circulation additives are fibers, flakes, granular masses, or mixtures. They include ground pecan hulls, redwood and cedar shavings, hay, pig hairs, shredded leather, mica flakes, laminated plastic, cellophane, sugar cane hulls, ground coal, and ground tires. After the lost-circulation zone has been drilled, it can be isolated by running and cementing a string of protection casing into the well (chapter 19).

Formation damage

When drilling a well with overbalance, part of the drilling mud liquid with some very fine-grained particles, called *mud filtrate*, is forced into any permeable rock adjacent to the wellbore. The mud filtrate can decrease or destroy the permeability of a reservoir rock near the wellbore (*formation damage* or *skin damage*). Formation damage in a well can be treated by well stimulation such as acidizing (wash job) or hydraulic fracturing (chapter 24).

Formation damage can be prevented by circulating *brine* (very salty water) or an oil-based emulsion or synthetic-based drilling mud while drilling through the lost-circulation zone. It can also be avoided by drilling with a lightweight drilling mud that exerts less pressure than formation pressure (*underbalance drilling*). The well can be drilled faster using underbalanced drilling, but it will not prevent fluids from flowing out of the rocks and into the well. To maintain pressure control during underbalanced drilling, a *rotating control head* is used on the rotary table. It has a rotating inner seal assembly that fits around the kelly in a stationary outer housing. The rotating control head is designed to divert any returns coming up the well from the rig floor. Underbalanced drilling is usually done only during part of the entire drilling operation. The well has to be killed by filling with heavier drilling mud before tripping out when drilling with underbalance.

Corrosive gases

In some areas, *corrosive gases* such as carbon dioxide (CO_2) and hydrogen sulfide (H_2S) can flow out of the rocks and into the well as it is being drilled. These gases can cause *hydrogen sulfide embrittlement* and weaken the steel drillstring. To prevent corrosion, a drillstring made of more resistant and expensive steel can be used, and chemicals can be added to the drilling mud.

Abnormal high pressure

Unexpected abnormal high pressure in the subsurface can cause a *blowout*, an uncontrolled flow of fluids up the well. Natural gas flowing out the well can catch fire (plate 16–1), causing the loss of the drilling rig. *Abnormal high pressure* is fluid pressure that is higher than expected hydrostatic pressure for that depth (fig. 16–2). The drilling mud pressure may not be able to contain the formation fluids. Fluids flow out of the subsurface rocks into the well in what is called a *kick*. As the water, gas, or oil flows into the well, it mixes with the drilling mud, causing it to become even lighter and exert less pressure on the bottom of the well. The diluted drilling mud is called *gas-cut*, *water-cut*, and *oil-cut mud*.

A kick and possible blowout can be detected by several different methods during drilling. As subsurface fluids flow into the bottom of a well during a kick, more fluids will be flowing out of the top of the well than are being pumped into the well. The sudden increase of fluid flow out of the well and the rise of fluid level in the mud pit are detected by an instrument called a pit-volume totalizer. The *pit-volume totalizer* uses floats in the mud tanks to continuously monitor and record drilling mud volume. It sounds an alarm if the mud volume decreases due to lost circulation or increases due to a kick.

The drilling mud can also be continuously monitored for sudden changes in weight, temperature, or electrical resistivity that would indicate that the mud is being cut by subsurface fluids. The drilling rate of penetration also increases in undercompacted shales that can have abnormal high pressures.

Plate 16-1. Blowout on an offshore drilling rig. (Courtesy of American Petroleum Institute.)

Abnormal high pressures often occur in isolated reservoirs of limited extent. As a reservoir of large, aerial extent is buried in the subsurface, it reacts to increased overburden pressure by compacting. The reservoir compacts by decreasing porosity and squeezing some of the fluids out of the pore spaces. This maintains normal, hydrostatic pressure. If the reservoir is isolated and limited in extent, such as encased in shale or cut by sealing faults, it cannot compact because fluids cannot be expelled from the reservoir. The pressure on the overlying rocks (lithostatic pressure) is then transferred to the pressure on the fluids. Abnormal high pressure can be more than twice hydrostatic pressure (fig. 16–8).

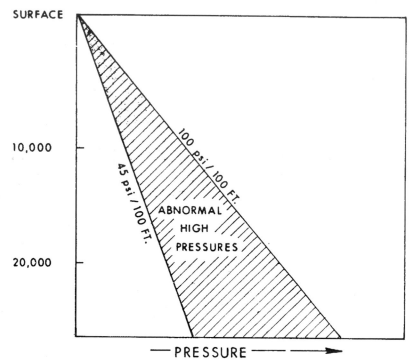

Fig. 16–8. Abnormal high pressure

Fluids cannot flow uncontrolled up the center of the drillstring during a kick because of the kelly cock (see chap. 15, fig. 15–20). Where fluids can flow up the well is along the outside of the drillstring in the space called the annulus. After a kick is detected, the well is *killed* (flow from the well is stopped) by throwing the blowout preventers. Heavier drilling mud (*kill mud*) is pumped into the well through the choke manifold, kill line, and a one-way valve in the blowout-preventer stack (*kill connection*) to circulate the kick out of the well.

Two methods to control the kick are used, depending on whether the heavy (kill) mud that is stored next to the rig is ready to be circulated. If it is already mixed and ready to use, the kick is controlled by the driller's method. If the kill mud is in dry sacks and needs to be mixed, the wait-and-weigh method is used.

In the *driller's method*, the first circulation by the mud pumps is used to replace the original drilling mud that was cut by kick fluids with original drilling mud without the kick fluids. The second circulation replaces the undiluted, original mud with kill mud. In the *wait-and-weigh method*, the well is shut in as the kill mud is being prepared. The mud pumps are started, and using a slower pumping speed, the original mud and kick fluids are replaced with the heavier kill mud during one circulation. The abnormal high-pressure zone is then drilled, and protection casing (steel pipe) is run and cemented into the well to isolate the zone (chapter 19).

About 50% of the blowouts occur during tripping out. This is because the drillstring displaces a volume of drilling mud in the well. As the drillstring is raised, the level of drilling mud falls in the well, and the pressure is decreased on the bottom of the well. If the level of the drilling mud is not maintained in the well, overbalance is lost, and a kick could occur. A *trip tank* (a 10- to 40-barrel volume steel tank that holds drilling mud on the drill floor) is used to automatically keep the well filled with mud during tripping out. Also, if the drillstring is pulled from the well too fast, it could suck gas out of the formation to start the kick. Blowout-preventer tests are periodically run on drilling rigs to test the equipment and crew response.

Dry Holes

There are two outcomes after drilling and testing a well. The well could be a *dry hole* that did not encounter commercial amount of gas and oil and is plugged and abandoned. The well could also encounter commercial amounts of gas and oil and be completed as a *producer*. *Risk* or *chance of success* is defined as the number of successful wells completed as producers divided by the total number of wells drilled. It is expressed as a percentage or decimal. The API reported that for all wells drilled in the United States during 2008, there was an average chance of success of 90%. The lowest was 39% for new-field exploratory wells and highest of 92% for developmental wells. Another way of expressing risk is *success ratio*, the number of wells drilled divided by those completed as producers and then multiplying by 100.

Mukluk was a very expensive dry hole drilled in 1983 in Alaska's Beaufort Sea, not far from the Prudhoe Bay oil field. The angular unconformity trap was located by seismic exploration, and it was thought to be another Prudhoe Bay oil field. The potential reservoir rock was the same Sadlerochit

Sandstone and the caprock rock was a shale. The leases sold for $1.5 billion. A gravel island had to be constructed in the Beaufort Sea to drill the well. The dry hole cost $120 million. Why the oil was not there may never be known for certain. Perhaps the shale caprock was not effective and the oil leaked out, or perhaps the trap was not in position when the oil migrated through the area.

Drilling Techniques

Vertical Well

Drilling contracts can have a clause that the well being drilled will be a *vertical well* or *straight hole* (fig. 17–1). The well cannot exceed a maximum deviation in degrees per 100 ft anywhere along the wellbore, and the well must fit entirely within a cone of specific degrees.

When rotary drilling rigs were originally introduced in the early 1900s, the drillers often could not drill the well straight down because of dipping beds of hard rocks such as limestone (fig. 17–2). If the bit hits a subsurface rock layer with a dip greater than 45°, the bit tends to be deflected downdip. If the hard rock layer dips less than 45°, the bit tends to be deflected updip. A well with an excessive angle that has not been drilled that way on purpose is called a *crooked hole*. An area that has dipping rock layers that cause crooked holes is called *crooked hole country*. A *slick bottomhole assembly* that has no stabilizers can be used to attempt to drill a straight hole.

Fig. 17–1. A straight hole

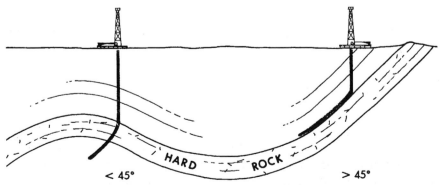

Fig. 17–2. Cause of crooked holes by dipping hard rocks

Directional Drilling

Modern rotary rigs can be controlled so that a straight hole is drilled. They can also drill a *directional* or *deviated well* out at a predetermined angle during *directional* or *deviation drilling* that ends up in a predetermined location called the *target* (fig. 17–3). The bottom of the vertical section where the well starts out at an angle is called the *kick-off point* (*KOP*), and starting a straight well out at an angle is called *kicking off the well*. If the well has been cased, a hole called a *window* is cut in the casing with a casing mill to kick off the well. The angle in which the well goes out from vertical at any location in the well is called the *deviation*. The well section below the KOP where the deviation becomes larger is called the *build* or *building angle section*. Using steel drillpipe, the maximum build angle in the build angle section is 20°/100 ft.

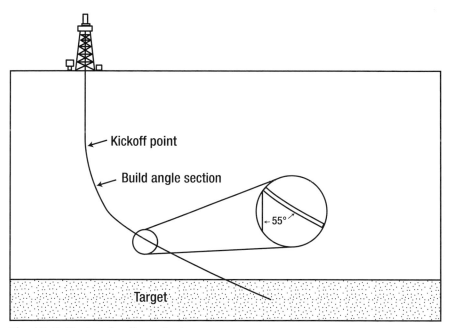

Fig. 17–3. Deviated well terminology

The first device used to kick off a well was a *whipstock* (fig. 17–4), a long steel wedge designed to bend the drillstring. The whipstock is run into the well on a drillstring and oriented by survey instruments. Weight is then applied to the drillstring to break a shear pin and separate the whipstock from the drillstring, which is then pulled out. A small-diameter bit is then run in the hole on a drillstring to drill a *pilot hole* out for 10 to 15 ft (3 to 4.6 m). The pilot hole is then surveyed. If it has the right orientation, the hole is then enlarged with a normal bit.

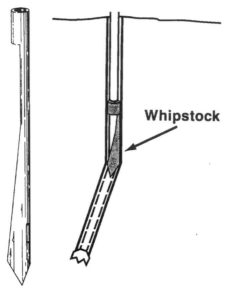

Fig. 17–4. Whipstock

The slant hole scandal occurred in the East Texas oil field during the late 1940s to the 1960s. It was discovered that many wells from adjacent nonproductive leases had been drilled with deviation drilling using whipstocks into the oil reservoir below adjacent leases. The Texas Rangers shut down 380 of these illegal slant holes after an estimated $100 million of oil had been stolen. No one was ever convicted.

In relatively soft sediments, a jet bit can be used to kick off the well. A *jet bit* is a tricone bit that has one large and two small nozzles. It is run into the well and then oriented with a surveying instrument. When the orientation is correct, mud is circulated at maximum possible flow rate without rotating the drillstring. The hydraulic action of mud jetting out of the large nozzle erodes the well out in that direction. The drillstring is then pulled, and the pilot hole is surveyed. If it is oriented correctly, the pilot hole is then drilled out with a normal bit.

A modern method used to kick off a deviated well is to run a *steerable downhole assembly* that is a combination of a bent sub, downhole mud or turbine motor, stabilizers, and a diamond or PDC (polycrystalline diamond compact) bit (fig.17–5). A *bent sub* is a short section of pipe with an angle of 0.5 to 3° in it. The *downhole mud* or *turbine motor* is driven by drilling mud flowing down the center of the drillstring. The mud strikes either a spiral shaft or blades in the motor, causing it to turn the drill bit. It can be turned on or off from the rig floor. A diamond or PDC bit is usually used with a downhole mud motor. Stabilizers are subs that centralize the assembly.

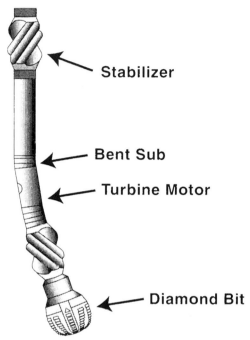

Fig. 17–5. Steerable downhole assembly

The steerable downhole assembly is run into the well and oriented in the right direction. The drillstring remains stationary as the mud motor is activated to kick off the deviated well.

The well can then be drilled out straight (*maintain angle*), drilled to increase the well deviation (*build angle*), or drilled to decrease the well deviation (*drop angle*). To maintain angle, the assembly is rotated from the rig floor similar to normal rotary drilling in the *rotating mode* (fig. 17–6a). To build or drop angle, the assembly is oriented in the right direction and not rotated. The downhole mud motor is activated to drill the well out in the direction the assembly is pointing in the *sliding mode* (fig. 17–6b) as the assembly slides along the bottom of the wellbore. The well is drilled more slowly in the sliding mode than in the rotating mode. Some steerable downhole assemblies have *adjustable bent subs* in which the angle in the bent sub can be adjusted from the surface as the assembly is in the well. Fiberglass drillpipe instead of steel can be used for build angles up to 100°/100 ft. It is, however, more expensive and cannot be used in high-temperature wells.

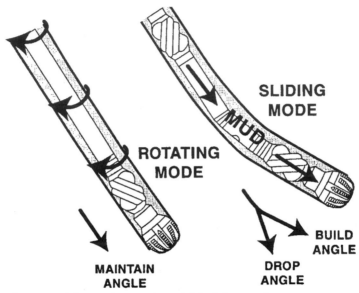

Fig. 17-6. (a) Rotating mode and (b) sliding mode

Another method to drill deviated wells is the *rotary steerable system*, which uses the *push-the-bit* technique. Along the sides of the downhole drilling unit are three or four nonrotating steering pads. The pads can be hydraulically expanded and contracted to force the bit off center to drill in that direction. The rotary steerable system has a higher rate of penetration than a steerable downhole assembly because it does not use sliding mode, it produces a smoother wellbore, and it provides for more constant steering.

The driller knows exactly which direction the bit is drilling during deviation drilling because of a *measurements-while-drilling* (MWD) system. It is a real-time system that senses the direction the bit is drilling using gyroscopes, magnetometers, and accelerometers just above the bit and transmits that information up the well to the rig floor.

Deviation drilling is common today for several reasons (fig. 17-7). Offshore production platforms are very expensive. An offshore field is best developed using one large production platform with numerous deviated wells that radiate out to the sides. Cognac, a production platform off the Mississippi River delta, has 62 deviated wells. Drilling offshore is considerably more expensive than drilling on land. An oilfield in very shallow waters can often be more economically developed by deviation drilling from the beach.

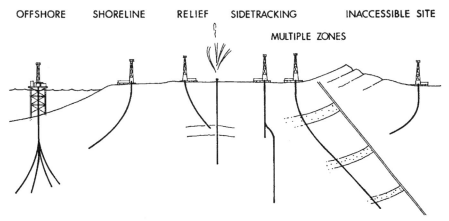

Fig. 17–7. Deviated well uses. (Modified from Beebe, 1961.)

If a well is on fire, the *wild well* can be brought under control by two methods. The fire is first extinguished, often by detonating an explosive on top of the well to remove oxygen. Then a valve can be lowered and attached to the wellhead and closed. A more expensive method is to drill a *relief well* at a safe distance from the wild well. The relief well does not have to intersect the wild well in the subsurface but just come close. It drills into the abnormal high-pressure zone that causes the blowout, and the pressure is relieved by producing the gas. After the pressure has been reduced, heavy drilling mud (*kill mud*) is then pumped from the relief well through subsurface rocks and into the wild well to control it.

After the Macondo well blew out in the Gulf of Mexico during 2010, a relief well successfully intersected it at a vertical depth in the Macondo well of about 13,000 ft (3,962 m) below the seafloor. The relief well had a total depth of 17,977 ft (5,479 m) below the seafloor. Because the Macondo well had steel casing in it, a downhole magnetometer in the relief well was used to locate and intersect the Macondo well. A kill mud was first pumped into the Macondo well, followed by cement.

Sidetracking, drilling a deviated well out from a straight hole, is common. If something breaks off or falls down a straight well and cannot be removed by fishing, the well can be drilled around the fish (*sidetracked*). A deviated well can be drilled to test several potential petroleum reservoirs rather than several straight holes being drilled to test each reservoir. Deviation drilling is also used to avoid a poor drilling location.

Several wells can be drilled from the same drilling pad using deviation drilling (fig. 17–8). The wells can be as close together as 10 ft (3 m). This leaves a significantly smaller footprint in environmentally sensitive areas. It also save money by greatly decreasing the distance the drilling rig has to be moved before drilling each well. After the wells have been completed and put into production, the drilling pad is then called a *production pad* and is often protected by a fence.

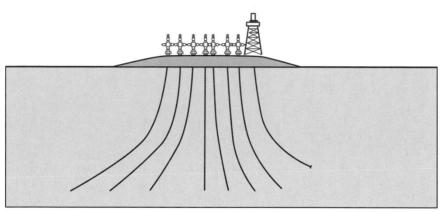

Fig. 17–8. Drilling from a drilling pad

Directional surveys that measure both angle (*deviation*) and compass orientation (*azimuth*) were originally run on deviated wells when they were first being drilled (fig. 17–9). The surveys were either single (one measurement) or multishot. The survey instrument contained a magnetic compass or gyroscope. The magnetic instrument was run when the downhole assembly was made of nonmagnetic drill collars called *K-Monel®*. The gyroscope instrument was used with magnetic drill collars or when there was magnetic iron such as casing in the area. Today, using the measurements-while-drilling system, a directional log is continuously made as the well is being drilled.

An *extended-reach well* is a deviated well that has one bend in the well and bottoms out several thousand feet horizontally from its surface location (see Introduction, fig. I-7). The world record for the *horizontal reach* from the surface location of the well to the bottom of the well on an extended-reach well is more than 7 miles (11 km).

A *horizontal well* is a deviated well that is drilled along the pay zone (*target*) parallel to the reservoir (fig. 17–10). There are two build angle sections separated by a *tangent section*. The *entry point* is where the well first penetrates the target. The horizontal part of the well is called a *lateral* or *horizontal section*. The *toe* is at the end and the heel is at the start of the lateral.

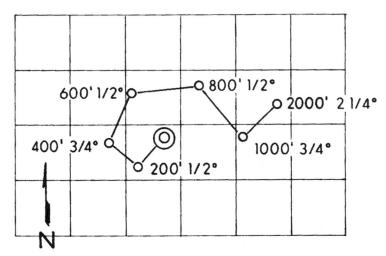

Fig. 17–9. Directional survey showing depth, deviation and location of the wellbore

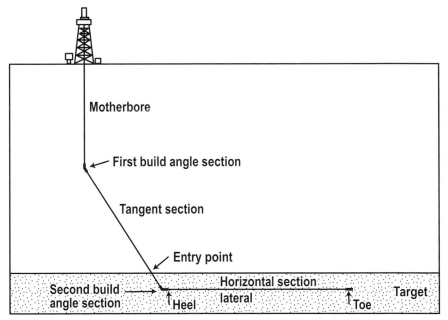

Fig. 17–10. Horizontal well terminology

Horizontal wells are described by their *build angle*, the rate of change in degrees per unit length as the well goes from vertical to horizontal, such as 8°/100 ft. The wells are classified as *short-radius*, *medium-radius*, and *long-radius horizontal wells*, depending on how sharp the build angle is. Horizontal wells produce gas and oil at a higher initial rate and can ultimately produce more gas and oil than a straight hole.

Horizontal wells have been most successful in drilling reservoirs with vertical and subvertical fractures such as the Austin Chalk (fig. 17–11). The Cretaceous age Austin Chalk is a very-fine-grained limestone that occurs in Texas and Louisiana (fig. 17–12a). It has high porosity and very low permeability because of the very small pores and pore throats but is productive where it is naturally fractured. The Giddings and Pearsall fields of south-central Texas were first discovered in the 1930s. In the Giddings field (fig. 17–12b), the Austin Chalk occurs at a depth of 6,500 to 12,000 ft (2,000 to 3,660 m). It is underlain by the Eagle Ford Shale, which is the source rock for the oil along with organic matter in the chalk itself. The chalk is overlain by shale caprock. It is a homocline, so there is no conventional trap. Vertical fractures in the chalk allowed the gas and oil to migrate up into the chalk. Horizontal wells are drilled to intersect the vertical factures and drain the Austin Chalk. Without the fractures, permeability in the Austin Chalk is less than 1 millidarcy (md). It was drilling in the Austin Chalk during the 1990s that the modern horizontal well drilling technology was developed.

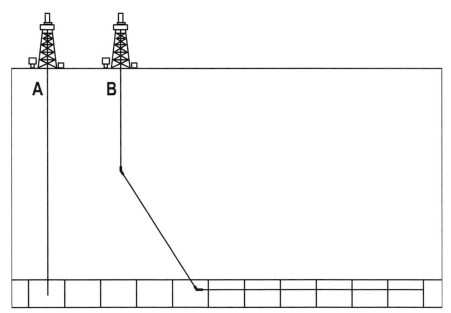

Fig. 17–11. Vertical and horizontal wells into vertical fractures

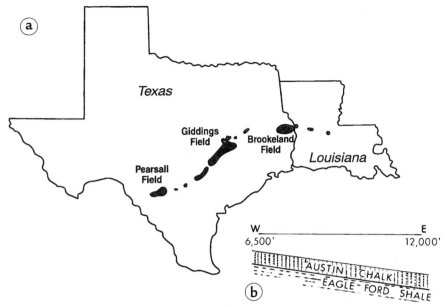

Fig. 17–12. (a) Map of Austin Chalk trend and (b) east-west cross section of Giddings field, Texas. (Modified from Galloway, Ewing, Garrett, Tyler, and Bebout, 1983.)

Horizontal wells are also used in low-permeability (tight) formations to increase ultimate recovery from the reservoir. They are also used to prevent coning, which produces excessive gas or water from above or below the oil reservoir (chapter 24). Horizontal drainholes are not much more expensive to drill than a comparable vertical well but are more difficult to log and complete. A lateral is any horizontal branch drilled out from the *motherbore*, the original vertical well, and several laterals can be drilled from one well (fig. 17–13).

Baker Hughes reported that in the United States in September 2001, 30% (589) of the drilling rigs were drilling vertical wells, 12% (231) were drilling directional wells, and 57% (1,134) were drilling horizontal wells.

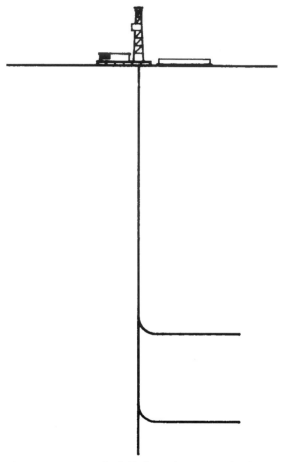

Fig. 17–13. Laterals drilled out from a motherbore

Well Depth

The depth of a well can be measured two ways (fig. 17–14). *Total depth* (*TD*), also called *measured, logged,* or *driller's depth,* is measured along the length of the wellbore. *True vertical depth* (*TVD*) is measured straight down and is less than total depth. It cannot be directly measured and has to be calculated.

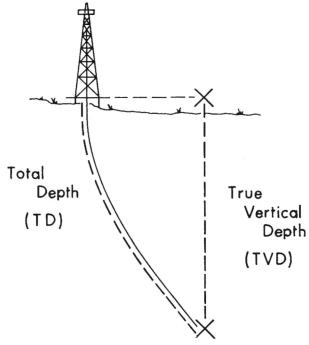

Fig. 17–14. Depth of a well

Air and Foam Drilling

Air drilling uses a rotary drilling rig with a skid- or trailer-mounted compressor that circulates air instead of drilling mud (plate 17–1). The rig and operations are almost exactly the same as a rotary drilling rig except without the circulating drilling mud system. Air is pumped down the drillstring and out the bit. It picks up well cuttings on the bottom of the well and returns up the annulus to blow out a *blooie* or *blooey line* into a pit adjacent to the rig. A *rotating head* is used on the top of the well to allow the kelly to rotate through it while maintaining a pressure seal around the kelly. The rotating head also diverts the air and cuttings coming up the well from the rig floor to the side of the rig.

Air drilling has a faster bit penetration rate than using drilling mud and helps avoid formation damage. It, however, does not build up a filter cake along the well walls to stabilize the well. Air drilling also cannot control formation fluids that flow into the well. Natural gas flowing into the well can cause a flammable mixture with the air.

Foam drilling is similar to air drilling but uses detergents in the air to form foam that better lifts water and well cuttings from the well. Soap and water are mixed and injected by a small pump into the air circulating into the well.

Plate 17–1. Truck-mounted air drilling rig. Note the smaller truck with an air compressor to the left. The blooie line is pointing out from the bottom of the drilling rig.

Testing a Well

Completing a well usually costs more money than drilling a well. Because of this, a well must be accurately tested after it has been drilled. Will this well produce enough gas or oil to make it worthwhile to complete? Old pictures and movies show gushers. Those gushers, however, were on old cable-tool drilling rigs that are rarely used today. Well testing is now based primarily on *well logs*, records of rocks and their fluids in the well.

Sample or Lithologic Log

A *sample, strip*, or *lithologic log*, recorded on a long strip of heavy paper, is a physical description of the rocks through which the well was drilled. At the top of the sample log is a *header* (fig. 18-1) with information about the well such as the operator, well name, and location. A *depth track*, usually along the left edge, shows depths in the well. In the next column, geological symbols for various rocks are used to identify the composition of each rock layer drilled (see chap. 8, fig. 8-8). The rock symbols may be colored yellow for sandstone, light blue for limestone, gray for shale, and black for coal. Next to the symbols, the rock is described. Rock texture, color, grain size, sorting, cementation, porosity, oil staining, and microfossil content can be noted there.

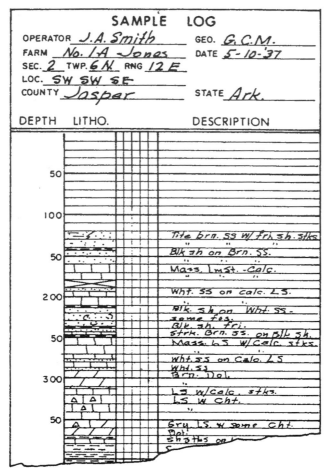

Fig. 18–1. Sample log

The source of information for the sample log is primarily from well cuttings, the small rock chips that are made by the drill bit and flushed up the well by the drilling mud. The coarser cuttings are caught on the shale shaker screen and sampled at regular intervals by the wellsite geologist, mud logger, or drilling crew. The well cuttings are typically sampled as a *composite sample* over each 10 ft (3 m) of depth. The sample interval can be shortened to a sample each foot as the reservoir rock is drilled. The well cuttings are washed to remove the drilling mud and stored in cloth or paper bags. Originally, there was a set of well cuttings from every well ever drilled. The cuttings are examined under a binocular microscope. If oil stains are present in the cuttings, they can be verified with ultraviolet light. Different °API gravity oils fluoresce with different colors.

The cuttings take time (*lag time*) to circulate from the bottom of the well to the shaker screens. Lag time is estimated by timing a tracer that is inserted into the circulating mud before it is pumped down the well. Engineering calculations using the flow from the mud pumps and the volume of the well can also be used to estimate lag time. As a general rule, in an 8-in. (20-cm) hole, it takes about 10 minutes for mud to circulate each 1,000 ft (305 m).

A *core*, a cylinder of rock drilled from the well, is the most accurate source of information about the reservoir. A *full-diameter core* ranges in diameter from 1¾ to 5¼ in. (4.4 to 13.3 cm) and can be up to 400 ft (122 m) long but is commonly 20 to 90 ft (6.1 to 27.4. m) long. To take a core, drilling must stop at the top of the subsurface interval to be cored. The drillstring is then pulled from the hole, and the drill bit is replaced with a rotary coring bit. The drillstring is then run back into the hole.

Some *rotary core bits* are solid metal with either industrial diamonds or tungsten steel inserts for cutting. Other rotary core bits use roller cones. The rotary coring bit cuts in a circle (fig. 18–2) around the outside of the well and is hollow. The core passes up into the hollow core bit and core barrel located above the bit (fig. 18-3). The *core barrel* consists of an inner and outer barrel separated by ball bearings. This allows the inner barrel to remain stationary to receive the core, while the outer barrel is rotated by the drillstring to cut the core. The inner barrel contains a core catcher with flexible fingers pointing upward and a check valve at the top to retain the core.

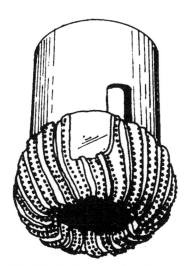

Fig. 18–2. Rotary coring bit

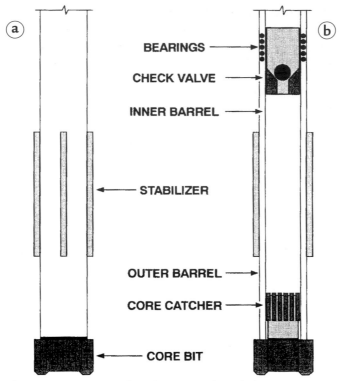

Fig. 18–3. Rotary core barrel: (a) outside and (b) cross section

During coring, drilling mud is circulated down the space between the inner and outer core barrel. After the core has been cut, the drillstring is raised to the surface and the core is removed from the core barrel. The cores are usually stored in cardboard boxes. The normal drill bit is then reattached, the drillstring is run into the well, and drilling is resumed. Because of the extra rig time involved, cores are expensive. The deeper the well is, the more expensive the core. This is why usually only the reservoir interval is cored, and in a field, only two or three wells are typically cored.

Subsurface rocks that are highly fractured, very porous, or unconsolidated are not usually retained in the core barrel. Loss of core can indicate good reservoir rock. *Oriented cores* are made with reference to geographic or magnetic north by cutting a groove along the length of the core as it is being drilled. A *native state core* is enclosed in a rubber sleeve as it is being drilled to retain all the fluids in the core under reservoir conditions. In the laboratory, the core is often *slabbed*, cut lengthwise to better examine the rocks. Plugs, small cores, are drilled from the core to measure porosity and permeability.

A faster and less expensive way to take samples is *sidewall coring*. After the well has been drilled, a sidewall coring device is run into the well until it is adjacent to the sample interval (fig.18–4). A *percussion sidewall coring tool* commonly has 30 small core tubes called *bullets* with explosive charges behind them. These are detonated, and the bullets are shot into the sidewall to take samples. The bullets are attached to the tool by wires so that the bullets and their samples are brought to the surface with the sidewall coring tool. The percussion method, however, can disturb the sediment grains and alter the porosity and permeability of the sample.

During *rotary sidewall coring*, the tube is drilled into the well wall to minimize sample alteration. The rotary sidewall coring tool is lowered into position in the well, and a small bit pivots out to drill the sample. The sample then falls into the tool. The tool can be moved to take another sample. The samples are kept separate by discs in the tool. A limitation of *sidewall cores* is that they are very small, 1 in. (2.5 cm) in diameter and 1¾ in. (4.4 cm) long.

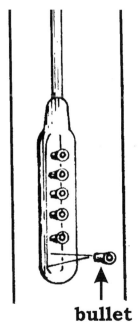

bullet

Fig. 18–4. Percussion sidewall coring tool

Many older sample logs were made by drillers with little or no training in geology. These are called *driller's logs* and display a considerable range of accuracy and usefulness. Drillers could usually distinguish between limestone, sandstone, and shale. Unfortunately, they would often use ambiguous terms such as "gumbo."

Drilling-Time Log

A *drilling-time log* is a record of the rate of drill bit penetration through the rocks (*rate of penetration*, or *ROP*). It is recorded in minutes per foot or meter drilled on a *geolograph* or *drilling recorder* on the drill floor. Because this log is recorded as the well is drilled, the drilling-time log is a real-time log. Changes in the subsurface rocks are recorded instantaneously.

The ROP depends on both drilling parameters and rock properties. Revolutions per minute, weight on bit, and bit type affect the drilling rate. If the drilling parameters are kept relatively constant, then rock properties cause the dominant variations in drilling rate. With a tricone drill bit, sandstones have the fastest drilling rate, shales are intermediate, and carbonates are slowest (fig. 18–5).

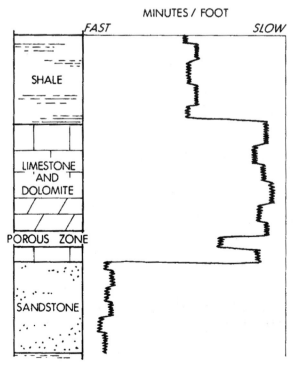

Fig. 18–5. Drilling-time log

A sudden change in the drilling rate is called a *drill* or *drilling break*. It occurs when the bit penetrates the top of a different rock layer. Drilling breaks on a drilling-time log are used to accurately determine the top and bottom elevations of subsurface formations. Because porous zones are less dense and easier to drill, drilling breaks can also be used to locate porous zones in a dense rock (fig. 18–5).

Mud Log

A *mud log* is a chemical and visual analysis of the drilling mud and well cuttings for traces of subsurface natural gas and crude oil as the well is being drilled. Any oil or gas above the normal expected background is called a *show*. The mud log is made by a service company in a *mud-logging trailer* at the wellsite (plate 18–1). The purpose of the mud log is to identify oil- and/ or gas-bearing rocks in the subsurface. Drilling mud circulating out of the well is sampled from a gas trap in the shale shaker, along with well cuttings from the shaker screens. The *mud loggers* are usually geologists. Typically, two geologists work 12 hours shifts at the mud-logging trailer, so that the mud log is maintained 24 hours a day.

Plate 18–1. Mud-logging trailer

The American Petroleum Institute has published a standard mud log format. A mud log (fig. 18–6) has a header at the top with operator, well name, location, elevation, and other information. A depth strip, showing depth in the well, runs down a column near the middle of the log, along with a sample log that was made by the mud loggers. On the left side is a drilling-time log recorded in ROP. The right side of the mud log shows the amounts of gas and oil detected in the drilling mud and well cuttings. A curve shows the total gas in the drilling mud in gas units measured by a gas detector. A more detailed chemical analysis (*show evaluation*) can be made of a gas show with a gas chromatograph. It measures the percentages of methane (C_1), ethane (C_2), propane (C_3), butane (C_4), and pentane (C_5). A black bar shows that the mud logger saw oil staining on the well cuttings in what is called a *show of oil*.

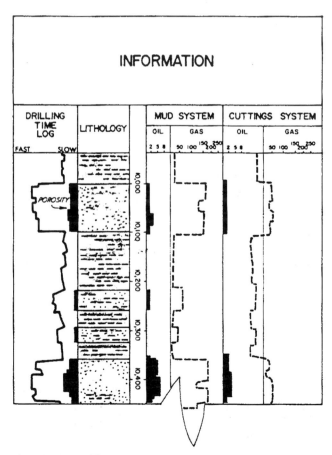

Fig. 18–6. Mud log

Wireline Well Logs

A well log made by running an instrument down a well on a wireline is called a *wireline well log*. They were invented in France in the mid-1920s and were first applied to the oil fields in California and Oklahoma in 1930. The first wireline well log was a resistivity log that was hand-recorded, point by point, on a strip of paper. Wireline well logs helped rotary drilling become more accepted. Because the early logs could be run in wells filled with drilling mud with filter cakes along their wellbore, they could be used to evaluate wells drilled with a rotary drilling rig.

To make a wireline well log after the well is drilled, the hole is first *conditioned* by circulating drilling mud, and the drilling equipment is pulled from the well. A logging truck (plate 18–2) with logging equipment is called out to the drillsite. A logging tool or sonde is run into the well (which is still filled with drilling mud) on a logging cable (fig. 18–7). The *logging cable* is wireline, an armored cable that has steel cables surrounding insulated conductor cables (fig. 18–8a). It is reeled out from a drum in the back of the logging truck.

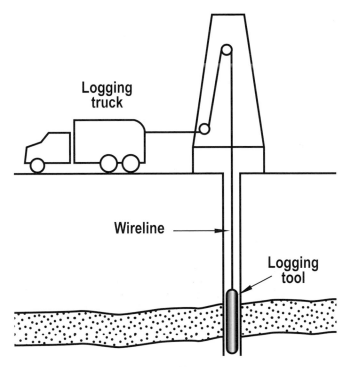

Fig. 18–7. Making a wireline well log

Plate 18–2. Logging truck. (Courtesy of Schlumberger.)

The *logging tool* or *sonde* is a cylinder (fig. 18–8), commonly 27 to 60 ft (8.2 to 18.3 m) long but sometimes up to 90 ft (27.4 m) long and 3 to 4 in. (7.6 to 10.2 cm) in diameter. The tool is filled with instruments. Several instrument packages such as formation density, neutron porosity, and gamma ray can be combined to form the logging tool. The tool has either one expandable arm or bow spring that puts the sensors in contact with the well walls or several expandable arms, bow springs, or contact pads that center the tool in the well (figs. 18–8b and 18–8c). As the logging tool is run back up the well, it can sense the electrical, acoustical, and/or radioactive properties of the rocks and their fluids and the geometry of the wellbore. In a directional well with a high deviation (>45°) or a horizontal well, the logging tool must be pushed down the well with tubing or drillstring or be pulled in with a small tractor that fits in the well. One trip down and up with a logging tool is called a *logging run*.

The data from the logging tool are transmitted digitally up the conductor cables to instruments in the logging truck where a logging engineer monitors the logging tool response and the data are recorded. The data are also processed later, and a cleaner log (*final print*) is made. The logging data are encoded and sent by radio telemetry back to a data center where only authorized persons can view the data on the Internet. On an offshore drilling rig, a small permanent cabin is used for logging. Logging a well can take from several hours to several days.

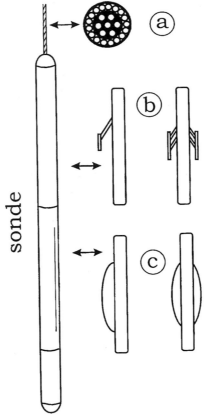

Fig. 18–8. Logging tool on a wireline: (a) cross section of armored cable, (b) logging tool with arm(s), and (c) logging tool with bow spring(s)

A wireline well log is commonly recorded on one of two similar formats (fig. 18–9). A *header* at the top of the log contains well information, followed by logging information. A *depth strip* or *track* runs down near the middle of both formats. The depth is measured from below kelly bushing (KB), drill floor (DF), or ground level (GL). On the left side of the depth strip on both formats is a graph called track 1. On the right side of the depth strip in both formats is another graph called track 2. On some logs, a third graph is located on the far right (track 3).

There is at least one line in each track that records some property of the rocks and their fluids in the well or the geometry of the wellbore. The caption at the top of each track (fig. 18–10) identifies the measurement being recorded. Each measurement line is also labeled in the track. There can be one, two, or three different measurements in each track. If more than one measurement is being recorded in the same track, a different line (heavy or light, solid or dashed) or color is used for each. When a measurement in a track is deflected from one side to the other, it is called a *kick*. An accurate scale for each measurement (either linear or logarithmic) is located on the top of each track. Two common vertical scales are used. A *correlation log* uses 2 in. to 100 ft (5 cm to 30.5 m), and a *detail log* uses 5 or more inches to 100 ft.

Most wireline well logs are *open-hole logs*, which can be run only in wells with bare rock walls. *Cased-hole logs* are less common and can be run accurately both in open holes and wells in which pipe (casing) has been cemented into the well. *Compensated logs* are measurements that have been adjusted for irregularities in the shape and roughness of the wellbore sides.

Certain measurements are usually recorded in specific tracks. Spontaneous potential, natural gamma ray, and caliper logs are usually recorded in track 1. Resistivity (electrical and induction), formation density, neutron porosity, and sonic logs are usually recorded in tracks 2 and 3.

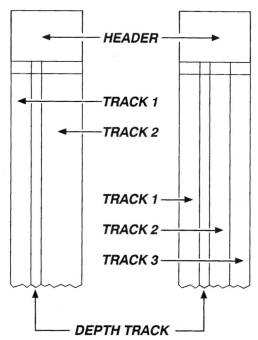

Fig. 18–9. Wireline well-log formats

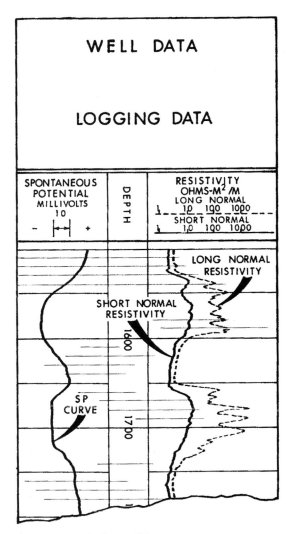

Fig. 18–10. Wireline well log

Electrical log

The first wireline log was an *electrical log* that measured resistivity. An electrical logging tool with electrodes in contact with the rocks along the wellbore was raised in the well (fig. 18–11). An electrical current was passed through the rocks, and the *resistivity* (*R*) of the rocks and their fluids to that current was measured.

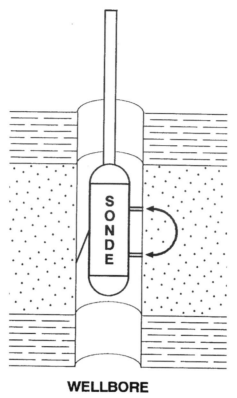

WELLBORE

Fig. 18–11. Making a resistivity measurement

Two resistivities were measured. *Short normal resistivity* was measured with electrodes closely spaced, 16 in. (40.6 cm). *Long normal resistivity* was measured with electrodes spaced far apart, 64 in (162.6 cm). Short normal resistivity was recorded in track 2 on the electrical log (fig. 18–12), with resistivity increasing to the right. Long normal resistivity was recorded in track 3, with resistivity also increasing to the right. Because of the short distance between the electrodes used in short normal resistivity, the curve sharply recorded the top and bottom of subsurface layers in the well. It was used to accurately determine the subsurface elevation of each rock layer.

An important factor in the resistivity of porous rock is the fluid in the rock. By measuring the resistivity of the rock, the fluid (water, gas, or oil) in the pores can be identified. However, when drilling with a rotary drilling rig, permeable rocks adjacent to the well are flushed with drilling mud. Because of the pressure on the drilling mud, some of the large solid particles in the mud are plastered against the well walls to form the mud cake, and some of the drilling mud liquid along with finer solid particles (mud filtrate) is forced into the rock. The area adjacent to the wellbore that is flushed with mud filtrate (*invaded zone*) goes from 0 to 100 in (0 to 254 cm) back from the wellbore, depending on the porosity and permeability of the rock.

Long normal resistivity puts the electrical current back behind the invaded zone to measure the true resistivity of the rock and determine the natural fluids in the pores of the rock. Saltwater conducts some electricity and has moderately low resistivity (fig. 18–13). Oil, gas, and fresh water, however, have very high resistivity. An oil or gas reservoir has a long normal resistivity kick to the right in track 3. Oil and gas cannot be differentiated on a resistivity log (R log). If there is no change in porosity, an oil-water or gas-water contact appears as a kick on the resistivity curve (fig. 18–14). Also, if the resistivity of the saltwater and the porosity are known, the oil saturation (percent oil in the pores) in a reservoir can be calculated from a resistivity log (fig. 18–15). The higher the oil saturation, the greater the resistivity. Electrical log resistivity has been replaced by modern induction and focused log resistivity measurements.

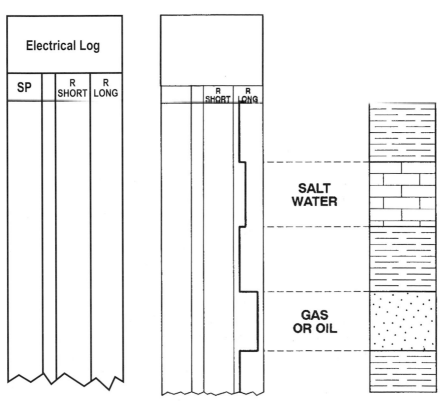

Fig. 18–12. Electrical log   **Fig. 18–13.** Resistivity responses

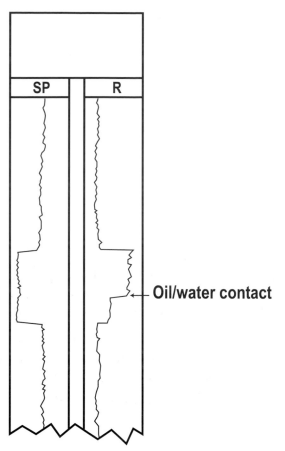

Fig. 18–14. An oil-water contact on a resistivity curve

A common measurement made with resistivity on an electrical log, both in the past and today, is *spontaneous* or *self potential* (*SP*). It is made with an electrode that is grounded on the surface and connected to another electrode in the logging tool. As the logging tool is run back up the well, the electrode is in contact with the rocks along the wellbore.

When two fluids of different salinities are in contact, a potential electrical voltage is created. A permeable reservoir rock drilled with a rotary drilling rig using drilling mud overbalance has an invaded zone flushed with mud filtrate adjacent to the wellbore. The mud filtrate usually has a different salinity than the water in the pores of the rock. This creates a potential electrical voltage along the top and bottom of the reservoir rock where it is in contact with shales (fig. 18–16). Spontaneous potential measures the magnitude of this voltage to identify potential reservoir rocks in the well.

It is recorded in track 1 with positive on the right and negative on the left (fig. 18–17). The SP curve deflects to the left to identify a potential reservoir rock and to the right for nonreservoir rocks such as shale, tight sand, or dense limestone.

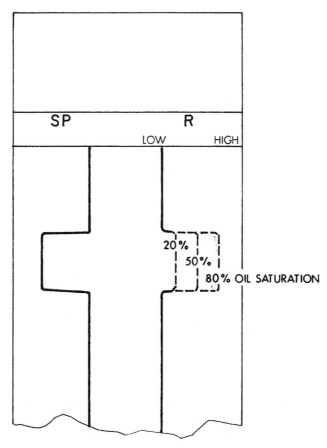

Fig. 18–15. Resistivity responses to different oil saturations

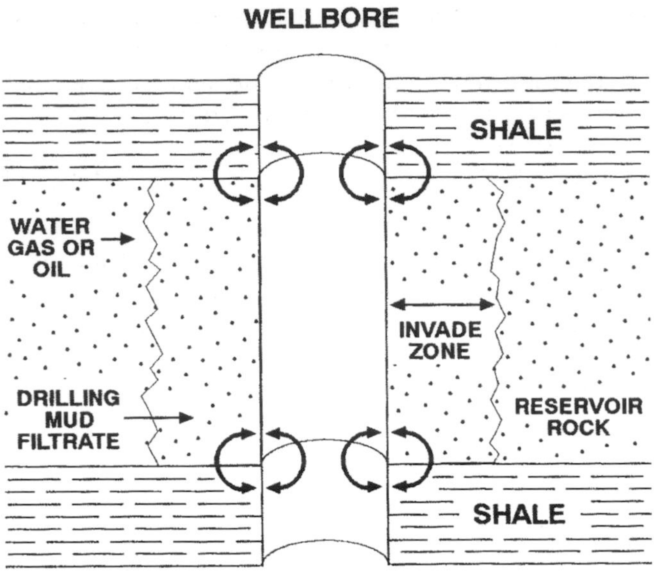

WELLBORE

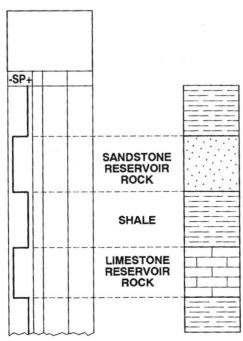

SHALE

WATER
GAS OR
OIL

INVADE
ZONE

DRILLING
MUD
FILTRATE

RESERVOIR
ROCK

SHALE

Fig. 18–16. Spontaneous potential

-SP+

SANDSTONE
RESERVOIR
ROCK

SHALE

LIMESTONE
RESERVOIR
ROCK

Fig. 18–17. Spontaneous potential responses

Tight sands, dense limestones, and shales have a characteristic signature on SP and R logs (fig. 18-18). Shales cause both curves to deflect to the center of the log with the SP deflecting to the right and the R deflecting to the left. Tight sands and dense limestones also cause the SP to deflect to the right, but they have high resistivities and cause R to deflect to the right. There is no way to distinguish between a tight sand and dense limestone on these logs.

An important use for electrical logs is the correlation of subsurface rocks (fig. 18-19). A *pick* is the top or bottom of a sedimentary rock layer on a wireline well log. It is located at the deflection of an SP, *R*, or other measurement. Lines are drawn along the same picks to correlate between well logs. The wireline well log is often the only well log available for correlation.

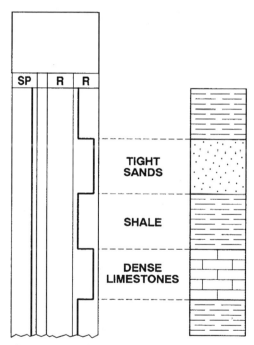

Fig. 18-18. The responses of a tight sand, dense limestone, and shale on SP and R

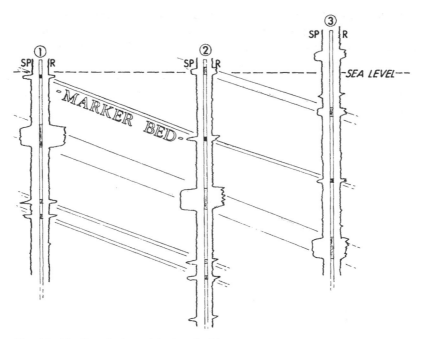

Fig. 18–19. Correlation with electrical logs

Induction log and laterolog

Resistivity is important because it is the only common log measurement that identifies the fluid in the pores of the rock. In the 1950s, wells were being drilled with oil-based drilling mud to prevent formation damage. A normal resistivity measurement could not be run in those wells because oil does not conduct electricity.

For the past several decades, resistivity has been recorded with induction logs and laterologs. An *induction log* is run in wells drilled with freshwater-based, oil-based, or synthetic-based mud or air. The induction-logging tool uses a transmitter coil to induce electrical currents into the formation adjacent to the wellbore. This creates a magnetic field that is recorded on a receiver coil on the tool to measure resistivity. Modern induction logs can measure formation resistivity in three dimensions adjacent to the wellbore.

A *laterolog* or *guard log* is run in wells drilled with saltwater-based mud. The logging tool focuses an electric current back into the formation adjacent to the wellbore. An electrode in the tool measures the formation resistivity.

Both induction logs and laterology are recorded in track 2. A logarithmic (log) scale is used with very low resistivity on the left and resistivity increasing very rapidly to very high resistivity on the right. A dual induction log measures shallow, medium, and deep induction resistivity (fig. 18–20), reflecting the depth in the formation back from the wellbore where the measurement was made. These resistivities show how far back into the reservoir the invade zone goes, indicating the permeability of the reservoir. If the well was drilled with freshwater-based drilling mud, the invade zone will have very low resistivity compared to the natural reservoir containing brine. A dual laterolog measures shallow and deep resistivity.

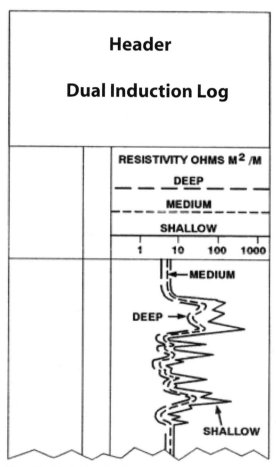

Fig. 18–20. Induction log

A *microlog* measures the resistivity an extremely short distance back beyond the wellbore walls. It uses a pad with closely spaced electrodes that is pressed against the wellbore wall. A permeable formation will have a thicker mudcake along the wellbore because more drilling mud filtrate was forced back into the permeable formation than nonpermeable formations. The microlog detects the thicker mudcakes that indicate permeable formations. It is also known as microlaterolog, proximity log, microspherical log, and microcylindrical log.

Gamma ray log

A *gamma ray* or *natural gamma ray log* (GR) uses a scintillation counter to measure the natural radioactivity from potassium, thorium, and uranium in the rocks along the wellbore. Most shales and some sandstones are radioactive. The GR log is plotted in track 1 (fig. 18–21) with low radioactivity to the left and high radioactivity to the right. Many shales, along with some sandstones, are "hot" and deflect to the right. Sandstones and limestones, potential reservoir rocks, deflect to the left. A natural GR log is relatively inexpensive and can be run accurately in both an open hole and cased hole. A *spectral GR log* is a type of gamma ray log that also identifies the source of the radiation (potassium, thorium, and uranium). This can be used to distinguish radioactivity from shales and radioactive sandstones.

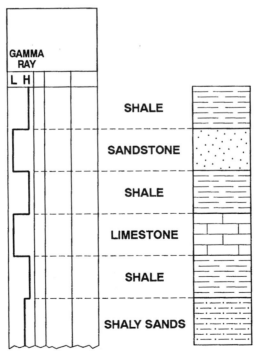

Fig. 18–21. Gamma ray (GR) log responses

In most logs, either an SP or GR curve will be located in track 1. Both logs are used to locate potential reservoir rocks that have characteristic kicks to the left. The *shaliness*, an estimate of the shale content, of a rock such as limestone can also be estimated with a gamma ray log.

Radioactive logs

A *radioactive log* is made by running a radioactive source into the well. The radioactive source is stored in a compartment in the back of the logging truck. The logging engineer uses a metal pole to remove the radioactive source and insert it into the logging tool and run into the well. Two types—neutron porosity and formation density logs—are commonly run today.

Neutron log. The *neutron* or *neutron porosity log* (*NL*) is used to measure the porosity of rocks in the well. The tool has a radioactive source that bombards the rocks adjacent to the well with high-speed atomic particles (neutrons) as the tool is raised in the well. If a high-speed neutron collides with a large rock atom, the atom will bounce the high-speed neutron back with almost no loss of energy. If the high-speed neutron collides with a hydrogen atom (a very small atom), the hydrogen atom absorbs some of the neutron's energy. The neutron will bounce back as a slow-moving neutron. The slow-moving neutron can be captured by another atom in the rock, causing that atom to emit gamma rays. The more hydrogen atoms in a rock, the more slow-moving neutrons and gamma rays the rock will produce when bombarded by fast-moving neutrons. The fewer hydrogen atoms in a rock, the more fast-moving neutrons will bounce back as the rock is bombarded.

Hydrogen atoms occur in water, gas, and oil in the pores of a subsurface rock. Each rock is bombarded with a certain number of high-speed neutrons. Either the number of gamma rays or slow neutrons is counted. The more porous a rock, the more slow neutrons and gamma rays are emitted and counted.

The log is calibrated and recorded as percent porosity in track 2 with low % porosity on the right side and high % porosity on the left (fig. 18–22). The calibration assumes that there is a liquid (oil or water) in the pores. If a gas occupies the pores, the neutron log will record a low porosity value. The neutron log is accurate in both open holes and cased holes. The compensated neutron porosity log (CNPL) is adjusted (compensated) for wellbore irregularities.

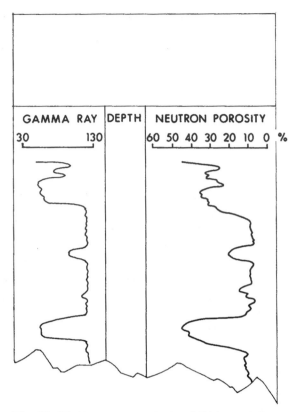

Fig. 18–22. Neutron porosity log (NL) in track 2

Formation density log. The *formation density log* (*FDL*) or *gamma-gamma* (*GG*) is another type of porosity measurement log. A radioactive source bombards the rocks in the well with gamma rays as the tool is being raised. If a gamma ray collides with a large rock atom that has a high electron density, part of the gamma ray energy is absorbed, and the weakened gamma ray is scattered back. Gamma rays are not significantly affected by small atoms such as hydrogen. Dense rocks have more rock atoms per unit volume than porous rocks. The denser and less porous a rock, the more gamma ray energy is absorbed and fewer scattered gamma rays are returned to the detector in the logging tool. The formation density log measures the density of the subsurface rock, and density is a good indication of the lithology of the rock. Table 18–1 shows the density of various sedimentary rocks with no porosity (matrix density). The more porous a rock, the more the gamma rays are able to return by scattering through the rock and back to the detector.

Table 18–1. The densities of some common sedimentary rocks with no porosity

	density (gm/cc)
salt (halite)	2.16
quartz sandstone	2.65
limestone	2.71
dolomite	2.87

(modified from Schlumberger, 1987)

The porosity of the rock (sandstone or limestone) can be computed from the density (fig. 18–23). The FDL is recorded in track 2 in one of three ways. First, it can record bulk density in units of grams per cubic centimeter. Second, it can record porosity that was usually calculated assuming that it is a limestone and is labeled limestone matrix. If necessary, a geologist or logging engineer can change the matrix to sandstone or dolomite. There will be a percent porosity scale with 0% on the right side of the track. If the rock is a sandstone and the assumed matrix is limestone, the actual porosity is slightly lower than the porosity shown on the porosity scale. Third, both the bulk density and porosity can be displayed in track 2 (fig. 18-24). The *formation density compensated log* (FDL) has been adjusted for wellbore irregularities.

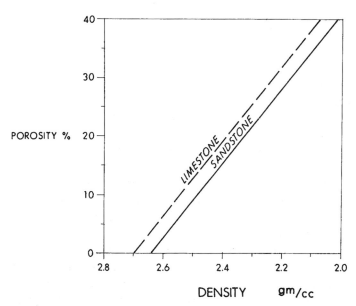

Fig. 18–23. The relationship between rock density and porosity. (Modified from Gearhart-Owen Industries *Chart Book.*)

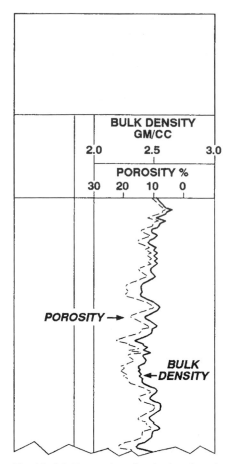

Fig. 18–24. Formation density log (FDL)

The porosity logs—neutron and formation density—are best calibrated if a core or two is available from a well or wells that have been logged in a field. The porosity values from these logs are adjusted to match the accurate porosity measurements made from the cores.

Gas effect

There are two states of matter that can occupy the pores of a subsurface rock, liquid (water and oil) and gas (natural gas). The neutron porosity is calibrated to measure porosity, assuming that a liquid is in the pores. It yields an inaccurate low-porosity reading on a gas reservoir. The FDL will give a more accurate but higher porosity calculation on a gas-filled rock. It is most affected by the density of the rock atoms, not the hydrogen atoms in the pores.

Natural gas is detected in subsurface reservoir rocks by running both porosity logs (formation density and neutron logs) in the well. They are plotted as porosity in track 2 on the log. If natural gas is present, the neutron log reads too low, and the formation density log reads too high (fig. 18–25). The crossover of the two curves is called the *gas effect*. A correction can be applied to the formation density and neutron logs to calculate the accurate porosity of the natural gas reservoir rock.

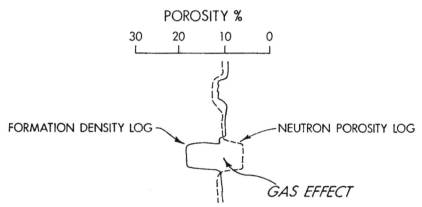

Fig. 18–25. Gas effect

Photo-electric factor log

The *photo-electric factor log* uses a logging tool with a radioactive source that emits gamma rays. Detectors on the tool are used to measure the absorption or capture of the gamma rays by the formation adjacent to the wellbore. This determines the photoelectric factor or index (Pe) of the formation that depends on the average atomic number of the formation. It is not significantly affected by porosity or fluid saturation. The measured formation photoelectric index is used to determine the mineral composition and lithology of the formation.

A *litho-density log* combines a photoelectric log with a formation density log. The photo-electric factor log is recorded on the left side of track 2 with the density log. The log combination is very useful in determining the exact formation lithology.

Caliper log

A *caliper log* (CAL) measures the diameter of the hole. The size of the hole depends upon the size of the drill bit, the strength of the well walls, and the thickness of the filter cake. Soft rocks, such as shale and coal, break off and sluff (cave) into the well, forming a wide hole. Strong rocks such as limestones, dolomites, and well-cemented sandstones have wellbores about the size of the drill bit. Salt layers can be dissolved by freshwater drilling mud, forming caverns that create serious drilling problems in certain areas of west Texas.

The caliper-logging tool has four arms that are expanded to touch the sides of the well. As the caliper tool is run up the well, the arms expand and contract to fit the well, and an electrical signal is generated to record the wellbore size. The CAL is recorded in track 1 (fig. 18–26). The units are inches or centimeters of diameter with a larger diameter wellbore to the right and smaller to the left.

Caliper logs are commonly run for three reasons. First, it is necessary to know the size of the hole for future engineering calculations. If the well is going to be cased (pipe cemented to the well walls) or plugged and abandoned, the volume of the well must be computed to order the right number of sacks of cement. Second, many of the other logs, called compensated logs, need to be calibrated for wellbore size to yield accurate results. Also, because permeable formations will accept more drilling mud filtrate during drilling, they will have a thicker mud cake (fig. 18–27). The caliper log can be used to locate thick filter cakes that form smaller diameter wellbores and identify permeable zones.

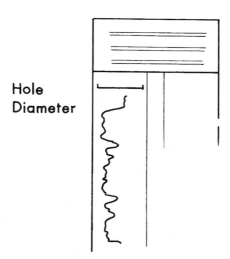

Hole Diameter

Fig. 18–26. Caliper log (CAL)

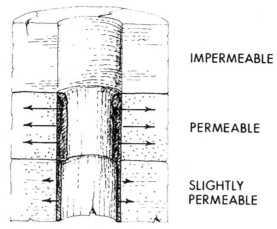

Fig. 18–27. Wellbore diameters

Sonic or acoustic velocity log

The *sonic log* (*SL*) or *acoustic velocity log* (*AVL*) measures the sound velocity through each rock layer in the well. The logging tool has a sound transmitter at the top of the tool and two sound receivers spaced along the tool (fig. 18–28). An impulse of sound is emitted by the transmitter and is recorded on the two receivers. The time it takes the sound to travel from one receiver through the rocks to the other receiver is recorded in units of microseconds per foot. This velocity is called the *interval transit time* or Δt of the rock.

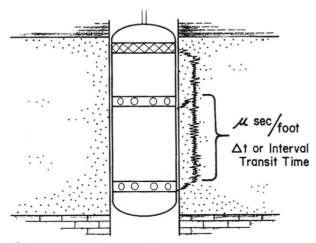

Fig. 18–28. Making a sonic log measurement

Table 18–2 shows the common ranges of sound velocities through sedimentary rocks, water, and natural gas. Of the common sedimentary rocks, shales have the lowest sonic velocities, sandstones have higher velocities, and limestone and dolomite have the highest. There is a wide range of sonic velocities in each type of sedimentary rock because sound velocity through gas and liquid is less than through solids such as rocks. The more porous a rock, the more gas or liquid it contains and the slower its sonic velocity will be.

Table 18–2. Typical sonic log velocities

	velocity (ft/second)	(m/second)	Δt(microsecond/ft)
shale	7,000 to 17,000	2,134 to 5,182	144 to 59
sandstone	11,500 to 16,000	3,505 to 4,877	87 to 62
limestone	13,000 to 18,500	3,962 to 5,639	77 to 54
dolomite	15,000 to 20,000	4,475 to 6,096	67 to 50
natural gas	1,500	456	667
water	5,000	1524	200

(Modified from Gearhart-Owens Industries *Chart Book.*)

The sonic log is plotted as interval transit time in track 2 or 3 (fig. 18–29). Fast interval transit time is on the right and slow on the left. The interval transit time gives a good indication of the rock composition. If the composition of the rock is known, the porosity of the rock can be computed from the interval transit time of the rock (fig. 18–30). Porosities calculated from sonic logs usually do not include any porosity formed by fractures in the rock. Fractures greatly attenuate (decrease) the amplitude of the sonic waves through rock. A type of sonic log, the *sonic amplitude log*, measures the attenuation of the sound and detects the presence of fractures.

Dipmeter

The *dipmeter* or *dip log* is a logging tool used to determine the orientation of rock layers in a well. The dipmeter consists of four arms, each with either two closely spaced electrodes or pads of electrodes that record resistivity. The arms expand in the well to touch the sides of the well. As the dipmeter is brought up the well, the electrodes on each arm are in contact with the rock layers (fig. 18–31). If the rock layer is dipping, different arms will contact the top and bottom of the layer at different depths. The orientation of the dipmeter in the well is known from a gyroscope. The sequence of contacts between individual arms and each layer can be used to compute the dip of the layer. If the layer is horizontal, all arms of the dipmeter contact the top and bottom of the layer at the same time.

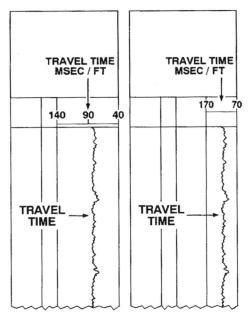

Fig. 18–29. Sonic logs

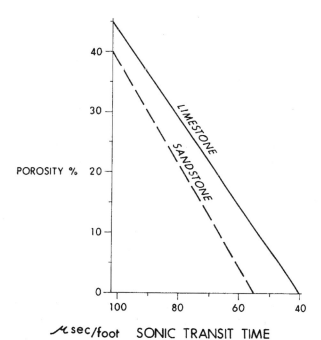

Fig. 18–30. The relationship between interval transit time and porosity. (Modified from Gearhart-Owen Industries *Chart Book.*)

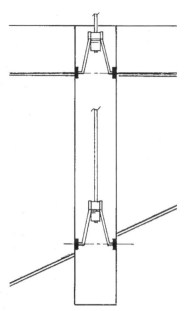

Fig. 18–31. Making a dipmeter measurement

Two ways to present dipmeter data are tadpole and stick plots. On a *tadpole plot* (fig. 18–32), dip is plotted on the horizontal axis with zero dip on the left. Depth in the well is the vertical axis. A small circle on the log gives the depth and dip of each measurement. A small line, similar to a tadpole tail, is oriented in the compass direction of the dip with north at the top of the log. A *stick plot* uses lines (sticks) to show the dip measurements (fig. 18–33). Depth is recorded on the vertical axis with the well represented by a vertical line. The depth of each dip measurement is where the stick intersects the well. The angle on the stick is the dip measurement.

Nuclear magnetic resonance log

A *nuclear magnetic resonance (NMR) log* uses a magnetic field from the tool that is turned on and off to orient and then relax hydrogen atom protons in the rock. The relaxing protons give off a radio-frequency signal that is recorded and used to directly measure or calculate an estimate of oil and water saturation, total and effective porosity, formation permeability, volume of the fluid that will flow (movable), and the irreducible water saturation along with locating gas and residual oil.

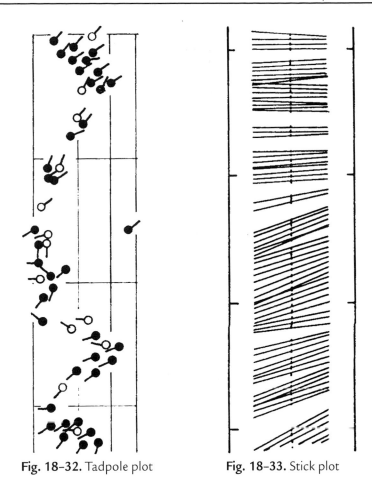

Fig. 18–32. Tadpole plot   **Fig. 18–33.** Stick plot

Wellbore imaging logs

Wellbore imaging logs make a picture of the rocks along the wellbore using either resistivity or ultrasonic reflections. The resistivity tool uses a large number of small, button electrodes that are closely spaced on pads mounted on four arms. The resistivity picture of the wellbore images sedimentary rock layers including thin shale layers, unconformities, fractures (both natural and induced by drilling), and vugs. It can be used to calculate the strike and dip of rock layers and fractures. The ultrasonic tool uses sound waves with frequencies higher than a human can hear. An ultrasonic transmitter in the tool bounces ultrasound waves off the wellbore, which are recorded on a receiver in the tool to image the wellbore.

Table 18–3 summarizes the uses of modern wireline logs.

Table 18–3. The names, track recorded in, and uses of common wireline well logs

Name	Track	Uses
spontaneous potential (SP)	1	· identify permeable formations
gamma ray (GR)	1	· measure natural radioactivity · identify potential reservoir rock · correlate formations · determine formation elevations
resistivity (R) induction or laterolog		
shallow resistivity	2	· determine formation elevations
medium resistivity	2	· depth of invade zone
deep resistivity	2	· identify fluid in pores
microlog, microresistivity, or minilog	2	· locate mudcake · identify permeable rocks
neutron or neutron porosity (N or NL)	2	· determine formation porosity
formation density log (FDL), density (D), or gamma-gamma (GG)	2	· determine formation density and calculate porosity
sonic log (SL) or acoustic velocity (AVL)	2 or 3	· determine formation sound velocity and calculate porosity · determine lithology
photo-electric factor (PEF)	2	· determine formation mineral composition and lithology
caliper (CAL)	1	· measure wellbore diameter · calculate wellbore volume

Computer-generated log

A *computer-generated log* is a combination of two or more logging measurements that have been calculated by a computer on the logging truck at the wellsite or in a computing center. The wellsite computer logs are called *quick-look logs* and can show water saturation; porosity; percentage of limestone, sandstone, and shale; the presence of fractures; and true vertical depth. *Computer center logs* are similar to quick-look logs but are more accurate and detailed.

Measurements-While-Drilling and Logging-While-Drilling

Wireline well logs are run after the well has been drilled. In the 1980s, sensors for the bottom of the drillstring and a data-transmitting process were developed to give real-time logs as the well is being drilled called *measurements-while-drilling* (MWD) and *logging-while-drilling* (LWD). MWD measures drilling and well parameters and LWD measures rock and fluid parameters.

The sensors are located just above the drill bit on the drillstring. The power to the sensors is supplied either by a turbine driven by the circulating drilling mud or electrical batteries. The data are transmitted to the surface by *fluid pulse telemetry*. The data are coded digitally in pressure pulses that are sent up the well through the drilling mud. They are recorded on a pressure transducer on the surface where they are decoded by a computer.

LWD makes a resistivity, natural, and spectral gamma ray, formation density, neutron porosity, nuclear magnetic resonance, and caliper log of the well. MWD is very useful in drilling deviation and horizontal wells. It uses gyroscopes, magnetometers, and accelerometers in the downhole tool to measure the orientation of the drill bit (*tool face*) and the direction (azimuth and deviation) in which the well is being drilled. It also measures torque, weight on bit, and well temperature and pressure.

Geosteering is the drilling of a horizontal well while continuously adjusting the direction of the bit to keep the well within the target formation. An *LWD* system is used to sense the target formation top or bottom, and an MWD system shows the orientation of the bit. A steerable downhole assembly is used to adjust the direction the well is being drilled to keep the well within a target formation that can be as thin as 7 feet (2.1 m).

Drillstem Test

A *drillstem test* is a temporary completion of a well. As the well is drilled and logged, it is kept filled with drilling mud. Drilling mud pressure keeps any fluids back in the pores of the rock adjacent to the wellbore. If the logs indicate a potential reservoir, a drillstem test can be run to further evaluate that reservoir. A drillstem, usually made of drillpipe, is run in a well that is still filled with drilling mud. The drillstem has one or two packers, perforated pipe, pressure gauges, and a valve assembly (fig. 18–34).

Packers are cylinders made of a rubber-like material that can be compressed to expand against the well walls to seal that portion of the well and prevent any vertical flow of fluid in that section of the well (fig. 18–35). If the formation is located on the bottom of the well, only one packer is used (fig. 18–34). If the formation is located above the bottom of the well, two packers (*straddle packers*) are used (fig. 18–35). Once seated, the packer(s) eliminate any drilling mud pressure on that formation. The water, gas, or oil can then flow out of the formation and into the well. A valve is opened on the drillstem, and the formation fluids flow into and up the drillstem.

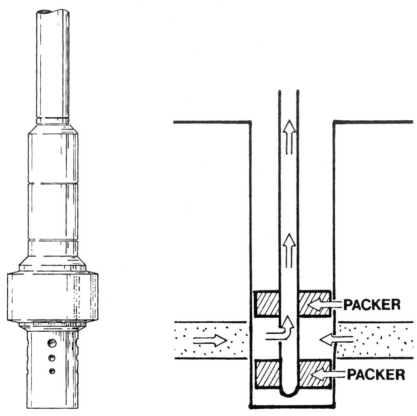

Fig. 18–34. Drillstem tool with one packer

Fig. 18–35. Drillstem test with straddle packers

If gas is present, it will flow up the drillstem and onto the surface where it is measured and *flared* (burned). Sometimes oil has enough pressure to flow to the surface during a drillstem test. Usually, however, the oil fills the drillstem only to a certain height that is measured. During the drillstem test, pressure on the fluid flowing into the drillstem is continuously measured. The valve on the drillstem is opened and closed several times, and fluid pressure buildup and drop-off are recorded on a chart of pressure versus time called a *pressure buildup curve* (fig. 18–36).

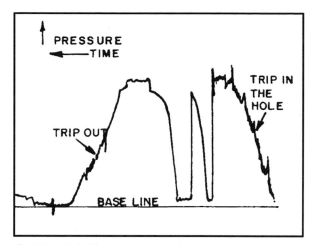

Fig. 18–36. Drillstem test record

Engineers use the pressure records to calculate formation permeability, reservoir fluid pressure, and the extent of any formation damage. The test can take 20 minutes to three days. Longer tests are more accurate but more expensive because the test involves rig time. The test can be run in either an open-hole or a cased well that has been perforated. It should not be run in a formation with unconsolidated sands that could collapse into the well during the test and trap the equipment. Drillstem tests are usually run only on exploratory wells because of the rig time expense.

Repeat Formation Tester

A *repeat formation tester* (*RFT*) or *wireline formation tester* (*WFT*) is a wireline tool that is several tens of feet long and used to sample reservoir fluids and measure formation pressure versus time. It can test several levels in a well. At the zone to be tested, a backup shoe from the tool is pressed against the well wall to force a rubber pad with a valve against the opposite wall. The valve is opened, and formation fluids can flow into the tool as the pressures are measured. The pressure records are used to calculate formation permeability. Two large sample chambers, each holding several gallons, are used to obtain a sample of formation fluids at that level. The samples are used to measure water resistivity (R_w) and saturation (S_w). This can be repeated at several locations in the well.

Completing a Well

After a well has been drilled and tested, there are two options. The well is either plugged and abandoned as a dry hole and the drilling rig is released, or the well is completed as a producer by setting pipe.

The *Oil & Gas Journal* estimated that 44,714 wells will be completed in the United States and 11,409 in Canada in 2011. In the United States, Texas will have the most well completions (34%).

Casing

A well is always cased (*set pipe*) to complete the well. *Casing* is relatively thin-walled steel pipe. Numerous joints of the same size casing are screwed together to form a long length of casing, called a *casing string*. The casing has an outer diameter of at least 2 in. (5 cm) less the wellbore diameter. The casing string is run into the well and cemented to the sides of the well (see Introduction, fig. I–10) in a *cement job*.

Casing stabilizes the well and prevents the sides from caving into the well. It protects freshwater reservoirs from the oil, gas, and saltwater brought up the well during production. Casing also prevents the production from being diluted by waters from other formations in the well. The drilling rig is usually used to run the casing. If a very large and expensive drilling rig was used to drill the well, it can be released, and a smaller, less expensive *completion rig* can be used to run the casing.

Casing is seamless pipe made to API standards. A *joint* of casing is commonly about 30 ft (9.1 m) long, though the length can be as short as 16 ft (4.9 m) or as long as 42 ft (12.8 m). Diameters range from 4½ to 36 in. (11.4 to 91.4 cm) but are commonly 5½ to 13¾ in. (14 to 34.9 cm). Casing is graded by the API for (1) outer diameter and wall thickness, (2) weight-per-unit length, (3) type of coupling, (4) length, and (5) grade of steel.

The end of each casing joint has male threads that are protected by a plastic or metal cap called a *thread protector* until the casing is ready for use. A *collar* or *coupling*, a short section of cylindrical steel pipe with female threads on the inside and a diameter slightly larger than the casing, is used to connect joints of casing.

Before a cement job, the well is conditioned by running a drillstring with a used bit into the well. Mud is then circulated for a period of time to remove any remaining cuttings. In order to scrape the filter cake off the well sides to prepare it for the cement, *wall scratchers* with protruding wires (fig. 19–1) are attached by collars or clamps to the casing string and run up and down or rotated in the well.

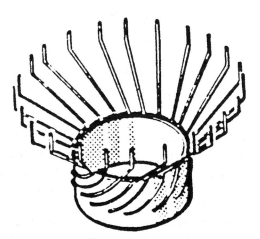

Fig. 19–1. Wall scratcher

A service company *casing crew* with one to five members usually cases the well. They have specialized equipment such as hydraulic casing tongs and a casing stabbing board. A *guide shoe*, a short section of rounded pipe with a hole in the end, is screwed into the end of the casing string to guide the casing string down the well. Spring-like devices called *centralizers* (fig. 19–2) are attached to the outside of the casing as it is run into the well to position the string in the center of the well.

Fig. 19–2. Centralizer

The casing is stored horizontally on the *pipe rack* on the ground next to the rig (plate 19–1). It is pulled a joint at a time up the pipe ramp, through the V-door in the derrick or mast and onto the drill floor (plate 19–2). Each casing joint is lifted by *casing elevators* on the traveling block and guided (*stabbed*) into the casing string already being suspended in the well. A thread compound is applied to the threads to make a tight seal. *Casing tongs* hung by cable above the drill floor are used to screw the casing joint onto the string. To assist in stabbing the joint of casing, a derrick operator from the casing crew stands on the *stabbing board*, located in the derrick or mast. The casing string is run into the well (plate 19–3) and finally *landed* by transferring the weight of the string to the casing hangers in the casinghead on the top of the well. *Casing hangers* use either slips (wedges) or threads to suspend the casing in the well.

Plate 19–1. Casing racked in front of a well. (Courtesy of Parker Drilling.)

Plate 19–2. Casing being run up the pipe ramp onto the drill floor. (Courtesy of API.)

The service company that provides the cementing equipment (plate 19-4) prepares the wet cement called *slurry* for the cement job by mixing sacks of dry cement and water together at the wellsite in a *hydraulic jet mixer*, *recirculating mixer*, or *batch mixer*. The cement is Portland cement made to one of eight API classes with additives for various situations. Common *cement additives* that come in sacks include *accelerators* that shorten cement setting time, *retarders* that lengthen the time, *lightweight additives* that decrease the slurry density, and *heavyweight additives* that increase the density. Other additives are used to change the cement's compressive strength, flow properties, and dehydration rate; to cause the cement to expand; to reduce the cost of the cement (*extenders*); and to prevent foaming (*antifoam*). *Bridging materials* can be used to plug lost circulation zones.

The *thickening* or *pumpability time* of the cement is the time the cement is fluid and can be pumped before it sets. Special cements are available for very corrosive environments and for *setting* (hardening) at deep depths under high temperature.

Plate 19–3. Casing being run into the well. (Courtesy of API.)

Plate 19–4. Cement job. (Courtesy of Halliburton.)

After the casing has been run into the well to just above the bottom, an L-shaped fitting (*cementing head*) is attached to the top of the wellhead to receive the slurry through a line from the pumps. It contains two wiper plugs. A *wiper plug* (fig. 19–3), made of cast aluminum and rubber, is designed to wipe the inside of the casing and separate two different liquids to prevent mixing.

Fig. 19–3. Wiper plug

The first wiper plug pumped down the casing is the *bottom plug*. It separates the drilling mud from the cement slurry. The cement slurry is pumped down the casing until the bottom plug is caught in a float collar located one or more casing joints above the guide shoe (fig. 19–4). The *float collar* is a short pipe with the same diameter as the casing. It contains a constriction to stop the wiper plug and also acts as a one-way valve. This allows the casing string to be run in the well without the drilling mud flowing up the inside of the casing. Because of this, the casing string floats in the drilling mud to partially support its weight. It does, however, allow liquid cement to be pumped down the casing.

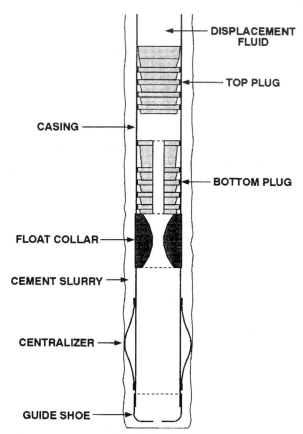

Fig. 19–4. Bottom of well during a cement job

After the bottom plug has landed in the float collar, the pump pressure is increased until the cement slurry ruptures a diaphragm in the bottom plug and flows through it. The slurry flows out the guide shoe and up the outside of the casing string. After a predetermined volume of slurry has been pumped down the well, the *top plug* is pumped down the casing, followed by a displacement fluid that is usually drilling mud. When the top plug hits the bottom plug and all the slurry has been displaced between them, the pumps are shut down, and the cement is allowed to set (*waiting on cement*, or *WOC*) for 8 to 12 hours. The wiper plugs, guide shoe, and cement on the bottom are then drilled out. A temperature log can be run to locate the top of the setting cement behind the casing by the heat given off by the setting cement.

Multistage cementing is used on long casing strings when the required pump pressures would be too high if the entire length of casing was cemented at once. Two or more sections of a single string are cemented separately. First, the lower section is cemented. Next, holes in a coupling located on the casing string are opened for the cement to flow through, and the section above the coupling is cemented. This can be repeated several times at various levels up the casing string.

A well is drilled and cased in stages called a casing program. During the *casing program*, the well is drilled to a certain depth and then cased, drilled deeper and cased again, drilled deeper and cased again. Each time the well is drilled deeper, a bit is used that is at least ½ in. (1.3 cm) smaller in diameter than the casing. The casing program defines the grades, lengths, and sizes of casing that are going to be used before the well is drilled. The casing is ordered and delivered to the drillsite as the well is being drilled.

A well commonly contains three or more concentric casing strings (fig. 19–5). Shallow wells can have two or even just one casing string, whereas deeper wells have more casing strings. Each casing string runs back up to the surface. The largest diameter and shortest length string is on the outside. The smallest diameter and longest length string is on the inside. The outside string is cemented first and the inside string last.

Conductor pipe is the largest diameter casing string (30 to 42 in. or 76.2 to 106.7 cm offshore and 16 in. or 40.6 cm onshore) and is often several hundred feet long. Either a hole for the conductor pipe is drilled into hard rock, or the conductor pipe is pile-driven into soft ground. Conductor pipe serves as a route for the drilling mud coming from the well to the mud tanks, prevents the top of the well from caving in, and isolates any near-surface, freshwater and gas zones. The blowout preventers are attached to the top of the conductor pipe if shallow gas is present.

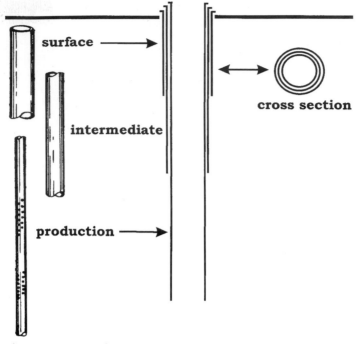

Fig. 19–5. Types of casing

The next casing string is *surface casing,* often 13¾ in. (34.9 cm) in diameter and several hundred to several thousand feet in length. It prevents soft, near-surface sediments from caving into the well. It also protects freshwater reservoirs from further contamination by drilling mud as the well is being drilled deeper. The length of surface casing in a well is often mandated by the government.

Protection or *intermediate casing* can be set to isolate problem zones in the well such as abnormal high pressure, lost circulation, or salt. It is typically 8⅝ in. (21.9 cm) in diameter. In some wells there is no intermediate casing string.

The final string of casing is the *production* or *oil string* that runs down to the producing zone. It is typically 5½ in. (14 cm) in diameter.

In New Mexico, special care must be taken during drilling into the Pennsylvanian age Morrow sands for gas. Drilling into the Morrow sands with heavy drilling mud causes extensive formation damage. Overlying the Morrow sands, however, is the Atoka Formation, which has abnormal high pressure. Heavy drilling mud must be used when drilling there to prevent blowouts. After the Atoka Formation is drilled, a string of intermediate or protection casing is cemented into place to seal off the formation. Lighter drilling mud is then used to drill into the Morrow sands and complete the well.

Instead of a casing string, a liner string can be set on the bottom of the well to save money. A *liner string* is similar to a casing string and is made of liner joints, which are the same as casing joints. A liner string, however, never runs all the way up the well to the surface as a casing string does. It is suspended in the well by a *liner hanger* using slips in a casing string, and it can be cemented in. A liner string is often used in deep wells to save the cost of running a long production casing string.

The casing and cement are pressure-tested during a *mechanical integrity test (MIT)*. The well is shut in, and a liquid (water or drilling mud) is pumped into the well until the maximum casing pressure is reached. The pumps are then stopped, and the pressure is monitored. A pressure drop indicates a leak in the casing.

After the Yates field in west Texas (see chap. 10, fig. 10–37) was developed in the late 1920s, several new oil seeps were observed in the adjacent Pecos River. It became apparent that the substandard casing used to complete the wells was leaking. Until the casing was repaired, an estimated 70 million bbl (11 million m^3) of oil was lost.

Bottom-Hole Completions

The bottom of the well is completed with either an open-hole or cased-hole completion. An *open-hole, top set,* or *barefoot completion* (fig. 19–6) is made by drilling down to the top of the producing formation, and then casing the well. The well is then drilled through the producing formation, leaving the bottom of the well open. This completion is used primarily in developing a field with a known reservoir and reduces the cost of casing. An open-hole completion, however, cannot be used in soft formations that might cave into the well. Because the casing is *"set in the dark"* in an open-hole completion before the pay is drilled, the casing cannot be salvaged if the pay proves to be unproductive.

If the producing formation is composed of unconsolidated sands that can cave into the well (a *sand control problem*), a gravel pack completion can be used. A *gravel pack completion* starts similar to an open-hole completion with casing set at the top of the producing formation, and the producing formation is then drilled through. A tool called an *underreamer* is then run on a drillstring down the hole. The underreamer (fig. 19–7) has arms on it that are expanded. The underreamer is then rotated to ream out a cavity in the formation.

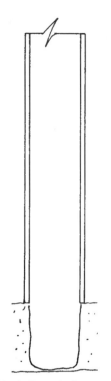

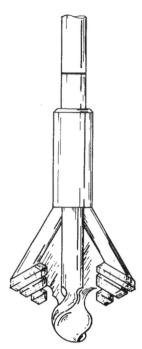

Fig. 19–6. Open-hole completion   **Fig. 19–7.** Underreamer

The cavity is then filled with very well-sorted, coarse sand, called a *gravel pack*. It is pumped down the well suspended in a carrier fluid that leaks off into the formation, leaving the gravel pack on the bottom of the well. The gravel pack is very porous and permeable. A section of slotted or screen liner is then run into the gravel pack (fig. 19–8). A *screen liner* has holes in the liner wall. Wire wrapped around the liner prevents loose sediments from flowing through the holes into the liner. A *slotted liner* has several long openings (slots) cut into it to allow fluids, but not sediments, to flow into the liner. A *prepacked slotted liner* is a slotted liner filled with a gravel pack that is held together with a resin coating.

Another completion technique is an *uncemented slotted liner* that is set in an open hole. It is used if there is a sloughing shale problem.

A *set-through completion*, in which a liner string (fig. 19–9a) or a casing string (fig. 19–9b) is cemented into the producing reservoir, is commonly done today. Holes called *perforations* are then shot through the liner or casing and cement and into the producing formation. A *perforating gun* is run into the well on an e-line, a braided wire cable with a core of insulated electrical wires, or tubing string to make the perforations. The original perforating guns used steel bullets, but *jet perforators* that use shaped explosive charges are commonly used today. When electrically detonated, the cone-shaped explosives produce extremely fast jets of gases that blow the perforations into the casing, cement, and producing formation. The explosives can be shaped to give either a maximum diameter or maximum length of penetration.

Perforating guns are either *expendable*, which disintegrate and leave debris in the well, or *retrievable*, which can be removed from the well. Perforations are described in shots per foot (*density*) and angular separation (*phasing*), such as 60°. Perforated completions are commonly used for multiple completions.

It is difficult to run casing through the build angle section on horizontal wells. Because of this, most of these wells are completed either as openhole, or with a slotted or cemented liner that has been perforated.

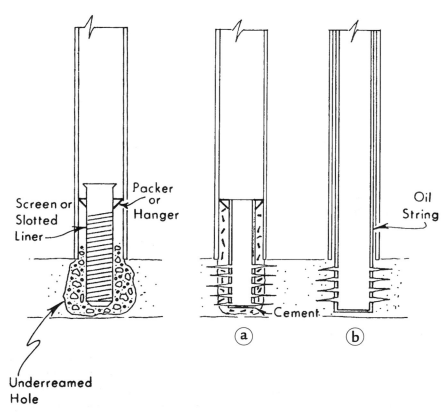

Fig. 19–8. Gravel pack completion

Fig. 19–9. (a) Perforated liner completion and (b) perforated casing completion

Expandable Casing and Liner

Expandable casing or *liner* is run in a well, then its diameter is permanently enlarged up to 30% by hydraulically pumping an expander plug through it. Elastomer can be bonded to the outside of the casing to form a seal. It is used to reduce the tapering effect of the diameter of casing and liner strings becoming smaller and smaller as they goes deeper into a well. Expandable casing can also be used to isolate water, fracture, lost circulation, and sloughing shale zones.

Expandable casing and liners have led to the construction of monobore wells. A *monobore well* has casing that is very similar in diameter from the top to the bottom of the well. A problem with deep wells and long deviated or horizontal wells is that the casing becomes smaller and smaller with depth. A monobore well eliminates that problem by allowing a larger-diameter production tubing string and higher production rates. It reduces well costs by requiring a smaller wellhead and smaller diameter surface casing, along with easier well servicing and intervention.

Tubing

Small-diameter pipe called *tubing* is run into the well to just above the bottom to conduct the water, gas, and oil (produced fluids) to the surface (fig. 19–10). Tubing is special steel pipe that ranges from $1\frac{1}{4}$ to $4\frac{1}{2}$ in. (3.2 to 11.4 cm) in diameter and comes in lengths of about 30 ft. (9.1 m) long (plate 19–5). The API grades tubing according to dimensions, strength, performance, and required threads. The thicker section with male threads on the ends of each tubing joint is called the *upset*. Tubing joints are joined together with *collars* (short steel cylinders with female threads) to form a tubing string. A *tubing* or *completion packer* is used near the bottom of the tubing string (fig. 19–10). The packer is made of hollow rubber that is compressed to seal the casing-tubing annulus. It keeps the tubing string central in the well and prevents the produced fluids from flowing up the outside of the tubing string.

Tubing protects the casing from corrosion by the produced fluids. Because the casing has been cemented in the well, it is very difficult to repair the casing. The tubing string, however, is suspended in the well and can be pulled from the well to repair or replace it during a workover. A *completion fluid*, commonly treated water or diesel oil, can be used to fill the annulus between the tubing and casing string to prevent corrosion.

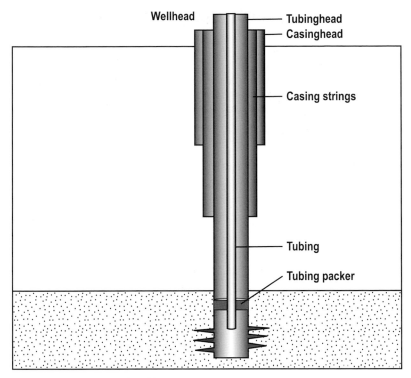

Fig. 19–10. Completed well

Plate 19–5. Tubing being run into a well

If the well needs to be pumped, a downhole pump is put on the end of the tubing string. A *tubing anchor* that uses slips can be used to secure the tubing to the casing on the bottom of the well. A *seating nipple* is a special joint of tubing used near the bottom of the tubing string. It has a narrow inner diameter designed to stop any equipment that is dropped or run down the tubing string. A *tubingless completion*, used on very high-volume wells, brings the produced fluids up the casing string without using tubing.

On offshore wells, a *subsurface safety valve* is used in the tubing to stop the flow during an emergency. The valve is held open by pressure. A drop in pressure automatically closes the valve. A *surface-controlled subsurface safety valve* is operated with hydraulic lines that run down the well.

Wellhead

The *wellhead* is the permanent, large, forged or cast steel fitting on the surface of the ground on top of the well (fig. 19–10). It is usually welded to the conductor pipe or surface casing. The wellhead consists of casingheads and a tubinghead. The larger, lower *casingheads* seal off the annulus between the casing strings and contain the casing hangers for the top of each casing string. *Casing hangers* are used to suspend the casing strings in the well. They either screw onto the top of each casing string or are held by slips (wedges). There is a casinghead and casing hanger for each casing string.

The casinghead for the surface casing is the largest in diameter and is located on the bottom. The deepest and longest string casing is hung in the uppermost casinghead with the smallest diameter. Each casinghead is bolted to the one below and has a gas outlet to provide pressure relief. Each time the well is drilled deeper and cased, the blowout-preventer stack is removed and reinstalled (*nippled up*) on the new casinghead. A *unitized wellhead* uses only one casinghead for all the casing strings.

The smaller *tubinghead*, located on the casingheads, suspends the tubing string down the well and seals the casing-tubing annulus. A *tubing hanger* uses slips or a bolted flange to hold the tubing string and fits as a wedge in the tubinghead.

Wellhead equipment is the equipment attached to the top of the tubing and casing. It supports the strings, seals the annulus between the strings, and controls the production. Wellhead equipment includes the casingheads, tubinghead, Christmas tree, stuffing box, and pressure gauges.

Chokes

A flowing well is seldom produced at an unlimited rate. It could result in a too-rapid depletion of reservoir pressure and a decrease in ultimate production. An excessive production rate creates a large pressure drop between the reservoir and wellbore, causing gas to bubble out of the oil and block the reservoir rock pores adjacent to the wellbore. Flow rates are limited by surface or subsurface *chokes*, which are valves that cause the fluid to flow through a small hole called an *orifice*. The smaller the orifice, the lower the flow rate. Chokes can be either *positive* with a fixed orifice size or *adjustable* to permit changing the flow rate.

Surface Equipment

Gas wells flow to the surface by themselves. There are some oil wells (4% in the United States) in which the oil has enough pressure to flow up the tubing string to the surface. For gas wells and flowing oil wells, a structure of pipes, fittings, valves, and gauges are welded to the wellhead to control the flow (fig. 19-11). This plumbing is called a *Christmas* or *production tree* (plate 19-6). All Christmas trees have a *master valve* sticking out of the lower part to turn the well off during an emergency. The plumbing going off the side of the Christmas tree to the flowline is called the *wing*. If there is only one producing zone in the well, it is a *single-wing tree*. Two producing zones require a *double-wing tree* with two wings on opposite sides to keep the production separate. On the wing is a *wing or flow valve* to turn flow on and off through that flowline. A *swab valve* on the upper part of the tree is used to open the well to allow wireline equipment to be lowered down the well during a workover. A *pressure gauge* at the top of the tree measures tubing pressure. Most Christmas trees are machined out of a solid block of metal.

In most oil wells (95% in the United States), the oil does not have enough pressure to flow all the way up the tubing to the surface. The produced water and oil will fill the well only to a certain level and has to be lifted to the surface by one of several methods called *artificial lift*. Even with a flowing oil well, as more fluids are produced from the subsurface reservoir, the pressure on the remaining oil can decrease until it no longer flows to the surface. When this happens, the Christmas tree has to be removed and a surface pumper installed in a process called *putting the well on pump*.

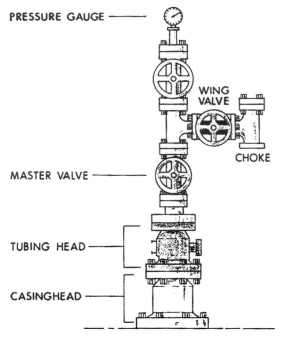

Fig. 19–11. Single-wing Christmas tree. (Modified from Baker, 1983.)

Plate 19–6. Christmas tree surrounded by a scaffolding for service

Sucker-rod Pump

The most common artificial lift system is a sucker-rod pump. A *sucker-rod* or *rod-pumping system* uses a sucker-rod pump on the bottom of the tubing string, a beam-pumping unit on the surface, and a sucker-rod string that runs down the well to connect them (plate 19–7). The *sucker-rod pump* is built to API specifications and has a standing valve and traveling valve (fig. 19–12). The *traveling valve* moves up and down, and the *standing valve* remains stationary. Both valves allow fluid flow in only one direction. They consist of a ball, a seat (a plate with a hole), and a cage to hold the ball over the seat. The steel ball allows the oil to flow up but not back down through the valve. Fluid flowing upward lifts the ball off the seat and opens the valve (fig. 19–13a). Fluid cannot flow down because gravity holds the ball in the seat (fig. 19–13b). Each upward stroke of the traveling valve lifts the oil and water up the tubing. There are commonly 10 to 20 strokes per minute.

Plate 19–7. Artificial lift with a sucker-rod pumping unit

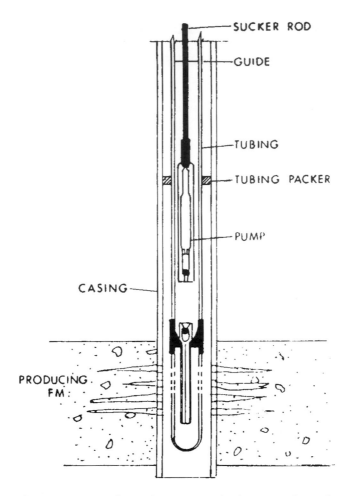

Fig. 19-12. A sucker-rod pump on the bottom of a well. (Modified from Baker, 1983.)

Three types of sucker-rod pumps are the insert or rod pump, tubing pump, and casing pump. The *insert* or *rod pump* (the most common) is run in the well as a complete unit on the sucker-rod string through the tubing string. It is usually held in place by a bottom anchor. The insert pump is the smallest of the pumps and has the lowest capacity. The *tubing pump* is run as part of the tubing string. The plunger and traveling valve are run on the sucker-rod string. A *casing pump* is a relatively large version of an insert pump that pumps the produced fluids up the casing. It is held in position by a packer and has a much larger volume than an insert pump.

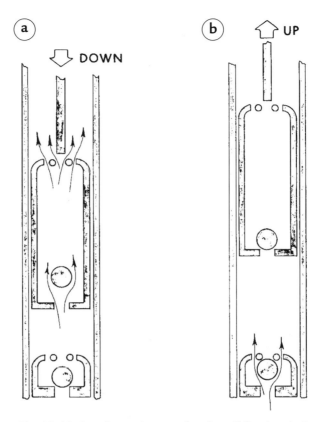

Fig. 19–13. A sucker-rod pump showing oil flowing during the (a) downstroke and (b) upstroke

A *beam-pumping unit* on the surface is used to active the sucker-rod pump (fig. 19–14). It is mounted on a heavy, steel I-beam base or a concrete base. The beam-pumping unit has a steel beam (*walking beam*) that pivots up and down on bearings on top of a *Samson post*. It is usually driven by an electric motor but could also be driven by a motor that uses natural gas produced from the well.

If an electrical motor is used, it can have a timer to periodically turn the pumping system on and off. This conserves electricity and prevents the pump from working when there isn't enough liquid in the well. The motor is connected by a V-belt drive to a *gear reducer* that decreases the rotational speed. The gear reducer rotates a *crank* on either side. Two steel beams (*pitmans*) and a cross bar (*equalizer*) connect the cranks on the gear reducer to the walking beam and cause the end of the walking beam to rise and fall.

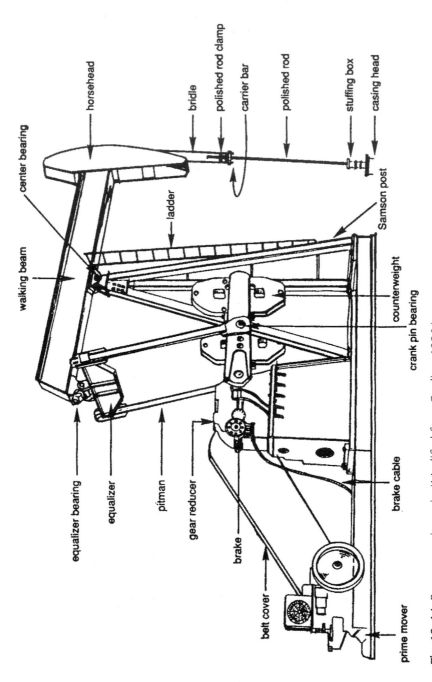

Fig. 19-14. Beam-pumping unit. (Modified from Gerding, 1986.)

horsehead

center bearing

bridle

polished rod clamp

carrier bar

polished rod

stuffing box

casing head

walking beam

ladder

Samson post

counterweight

crank pin bearing

equalizer bearing

equalizer

pitman

gear reducer

brake

belt cover

prime mover

brake cable

Attached to the opposite side of the walking beam is the *sucker-rod string* that runs down the center of the tubing string to the downhole pump. Sucker rods (fig. 19–15a) are solid steel alloy rods between ½ and 1¼ in. (1.3 to 3.2 cm) in diameter and are usually 25 ft (7.6 m) long. The API grades sucker rods according to their alloy composition, recommended well depth, maximum allowable stress, and environment. The rods have male threads on both ends. They are connected together with short steel cylinders with female threads on the inside called *sucker-rod couplings* (fig. 19–15b). Flat areas (*wrench flats*) allow a wrench to grip the coupling without harming it. Light-weight sucker rods are made of fiberglass.

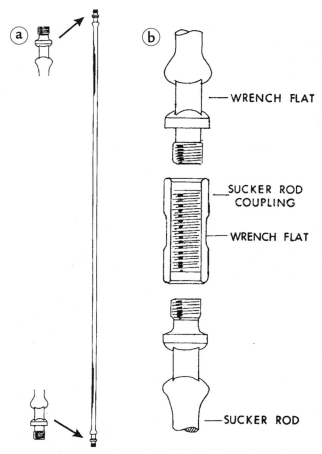

Fig. 19–15. (a) Sucker rod and (b) sucker-rod couple

The sucker-rod string is kept central in the tubing string by *sucker-rod guides* (fig. 19–12) made of rubber, plastic, nylon, or metal. They move up and down with the sucker-rod string as produced fluids flow up the tubing through slots in the guides. Shorter lengths of sucker rods (*pony rods*) can be used to adjust the length of the sucker-rod string. The string can be either *untapered* (all one diameter) or *tapered* with a decrease in rod diameter down the well.

Counterweights of steel are used to balance the weight of the sucker-rod string on the walking beam of a beam-pumping unit during the upstroke. Two rotating counterweights are located on both sides of the rotary crank on a *crank-balanced pumper* (fig. 19–14). The counterweights on a *beam-balanced pumper*, used for shallow wells, are located on the walking beam opposite the wellhead (fig. 19–16). They are adjustable by moving them along the walking beam.

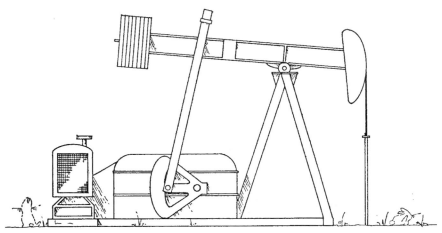

Fig. 19–16. Beam-balanced rod pumping unit

A large, curved steel plate called a *horsehead* is used on the well side of the walking beam to keep the pull on the sucker-rod string vertical (fig. 19–14). Two wire ropes (*bridles*) and a steel bar (*carrier*) are used to connect the horsehead to the top of the sucker-rod string.

A *polished rod* is used at the top of the sucker-rod string. It is a smooth length of brass or steel rod with a diameter of 1¼ to 1½ in. (3.2 to 3.8 cm) and ranges from 8 to 22 ft (2.4 to 6.7 m) long. The polished rod moves up and down through the *stuffing box*, a steel container on the wellhead that contains flexible material or packing such as rubber that provided a seal around the polished rod.

A variation is the *air-balanced beam-pumping unit*, which uses a piston and rod in a cylinder filled with compressed air to balance the weight of the sucker-rod string (fig. 19–17). It is more compact and lighter than the crank- and beam-balanced units. Another type, the *Mark II*, uses a lever system to counterbalance the sucker-rod string weight (fig. 19–18).

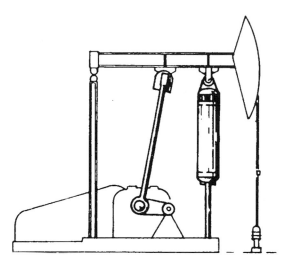

Fig. 19–17. Air-balanced beam-pumping unit

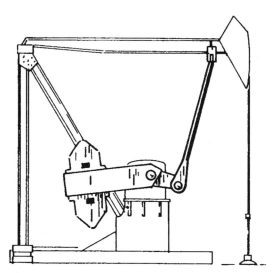

Fig. 19–18. Mark II rod pumping unit

Gas Lift

In *gas lift*, another type of artificial lift, a compressed, inert gas (*lift gas*) is injected into the annulus in the well between the casing and tubing (fig. 19–19). The lift gas is usually natural gas that was produced from the well. *Gas lift valves* are pressure valves that open and close and are spaced along the tubing string. They allow the lift gas to flow into the tubing where it dissolves in the produced liquid and also forms bubbles. This lightens the liquid density, which, along with the expanding bubbles, forces the produced liquid up the tubing string to the surface where the gas can be recycled.

The advantages of gas lift are that there is very little surface equipment and there are few moving parts. Gas lift is a very inexpensive technique when many wells are serviced by one central compressor facility. However, it is effective only in relatively shallow wells. Offshore oil wells that need artificial lift are usually completed with gas lift. Gas lift is either *continuous* or *intermittent* (periodically on and off) for wells with low production.

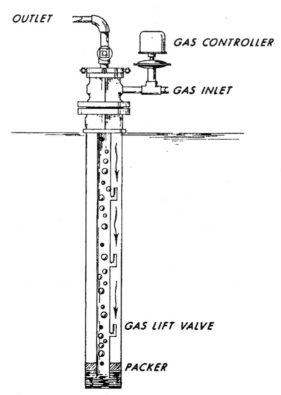

Fig. 19–19. A gas lift well

Electric Submersible Pump

An *electric submersible pump* (*ESP*) or *sub pump* uses an electric motor that drives a centrifugal pump with a series of rotating blades on a shaft on the bottom of the tubing (fig. 19–20). An armored electrical cable runs up the well, strapped to the tubing string. The electricity comes from a surface transformer. The downhole electric motor has a variable speed that can be adjusted for efficiently lifting different volumes of liquids. Electric submersible pumps are used for lifting large volumes of liquid up the well and for crooked and deviated wells. A gas separator is often used on the bottom of the pump to prevent gas from forming in the pump and decreasing the pump's efficiency.

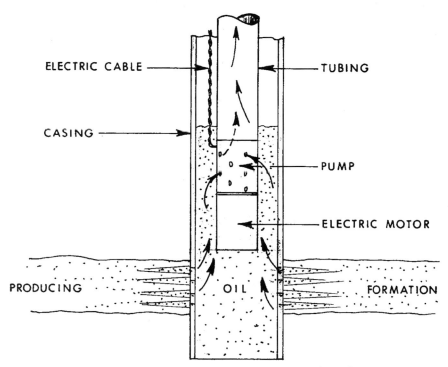

Fig. 19–20. An electric submersible pump on the bottom of a well

Hydraulic Pump

A *hydraulic pump* is identical to a sucker-rod pump except it is driven by hydraulic pressure from a liquid pumped down the well. It uses two reciprocating pumps. One pump on the surface injects high-pressure *power oil* or *fluid* (usually crude oil from a storage tank) down a tubing string in the well. The power fluid drives a reciprocating hydraulic motor on the bottom of the tubing. It is coupled to a pump, similar to a sucker-rod pump, located below the liquid level in the well. The pump lifts both the spent power fluid and the produced fluid from the well up another tubing string. The power fluid causes the upstroke, and the release of pressure causes the downstroke. This type is called a *parallel-free pump*. In a variation, the *casing-free pump*, the power fluid is pumped down a tubing string, and the produced liquid is pumped up the casing-tubing annulus.

The stroke in a hydraulic pump is very similar to a sucker-rod pump stroke, except it is shorter. Hydraulic pumps can be either *fixed* (screwed onto the tubing string) or *free* (pumped up and down the well). They can be either *open*, with downhole mixing of power and produced fluids, or *closed*, with no mixing. Most hydraulic pumps are free and open.

Artificial lift in the United States consists of 82% beam pumper, 10% gas lift, 4% electric submersible pump, and 2% hydraulic pump.

Multiple Completions

Production from two or more zones in a well can be mixed (*commingled*) and brought up the same tubing string. Usually, however, the production is kept separate. If there are two producing zones, two packers and two tubing strings are used in a *dual completion* (fig. 19–21). Sometimes only one packer and tubing string are used in a dual completion, and the production from one zone is brought up the casing-tubing annulus. If it is a flowing well, there will be a double-wing tree (plate 19–8). If beam pumping is used, there will be two beam pumpers and sucker-rod strings per well. Triple completions are the most completions that can be made in one well.

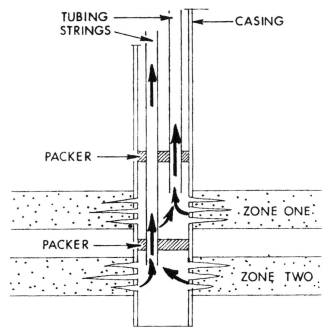

Fig. 19–21. The bottom of a dual completion well

Plate 19–8. Double-wing Christmas tree

Intelligent Wells

An *intelligent* or *smart well* is a well that has downhole sensors that can measure well flow properties such as rate, pressure, and gas/oil ratio. An adjustable choke on the bottom of the well can be either automatically or manually adjusted, usually by hydraulics, to obtain an optimum production rate.

Surface Treatment and Storage

Flowlines

A *flowline* is a pipe made of steel, plastic, or fiberglass that conducts the produced fluids from the wing on a Christmas tree or the *tee* on the wellhead of an artificial lift well (fig. 20–1) to the separation, treatment, and storage equipment. The flowline is located either on the surface or buried in the soil below the freeze line for protection from the weather. Each well can have its own separation and storage facilities, or a *central processing unit* (*CPU*) with shared separation and storage facilities can be used to service a group of surrounding wells. A gathering system of flowlines connects the wells to a CPU. *Headers* are larger pipes that collect the fluids from several smaller flowlines. A *radial gathering system* has the flowlines converging on a CPU (fig. 20–2a). An *axial* or *trunk-line* gathering system has flowlines emptying into several headers that flow into a larger trunk line, which takes the produced fluids to the CPU (fig. 20–2b).

Valves are gates on the flowlines that are designed to regulate the rate of flow through the line and open and close the line (fig. 20–3). They can be either manual or automatic. Valves are named by their construction (e.g., butterfly, needle, plug, ball, gate, bellows, or globe) or by their use (e.g., metering, check, safety, relief, regulating, pilot, or shut off).

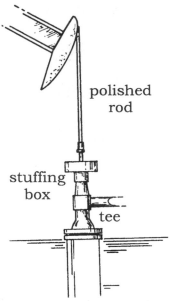

Fig. 20-1. Wellhead with tee for flowline

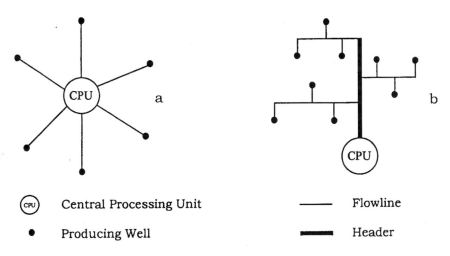

Fig. 20-2. Flowline patterns (a) radial and (b) axial or trunk line

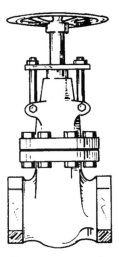

Fig. 20–3. Gate valve

If both natural gas and water vapor flow through the flowline, the expanding gas cools with dropping pressure. This can chill the water to form a hydrate that blocks the flowline. *Hydrates* are similar in appearance to snow and are methane molecules (CH_4) trapped in the ice crystals. They form between 30° and 70°F (–1° to 20°C), depending on the pressure. Drying the gas before it enters the flowline can prevent the formation of hydrates. Heaters can be installed on the flowlines or chemicals such as glycol or methanol can be added to the produced fluids to prevent the formation of hydrates.

Separators

Most oil wells produce saltwater along with gas bubbling out of the oil. They are separated in a long cylindrical steel tank called a *separator*. The separator can be either *vertical* (up to 12 ft or 3.7 m high) or horizontal (up to 16 ft or 4.9 m in length). Vertical separators take up less surface space, but horizontal separators have longer *retention times*, the time that the produced fluid is in the separator. The separator is either a *two-phase separator* that separates gas from liquid (fig. 20–4) or a *three-phase separator* that separates gas, oil, and water.

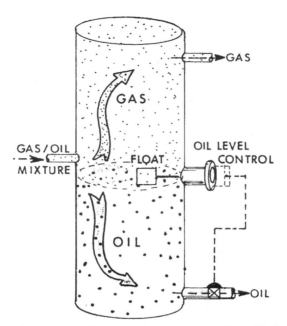

Fig. 20–4. Vertical two-phase separator. (Modified from Baker, 1983.)

If gravity readily separates the produced oil and water, the water is called *free water*. In contrast, an *emulsion* has droplets of one liquid that are completely suspended in another liquid. A *water-in-oil* or *reverse emulsion* that has droplets of water suspended in oil is the most common emulsion produced from a well. Less common is an *oil-in-water emulsion* that has droplets of oil suspended in water. The *tightness* of an emulsion is the degree to which the droplets are held in suspension and resists separation. An emulsion can be *tight* and resist separation or *loose* and readily separate.

On each separator there is an inlet for fluids from the flowline and separate outlets at different elevations for each of the separated fluids. Every separator has a *diffuser section* that makes an initial separation of the gas and liquid from the inlet. The gas rises to the *gas-scrubbing section* at the top of the separator, where most of the remaining liquid is removed from the gas before it goes out the *gas outlet*. The liquid falls to the bottom, where the *liquid-residence section* removes most of the remaining gas from the liquid before it goes out the *liquid outlet*.

All separators use gravity to separate gas, oil, and water. In the diffuser section of a vertical separator, the fluids from the inlet are spun around the shell of the vessel to let centrifugal force help the initial separation. In a horizontal separator, the fluids from the inlet strike a flat, metal plate or angle irons to slow and divert the flow direction to help the initial separation.

In the gas-scrubbing section, *mist extractors* are often used to coalesce and remove liquid droplets before the gas flows out the gas outlet. A wire mesh pad type of mist extractor uses finely woven mats of stainless steel wire packed in a cylinder. *Vanes* that are parallel metal plates with collection pockets for the liquid can also be used. The liquid-residence section is often a relatively empty chamber that lets gravity make the separation. *Baffles*, flat plates over which the liquid flows as a thin film, can be utilized. A *weir* is a dam in the lower, liquid-residence section of the separator that impounds liquid behind it. It aids in separation by allowing only the lightest liquid, such as oil floating on water, to flow over the weir.

Separators have *liquid level controls*, which are floats on the oil-gas and water-oil surfaces that control the level of those fluids by opening and closing valves. Controls on a separator include liquid level, high- and low-pressure, high- and low-temperature, safety relief valve, and *safety head* or *rupture disc* that breaks at a set pressure. A *backpressure valve* on the gas outlet maintains gas pressure in the separator.

A *free-water knockout* (FWKO) is a three-phase, horizontal, or vertical separator used to separate gas, oil, and free water by gravity (fig. 20–5). Water is drawn out the bottom, oil from the middle, and gas from the top.

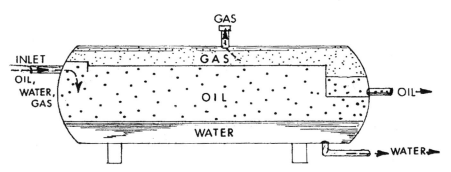

Fig. 20–5. Horizontal free-water knockout (FWKO). (Modified from Baker, 1983.)

A *double-barrel* or *double-tube horizontal separator* has two horizontal vessels mounted vertically (fig. 20–6). The produced fluid enters into the upper vessel where it flows over baffles to make an initial gas-liquid separation. The liquid then flows to the lower vessel to complete the oil-water separation. Oil-free gas flows out the upper barrel and gas-free oil flows out the bottom. A double-barrel horizontal separator can process a higher volume of produced fluids than a single horizontal separator.

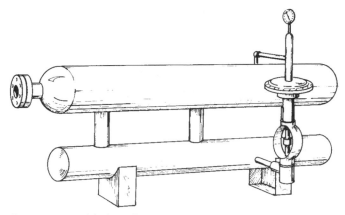

Fig. 20–6. Double-barrel separator

To separate or break an emulsion, the emulsion has to be heated. It is treated in a *heater treater*, a vertical or horizontal separator that has a *fire tube* in it (fig. 20–7) where natural gas is burned. The fire tube can be either in contact with the emulsion (*direct-fired*) or in contact with a water bath that transfers the heat to the emulsion (*indirect-fired*). Gravity then separates the heated emulsion.

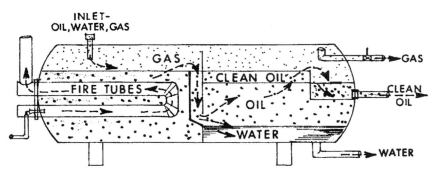

Fig. 20–7. Heater treater. (Modified from Baker, 1983.)

Electrode plates and electricity can also be used in an *electrostatic precipitator* to separate emulsions. If the emulsion is not very stable (loose), a large settling tank called a *gun barrel* or *wash tank* is used for gravity separation. A *demulsifier* is a chemical that can be injected into a treating vessel to help separate emulsions.

Separators are rated by *operating pressures* that are between 20 and 1,500 psi (1.4 and 105 kg/cm²). To maximize the retention of highly volatile components of oil, *stage separation* with several separators operating at decreasing pressures is used. The produced fluids flow first into a high-pressure separator and then through progressively lower pressure separators. The stock tanks are considered one last stage in the separation. *Three-stage separation* (fig. 20–8a) uses a high- and a low-pressure separator along with the stock tanks. *Four-stage separation* (fig. 20–8b) uses high-, medium-, and low-pressure separators and the stock tanks. *Retention time* varies from one minute for light oils to five or six minutes for heavy oils in three-phase separators.

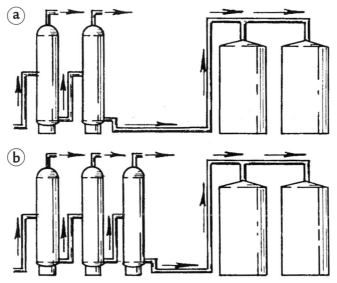

Fig. 20–8. Stage separation: (a) three-stage and (b) four-stage

Gas Treatment

Gas with minimal processing in the field so that it can be transported to a final processing plant is called *transportable gas*. Gas that has had final processing and is ready to be sold is called *sales-quality* or *pipeline-quality gas*. Natural gas is sold to a pipeline and must meet the specifications of the gas pipeline purchase contract. Pipeline-quality gas is dry enough so that liquids will not condense out in the pipeline and does not contain corrosive gases. It has a minimum heat content (calorific value) that is usually between 900 and 1,050 Btu/cf. The pressure must at least match pipeline pressure that is usually between 700 and 1,000 psi (49 to 70 kg/cm^2). A maximum dew point for both water and hydrocarbons is specified. *Dew point* is the temperature at which a gas becomes a liquid as the temperature decreases.

Impurities in natural gas are removed by *gas conditioning* in the field before the gas can be sold to a pipeline (plate 20–1). Removal of a liquid from a gas is called *stripping*.

Plate 20–1. Christmas tree and gas-conditioning equipment

Gas-conditioning equipment usually includes *dehydration* in which water is removed down to pipeline specifications that are often 7 lb of water vapor per MMcf. To remove the water, the gas can be bubbled up through bubble trays (see chap. 1, fig. 1–4) filled with a liquid desiccant such as glycol in a vessel called a *glycol absorber tower*. Triethylene glycol (TEG) is commonly used. When the glycol becomes saturated with water, the water is removed by heating in a vessel called a *reboiler*. The gas can also be dehydrated by passing it through solid desiccant beds of silica or alumina gel in a steel vessel called a *contactor*. Two contactors are used so that one can be kept on line while the other has water being removed from the desiccant by heating.

Corrosive gases such as carbon dioxide (CO_2) and hydrogen sulfide (H_2S), called *acid gases*, are removed by *sweetening*. The natural gas is passed through a *sweetening unit* that contains *iron sponge* (wood chips with impregnated iron) or chemical organic bases called *amines*. After the iron sponge or chemicals have become saturated, they are regenerated with heat to be reused. Pipeline specification often require the H_2S content to be reduced to 4 ppm by volume and CO_2 to 1 to 2% of the volume.

Solids in the gas are removed by filters that are metal cylinders packed with very fine glass fibers. They are designed to remove just solids or both solids and liquids on mist-extracting baffles.

If the gas is wet gas, the valuable hydrocarbon liquids are removed in a *natural gas processing plant* or *gas plant*. The gas plant uses either cooling or absorption to remove condensate, butane, propane, and ethane, called *natural gas liquids* (*NGLs*). In one cooling method, a heat-exchanger vessel (*chiller*) uses propane as a refrigerant to cool the wet gas. Expansion also causes the gas to cool. A *low-temperature separator* (*LTX*) *unit* passes the gas through an expansion or choke valve. *Expansion turbines* or *turboexpanders* in a *cryogenic* or *expander plant* cause the gas to rotate turbine blades to cool the gas.

Absorption of NGLs can also be done in an *absorption tower* as the gas is bubbled up through bubble trays (fig. 1–4) containing a light hydrocarbon liquid that is similar to gasoline and kerosene. The NGLs are then removed by distillation. A *hydrocarbon recovery unit* uses beds of silica, activated charcoal, or molecular sieves made of zeolites to remove the liquids.

The gas remaining after the NGLs have been removed is called *residue* or *tail gas* and is primarily composed of methane that goes to the pipeline. A *straddle plant* on a high pressure pipeline recovers condensate from the gas and returns the residual gas to the pipeline.

The saltwater and/or condensate that is separated from natural gas is stored in the field next to the gas wells in steel tanks. They are similar but smaller in size than oil stock tanks.

Compressors are used to increase gas pressure and cause the gas to flow. A *positive-displacement compressor* uses reciprocating pistons in cylinders. *Centrifugal compressors* are more common and uses spinning, turbine-like propellers on a shaft driven by a gas engine, gas turbine, or an electrical motor. The compressor is described by the *compression ratio* (*CR*). It is the gas volume before compression divided by the gas volume after compression, such as 10:1. *Multiple-stage compressors* are compressors that are joined together to compress gas in increasing increments. They are necessary for high pressures.

Storage and Measurement

Crude oil from the separator goes through a flowline to the *stock tanks* to be stored (plate 20–2). Stock tanks are made either of bolted or welded sheets of carbon steel in sizes holding 90 to several thousand barrels of oil, but commonly hold 210 bbl. They can be made to American Petroleum Institute (API) specifications.

Plate 20–2. Stock tanks

Bolted steel tanks are usually made of 12- or 10-gauge steel that is galvanized or painted. They come in many sizes and have the advantage that they can be assembled and repaired on the site. *Welded steel tanks* are prefabricated of ³⁄₁₆ in. (0.5 cm) or thicker steel and are transported to the site. The tank has either a flat or cone-shaped bottom to collect sediments.

Stock tanks can have a *vapor recovery system* to prevent loss of volatile hydrocarbons. The system is often a compressor vacuum line that takes the vapors from the stock tank to a suction separator where hydrocarbons liquids are separated from the gas. The liquids are then returned to the stock tank. The gas can be used in the field for an engine or sent to a pipeline.

A minimum of two and usually three or four stock tanks are connected by pipe in a *tank battery* (fig. 20–9) and are filled in sequence. The tank battery should hold a minimum of four days' production. When the tanks are filled, either a service company will truck the oil to the refinery or the oil is transferred into a pipeline. About 1 ft (0.3 m) above the bottom of each tank is an *oil sales outlet* that allows the oil to be drained from the tank.

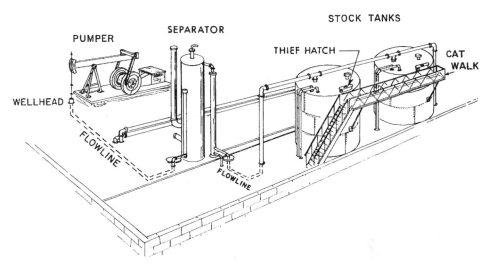

Fig. 20–9. Pumper, separator, and tank battery connected by flowlines

A steel stairway leads up to the *catwalk* that runs along the top of the stock tanks (plate 20–2 and fig. 20–9). This allows a person called a *gauger* to measure (*gauge*) the amount and quality of the oil in each stock tank. On the top of each tank is a *thief hatch* that can be opened (plate 20–3). The gauger lowers a *gauge tape* marked in ⅛- or ¹⁄₁₀-in. intervals with a brass weight on the end, down into the tank from a reel (fig. 20–10). A *gauge* or *tank table* is used to relate the height of oil to the volume of oil in that stock tank. The gauger also samples the oil by using a *thief*, a brass or glass cylinder about 15 in. (38 cm) long or with a 1-quart bottle with a stopper that is lowered in the tank. Samples can also be obtained from valves (*sample cocks*) that can be located on the sides of some tanks.

The temperature of the oil is measured and the sample is centrifuged in a glass tube with a *shake-out machine* (fig. 20–11) during a *shake-out test*. The heavier sediment and water are spun to the bottom of a glass container as the lighter oil rises. These impurities are called the *basic sediment and water* (*BS&W*) content of the oil. Generally, a pipeline or refinery will not accept oil with greater than 1% BS&W. Oil that is acceptable is called *clean* or *pipeline oil*.

Plate 20–3. Top of stock tanks showing open and closed thief hatches and catwalk

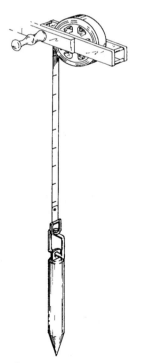

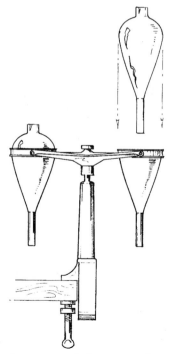

Fig. 20–10. Gauge tape   **Fig. 20–11.** Shake-out machine

Turbine meters can also be used to measure the volume of oil being transferred. They measure volume by the number of spins on the turbine shaft as the oil flows through the turbine and turns blades on the shaft. A *positive-displacement* meter can also be used to measure the volume of oil in separate, equal units.

The tank battery usually needs to be located next to a road to make it accessible to the tank trucks that take the oil to a refinery (plate 20–4). Wells in remote locations can be connected to an accessible separator and tank battery with a long production flowline.

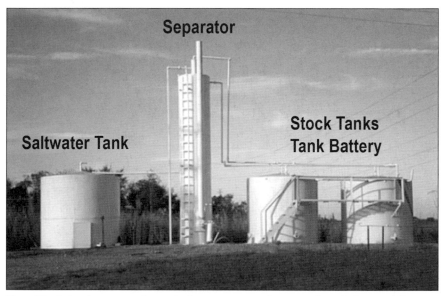

Plate 20–4. A saltwater tank, vertical separator, and two stock tanks in a tank battery

When the oil is transferred from a tank battery to a tank truck or pipeline, a receipt called a *run ticket* is made in triplicate by the gauger and witnessed by the pumper. A copy goes to the oil purchaser. It is the legal instrument by which the operator is paid. It includes the well number, operator, °API gravity of the oil, tank size, and gauge readings, BS&W content and temperature of the oil, seal number on the outlet, and time of transfer.

Natural gas from a gas well is also sampled, tested, and measured. Common gas tests include compression and charcoal tests and a fractional analysis. A *compression test* uses a truck with a compressor and refrigerant to remove and measure the amount of condensate in the gas. A *charcoal test* also measures the condensate content of the gas by passing a sample of the gas over activated charcoal to adsorb the condensate. The condensate is then removed by distillation and measured. A *fractional analysis* of natural gas is made by a laboratory. It is a chemical analysis of the gas sample by a gas chromatograph. The analysis includes a percentage of each hydrocarbon and water component, the gallons per 1,000 cubic feet (GPM) of propane and higher gases, and the heating value.

Gas volume is measured by a gas meter on the flowline. An *orifice gas meter* is commonly used. It measures the difference in gas pressure on gas flowing though an *orifice* (a round hole in a plate) on both sides of the orifice. The higher the flow rate, the greater the pressure drop across the orifice. The gas velocity can be recorded on a circular *gas meter chart* in a *charthouse* or *meterhouse*. The gas velocity has been calibrated to gas volume by a *meter prover*. Meter provers are used for both gas and liquid meters. They compare the volume of fluid flowing through the accurately calibrated meter prover with the same volume of fluid flowing through the meter being tested. The gas meter chart also shows the location of the meter, size of the orifice, date, and time. It is used to determine gas payments to the operator.

On many modern gas wells, equipment monitors the well's performance and the amount of gas being transferred (*electronic flow measurement*, or *EFM*). It can be powered by a solar panel on the well.

A *field superintendent* is in charge of a producing field and gives orders to *production foremen* who work for him or her. Production foremen direct the crews that work in the field. A *pumper* is responsible for keeping the production equipment in operating order. *Roustabouts* help the pumper with the general labor.

Modern Lease Operations

A *lease automatic custody transfer (LACT) unit* uses automatic equipment to measure, sample, test, and transfer oil to a pipeline in the field and eliminate some of the labor. A probe in the LACT unit records the temperature, °API gravity, and BS&W content of the oil. If the BS&W content is too high, it can automatically send the oil back through the separation system. A positive-displacement meter measures the volume of oil. The unit can also shut down the operation by closing a divert valve to the pipeline if there is too high a BS&W content. The LACT unit keeps records of all this.

Data collection and well control for both gas and oil wells has been modernized by the use of the Internet. The oil and water levels in tanks can be very accurately measured by an instrument in the tank top. It bounces pulses of electromagnetic microwaves off the liquid surfaces and times the echoes to determine the height of oil and water in the tank. Information such as electronic flow meter rates, compressor status, tank levels, run times, and casing and tubing pressures are measured automatically at the wells. The systems are often powered by solar panels at the wellsite. The data can be automatically or on command be sent digitally by telephone line or satellite to a server. These data are then made available with limited access only to people who work on those wells. The system can issue an *alarm call-out* if equipment such as a compressor is down or not working at optimum. Commands can be sent back to the well to activate or adjust equipment such as speeding up or slowing down a compressor, opening or closing a choke, or dumping a tank.

Offshore Drilling and Completion

Baker Hughes reported 338 active offshore drilling rigs in the world during 2010. During September 2011, 1% (19) of the United States rigs (1,958) were drilling inland waters, and 1.6% (32) were drilling offshore. Most of the offshore rigs (30) were drilling in the Gulf of Mexico. The API reported that the average cost of a United States offshore well in 2008 was $108,286,000 or $8,191 per foot. During 2010, offshore United States Gulf of Mexico averaged oil production of 1.7 million bbl/d (0.27 million m^3/d) for just less than one-quarter of the total United States daily oil production and average gas production of 6 Bcf/d (0.17 m^3/d) for about 10% of total United States daily gas production.

Mineral Rights

In an offshore area in the United States, the states own the mineral rights out to 3 nautical miles from the shoreline (fig. 21–1). Mineral rights for the states of Texas and Florida extend out to 3 leagues (9 nautical miles). The federal government owns the sea-bottom mineral rights from the state limit out to 200 nautical miles. This federal offshore land is called the *outer continental shelf* (*OCS*). Blocks on the OCS, often 9 square nautical miles in size, are offered in periodic closed-bid sales, going to the highest bonus offer. There is a one-sixth royalty and a 5-year primary term, except for deep-water leases, which have a 10-year term. In Canada, the offshore extent of mineral rights ownership varies considerably between different provinces. Every country that borders the ocean has an Exclusive Economic Zone, defined by the United Nations Law of the Sea, that includes the mineral rights out to 200 nautical miles.

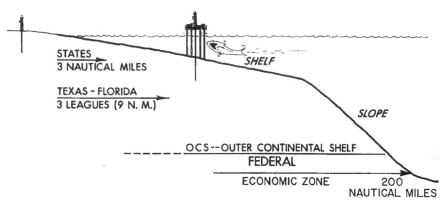

Fig. 21–1. Mineral rights in offshore US waters

In 1959, the Groningen gas field that has 73 trillion cu ft (12.1 trillion m^3) of recoverable natural gas was discovered on land in the Netherlands. The trap is a faulted dome. The reservoir rock is a Permian age sand dune sandstone (Rotliegend Formation), the caprock is Permian age salt (Zechstein Salt), and the source rock was the underlying Pennsylvanian age coals. The same rock layers underlie the North Sea, and geologists immediately realized that there could be gas and oil fields in the North Sea. In 1964, mineral rights in the North Sea were divided among the countries that border the North Sea by agreement (fig. 21–2).

Offshore Drilling

Before the offshore rig is positioned, a *subsea site* or *soil investigation* of the bottom slope, composition, and load-bearing capacity is made to make sure it can support the rig. Buoys or global positioning are then used to mark the site.

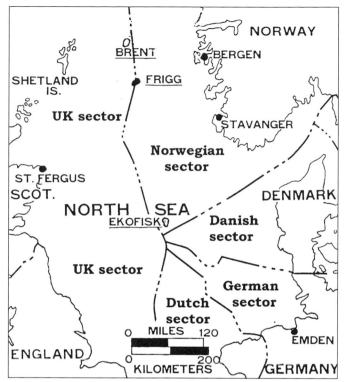

Fig. 21–2. North Sea mineral rights

Offshore rotary drilling rigs are the same as land rotary drilling rigs, except only the most modern rotary drilling rigs are used offshore because of the high drilling costs. The top drive on an offshore drilling rig moves up and down vertical rails to prevent it from swaying with any motion from waves. The drilling methods are the same as on land. There are usually three crews servicing an offshore drilling rig, two working on the rig and one off-duty ashore. An onshore and an offshore crew are rotated once every two weeks. The crews work 12-hour shifts from 11 a.m. to 11 p.m. and 11 p.m. to 11 a.m.

An offshore crew can have a driller, assistant driller, derrick operator, roughnecks, motor operator, diesel engine operator, pump operator, mud person, crane operator, and roustabouts. The *driller* handles the rotary controls, drawworks, and pumps, and the *assistant driller* handles the pipe and racking system and the iron roughnecks. The assistant driller can also relieve the driller when necessary. The *crane operator* operates the crane used to lift equipment and supplies from supply ships and barges onto the offshore rig. The crane operator is also usually the *head roustabout* in charge of the roustabouts, who handle the supplies and equipment. A tool pusher and company person will also be aboard. On some offshore rigs, a *pit watcher* may be responsible for the drilling mud and circulating equipment. There are often rig mechanics and electricians to maintain the rig.

On a large drillship, there can be quarters for 200 people. If the production platform is close to a port, the crew can be ferried out to the platform on fast crew boats for a crew change every two weeks. On many, however, the crew and other personnel are transported to and from the production platform by helicopters that land on the *helideck* (a flat platform).

On very modern offshore drilling rigs, the driller is called a *cyber driller* and sits in a climate-controlled, glass-enclosed cabin overlooking the drill floor. Computer monitors keep the driller informed of current drilling parameters. Weight on bit, revolutions per minute, and other drilling factors can be adjusted from the driller's station using a joystick and touch screen. Mechanical devices, such as a pipe-handling unit that consists of elevator and torque wrench systems to manipulate the drillpipe, are controlled by an assistant driller who also uses a joystick and touch screen.

Exploratory Drilling

In shallow, protected waters such as lagoons and canals up to about 25 ft (7.6 m) deep, the rig can be mounted on a *drilling barge*. Drilling barges have been very successful in canals on the Mississippi River delta marshes. A *posted barge* is a drilling barge designed to be sunk and rest on the bottom while drilling. The drilling deck is mounted on posts to keep it above the surface of the water. The API reported that there were 52 active drilling barges in the United States during 2008.

Offshore, a *mobile offshore drilling unit* (*MODU*) is used. Three types are jackup, semisubmersible, and drillship. A *jackup rig* usually has two barge-like hulls and at least three vertical legs through the hulls (fig. 21–3). The legs are either (1) *open-truss* with tubular steel members that are cross-braced or (2) *columnar* made of large diameter steel tubes. The *cantilevered jackup rig* is most common with the drilling rig mounted on two large steel beams that protrude over the edge of the deck. Occasionally the drilling rig is mounted on the deck over a slot or keyway in the deck.

The jackup rig is towed into position. While moving, both hulls are together, and it floats like a barge with the legs raised high. At the drillsite, the lower hull (*mat*) is jacked down and positioned on the seafloor. On each of the legs is a *jack house* that uses a rack-and-pinion arrangement powered by an electric or hydraulic motor to raise and lower the hulls. The upper hull is jacked up on the legs until usually about 25 ft (7.6 m) above the sea surface. The drilling rig on the upper hull is then secure above the waves, and the mat acts as a stable foundation, even with a soft bottom. If the ocean bottom is relatively hard, smaller cylinders (*cams*) with a point on the bottom can be used on the bottom of each leg instead of a mat. After drilling, the hulls can be joined again and the rig towed to another site. Legs on these rigs are constructed up to 550 ft (167.6 m) high. Jackups are generally used in water depths up to 400 ft (122 m).

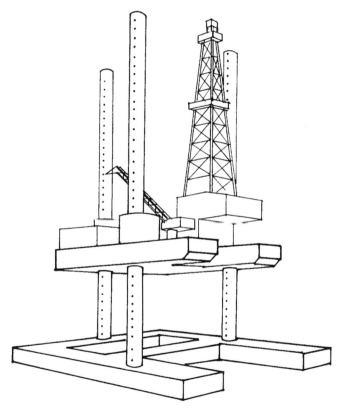

Fig. 21–3. Jackup rig. (Modified from Exploration Logging Inc., 1979.)

For deep-water drilling, a *floater* (i.e., a semisubmersible or drillship) is used. A *semisubmersible* (or *semi* as it is commonly called) is a floating, rectangular-shaped drilling platform (fig. 21–4). The most common type is the *column-stabilized semisubmersible*. Most of the rig flotation is in the ballasted pontoons, located 30 to 50 ft (9.1 to 15.2 m) below sea level when on station. Square or circular hollow structural columns connect the operating deck to the pontoons. Because most of the flotation is below sea level in the pontoons, the rig is very stable even during high seas and winds.

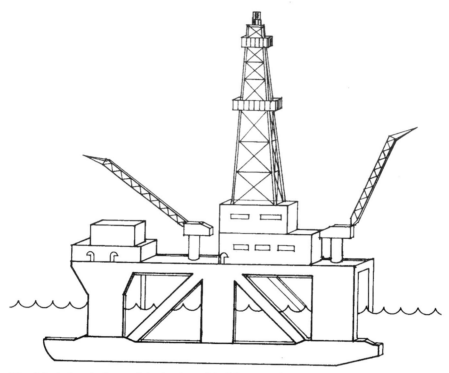

Fig. 21–4. Semisubmersible rig. (Modified from Exploration Logging Inc., 1979.)

The semi is towed to the drillsite. To move the semisubmersible, the pontoons are emptied, and the rig floats high in the *transit mode* for easier towing. A *ballast-control specialist* supervises the raising and lowering of the semi and keeps it stable. For long-distance transport, the semi can be carried on the deck of a special ship during a *dry tow*.

A *drillship* is a ship with a drilling rig mounted in the center (fig. 21–5). The ship steams out to the drillsite and then drills through a hole in the hull, called the *moonpool*. Drillships are very expensive. For efficiency, some modern drillships have the equipment and ability to drill two wells at the same time. There are two derricks with two traveling blocks and top drives, and the ship has two independent stations for each driller and assistant driller and two setback areas to rack the pipe.

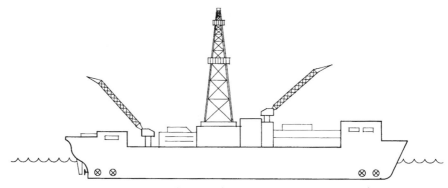

Fig. 21–5. Drillship. (Modified from Exploration Logging Inc., 1979.)

Drillships are not as stable in waves as a semisubmersible. The advantage of drillships is they can move faster to the next drillsite and have a larger load capacity than semis. The *Discoverer Enterprise* is a modern drillship with quarters for 200 people. It can drill, test, and complete a well in water depths up to 10,000 ft. (3,048 m) and can drill a well down to 35,000 ft. (10,668 m) below sea level.

Both semisubmersibles and drillships float over the drillsite. To keep the floater on station, either a mooring system or dynamic positioning is used. A *mooring system* uses a combination of steel cables and chains and lightweight polyester fiber that radiates out to eight or twelve anchors. The most common anchor used in deep water is a suction pile. A *suction pile* is a steel cylinder that is closed on the top. When placed on the ocean bottom, a pump on the top removes water from the suction pile to suck it into the ocean bottom. *Dynamic positioning* uses onboard satellite dishes to track navigational satellites and several computers to constantly recalculate the exact location of the floater. If the floater drifts off the drillsite, a computer engages the ship's propellers and puts it back on location. Floaters have propellers on the side of the ship (*bow* and *stern thrusters*) and can move both back and forth and sideways.

Offshore drilling is very expensive. The semisubmersible *Deepwater Horizon* that was drilling in 4,992 ft (1,522 m) of water and sank when the Macondo well blew out in the Gulf of Mexico was costing $1 million per day.

Spudding an Offshore Exploratory Well

On a well drilled by a jackup rig, several hundred feet of large-diameter (26 or 30 in.—66.0 or 76.2 cm) *conductor casing* is set into the sea bottom.

On a very soft sea bottom, the conductor casing is jetted into the bottom by pumping seawater through the center. On a harder bottom, the conductor casing is pile-driven into the bottom. On a very hard bottom, a hole is drilled, and then the casing is run into the hole and cemented. The conductor casing extends above sea level to just below the drilling deck. A smaller diameter hole is then drilled through the casing into the seafloor to several hundred feet below the bottom of the conductor casing. Surface casing is run into the hole and cemented. Next, a blowout-preventer (BOP) stack is bolted to the top of the surface casing. The rest of the well is then drilled and cased similar to a well on land.

On a well drilled by a floater (a semi or drillship), a *temporary guide base* or *drilling template* is installed on the sea bottom. The temporary guide base is a hexagonal-shaped steel framework with a hole in the center for the well. It is attached to bottom of a drillstring and lowered to the sea bottom. Four steel *guidelines* run from the sides of the temporary guide base up to the floater. They are used to lower and position other equipment onto and in the well.

The drillstring is then raised back up to the floater leaving the temporary guide base on the sea bottom. A *guide frame* is then attached to the bottom of the drillstring. It has two or four arms through which the guidelines run. The drillstring and guide frame are then lowered down the guidelines to the temporary guide base. A large-diameter hole (30 or 36 in.—76.2 or 91.4 cm) is drilled through the center of the temporary guide base to about 100 ft (30.5 m) below the seafloor. The drillstring and guide frame are then raised back to the floater.

The guide frame is then attached to the lowest joint on the *foundation pile*, the first casing string run into the well. A *foundation pile housing* and *permanent guide structure* are attached to the top foundation pile joint. The foundation pile is then run into the hole and cemented. The permanent guide structure is attached to the temporary guide base on the sea bottom. The hole is then drilled deeper, and a string of conductor casing is run and cemented into the hole.

A subsea BOP stack that can be activated from the floater is then lowered and locked onto the wellhead with a hydraulic wellhead connector. The drilling rig is then connected to the BOPs by a flexible metal tube (*marine riser*). The drillstring goes through the marine riser into the well. A *tensioner system* of wire rope and pulleys on the floater supports the upper part of the marine riser. The marine riser completes a closed system to circulate drilling mud down the drillstring and up the annulus between the drillstring and marine riser.

To compensate for the up-down motion (*heave*) of the floater on the surface, a *telescoping joint* that expands and contracts is used on the top of the marine riser. A *heave compensator* is also located between the traveling block and hook on the drilling rig. The compensator has pistons in cylinders that hold the hook and drillstring stationary as the floater heaves.

During an emergency such as severe weather, the BOPs can be closed and the marine riser disconnected from the BOP stack. The floater can then be moved off station to safety. After the emergency has passed, the well can be relocated and reentered.

Offshore wells have casing programs and bottomhole completions similar to wells on land. Offshore wells, however, can have more casing strings. The Macondo well that blew out in the Gulf of Mexico has nine casing strings starting at 36 in. (91.4 cm) and ending in 9⅞ in. (25.1 cm) diameter. Loose sand is often a problem (*sand control problem*) offshore, and gravel packs and prepacked slotted liners are common.

Developmental Drilling and Production

After a commercial offshore field has been discovered, it can be developed with a fixed production platform in shallow water or a floating tension-leg platform, compliant platform, or semisubmersible in deeper water. A *fixed production platform* has legs and sits on the bottom. One type of fixed platform is called a *gravity-base platform* because it has a large mass of steel-reinforced concrete on the bottom of the legs, and gravity holds it in position (fig. 21–6). The massive base has hollow cells that can be used for floatation when assembling and towing the platform into position. On location, the cells can be used for ballast or storage of crude and diesel oil. It is constructed in a sheltered, deep-water port along the shoreline and then towed into position. This type is used in areas of very rough seas.

A more common type of fixed production platform is the *steel-jacket platform*, named as such because it has legs (*steel jacket*) that sit on the bottom (plate 21–1 and fig. 21–7). It is constructed on land and either floated horizontally or carried on a barge out into position. It is then flooded and rotated vertically. Piles are driven into the sea bottom and bolted, welded, or cemented to the legs to hold it in position. A crane is used to lift the deck and modules such as power generation, crew quarters, and mud storage off barges up into position on the platform.

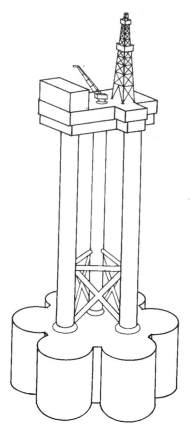

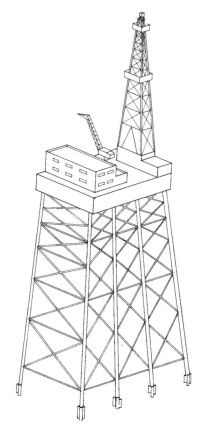

Fig. 21–6. Gravity-base production platform. (Modified from Exploration Logging Inc., 1979.)

Fig. 21–7. Steel-jacket production platform. (Modified from Exploration Logging Inc., 1979.)

Offshore platforms often have several *decks* (flat surfaces) on top of each other to serve various functions such as power and drilling (plate 21–1). Wellheads are usually located on the lower *cellar deck*. Separators, treaters, and gas compressors are located on the platform. The treated oil or gas is then usually sent ashore through a submarine pipeline. A crane is used to lift supplies and equipment aboard the platform from supply boats. Usually one or two derricks are left on deep-water platforms after the wells have been drilled to use for workovers.

Fixed production platforms can be used out to 1,500 ft. (457 m) water depth. In relatively shallow waters, there can be a separate *quarters platform* for the crew next to the *production platform* as a safety precaution. A *bridge* connects the platforms.

Plate 21–1. Fixed production platform. (Courtesy of American Petroleum Institute.)

A *tension-leg platform* (*TLP*) floats above the offshore field. It is held in position by heavy weights or suction piles on the seafloor (fig. 21–8). The weights or piles are connected to the tension-leg platform by hollow, steel tubes, 1 to 2 ft (0.3 to 0.6 m) in diameter, called *tendons*. The tendons pull the platform down in the water to prevent it from rising and falling with waves and tides. The tendons also allow the platform to move laterally within 8% of water depth. A *tension-leg well platform* is similar but has only wellheads and no production-treating facilities onboard. The produced fluids are sent by seabed pipeline to a production platform in shallow water for treatment.

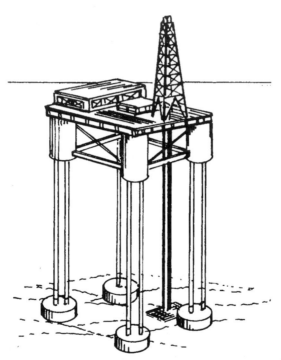

Fig. 21–8. Tension-leg platform

A *compliant platform* (fig. 21–9) is a relatively light production platform that is designed to sway with wind, waves, and currents. One type, a *guyed tower*, is attached to a pivot on the ocean bottom. Another type, a *spar*, is a floating production platform with one large-diameter or several smaller diameter, closed, vertical cylinders (fig. 21–10). The spar is designed to not rise and fall with the waves. Both the guyed tower and spar are held in position with a mooring system.

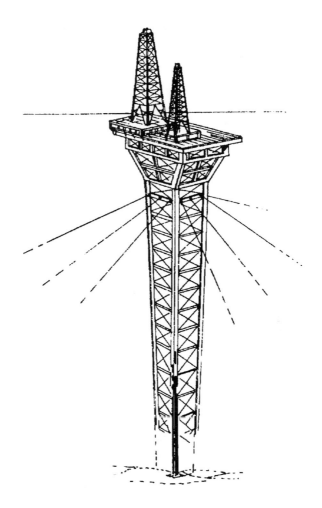

Fig. 21–9. Compliant tower

Semisubmersibles have proven to be very stable and are being used for deep-water production. The Na Kika production platform in the Gulf of Mexico (fig. 21–11) is a semi secured by 16 chain and wire rope lines to anchors that extend 6,600 to 8,300 ft (2,012 to 2,530 m) horizontally away from the platform in a water depth of 6,340 ft (1,932 m). It has quarters for a crew of 60, and the deck has separation, dehydration, and treating equipment for gas and oil. The platform services six small- to medium-size oil and gas fields by flowlines with wells that are up to 12 miles (19 km) from the platform. It will eventually recover 300 million bbl (48 million m^3) of oil. Na Kika was named after a South Seas octopus god.

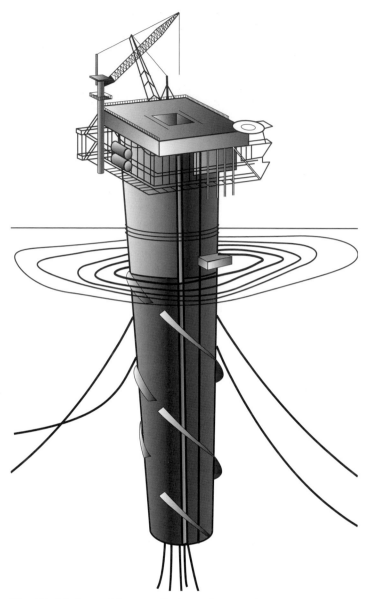

Fig. 21–10. Spar with one cylinder and a mooring system

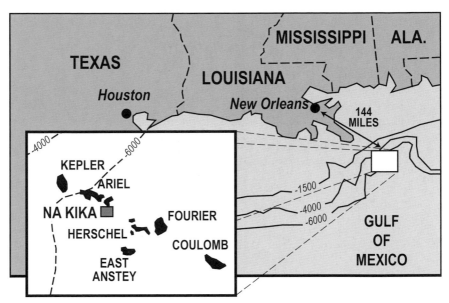

Fig. 21–11. Map of Na Kika platform and producing fields. (Modified from Rajasingam and Freckelton, 2004.)

The Independence Hub, a semi, is a gas production platform in 8,000 ft (2,440 m) water depth in the Gulf of Mexico (fig. 21–12). It is anchored in place 110 miles (177 km) southeast of the Mississippi delta with twelve 9-in. (23-cm) polyester ropes that are each 2.4 miles (3.9 km) long. The Independence Hub has a deep draft (105 ft or 32 m) and has two production decks with accommodations for 16. It processes 1 Bcf (28 million m³) of natural gas and 5,000 bbl (800 m³) of condensate per day. The 10 surrounding gas fields are located in 7,800 to 9,000 ft (2,377 to 2,743 m) of water depth and are connected to the hub with 8- to 10-in. (20- to 25-cm) diameter flowlines. None of the gas fields are economical by themselves. The treated gas and condensate goes ashore in a 24-in. (61-cm) diameter submarine pipeline that is 134 miles (216 km) long called the Independence Trail.

Wells, often numbering 32 to 40, are drilled through a *well template* on the ocean bottom, which is used to position and separate the wells. The well template is a steel frame with slots for each well. It supports the equipment necessary to drill and produce the wells. Each slot on the template locates and standardizes the instillation of a well. There are usually one or two extra slots left undrilled on the template for any further field development.

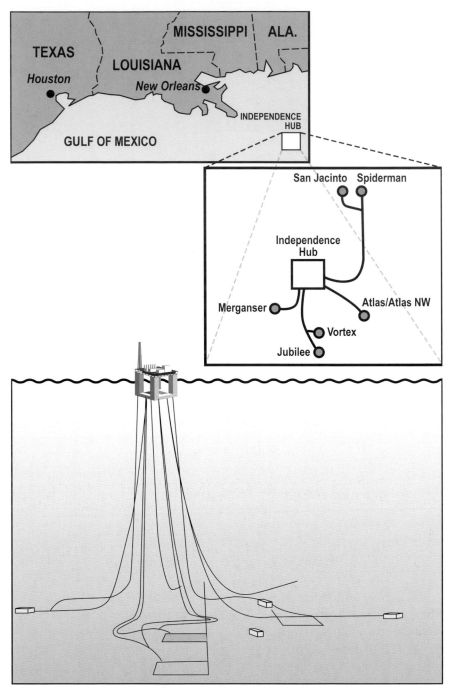

Fig. 21-12. Map of Independence Hub and producing fields and cross section showing mooring system. (Hilyard, 2008.)

The offshore field is developed by deviation drilling from one platform. If the offshore well flows, it is completed with a Christmas tree. An offshore oil well that needs artificial lift is usually completed with gas lift. Offshore wells are required by law to be equipped with *storm chokes*. The choke is installed on the bottom of the well and is closed either manually or automatically during an emergency.

Subsea Work

Installation and work on a well underwater can be done by a diver, a diver in a 1-atmosphere diving suit, and a remotely operated vehicle. A *saturation diver* breathing a helium and oxygen mix can work down to 1,000 ft (305 m). The *1-atmosphere diving suit* (*ADS*) is a hard diving suit with 1 atmosphere of air pressure in it and human-powered limbs. It can operate down to 2,300 ft (701 m). A *remotely operated vehicle* (*ROV*) is an unmanned submersible that can effectively operate down to 15,000 ft (4,572 m). It is connected to a mother ship on the surface by a cable (*umbilical*). A closed-circuit television camera on the ROV allows operators on the surface to see what is ahead of the ROV and to manipulate the ROV with thrusters and do work with manipulator arms. The ROVs used to work on offshore wells are very similar to those used to discover and explore sunken ships.

Subsea Wells

Subsea wells are used to develop a small offshore field that does not economically justify a production platform. The wellhead and production equipment such as Christmas tree or gas lift are on the bottom of the ocean. The wellhead has production sensors and the Christmas tree has hydraulically controlled valves that are remotely operated. ROVs are used to service subsea wells. The *subsea well* is drilled from a floater. The completion can be either *dry*, with an atmospheric chamber surrounding the equipment (fig. 21–13a), or *wet*, which is exposed to seawater (fig. 21–13b). In 2011, Shell Oil Co. drilled and completed a subsea well in 9,356 ft (2,852 m) water depth in the Gulf of Mexico.

The production from a subsea well can be taken by flowline to a *subsea manifold* where it is commingled with production from other subsea wells. It is then taken by a flowline to a production platform in shallow water, a tension-leg platform, a semisubmersible facility, a spar tower, or up a production riser to a floating production, storage, and offloading vessel (fig. 21–14) for processing. The *floating production, storage, and offloading (FPSO) vessel* is a converted tanker, a semisubmersible, or a specially built ship that contains separation and treating facilities on deck. It can be kept on position by a mooring system or by dynamic positioning. The treated oil is then transferred from the FPSO vessel to a *shuttle tanker* to be brought ashore. Ultra-deep-water wells are done with subsea completions.

Remote portions of an offshore field and several smaller fields that are not economical by themselves can be developed as subsea wells called *satellite wells*. They are all tied to a single production facility by flowlines.

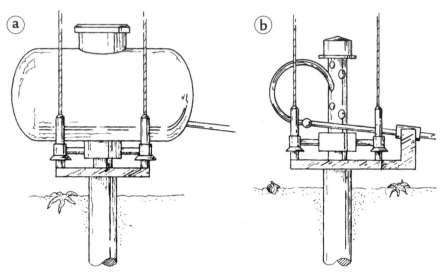

Fig. 21–13. Subsea wells: (a) dry and (b) wet

Unstable Sea Bed

The seabed is very unstable off deltas where landslides called *submarine mudflows* are common in the loose sediments (fig. 21–15). During a large hurricane off the Mississippi River delta in 1969, several offshore oil platforms failed. At first it was thought that the large hurricane waves and high winds were directly responsible for the failure of the rigs. It was later shown, however, that the hurricane waves caused large submarine mudflows that flowed down the Mississippi River delta bottom, knocking the legs out from under the rigs. Because of the instability of the sea bottom off the Mississippi River delta, dozens of rigs have failed and, in an average year, 110 pipelines fail.

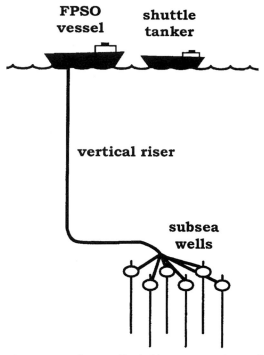

Fig. 21–14. Subsea wells tied in to an FPSO vessel

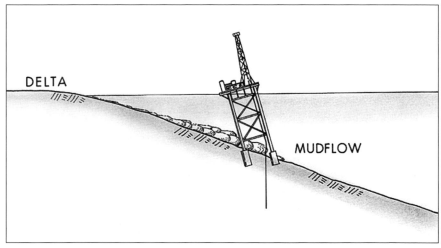

Fig. 21–15. Submarine mudflow

Workover

There are times in the life of a producing well in which the well must be *shut in* (production stopped) and remedial work done on the well to maintain, restore, or improve production by a *workover*. Workover includes both repairing mechanical problems and cleaning out the well. World Oil reported that there were 2,047 active workover and well servicing units in the United States and 655 in Canada during 2011.

Equipment

In the past, when a well was drilled and completed, the derrick was left standing above the well to raise and lower equipment into the well during a workover. Today, a service company does a workover with a mobile *production rig* that is either a workover rig for more extensive work or a smaller well-servicing unit. This eliminates the need for keeping the derrick.

A *workover rig* looks similar to a drilling rig. It can drill using a workstring of drillpipe or tubing and circulate with either a water-based or oil-based mud or foam made of air or nitrogen.

A *well-servicing* or *well service unit* uses hoisting equipment (winch, cable, and mast) mounted on a truck body or trailer. The winch is driven off the truck engine. There is a crown block, traveling block, and hoisting line similar to a small drilling rig. The hoisting line is either a *wireline* made of braided steel wire or a *slick line* of one solid steel wire wound around the drum in the winch. There can be either one or two drums. Elevators on the traveling block are used to clamp onto joints of tubing and sucker rods. Hydraulic slips and hydraulic tongs (both sucker-rod and tubing) are used to make up and break out connections. A swivel and rotating head with a small kelly can be used to rotate a workstring.

Two types of well-servicing unit masts are pole and structural. A *pole mast*, made of tubular steel, can both pivot and telescope up and down to several heights. It is either *single pole* or *double pole*, depending on the number of tubulars. A single-pole well-servicing unit (plate 22–1) is used for shallow wells and requires that any tubing or sucker rods that are pulled from the well be laid on the ground. A double-pole unit is more efficient and has either a single racking platform or both a rod basket and hanger and a tubing platform. A person can stand on the *racking platform* and rack the tubing or sucker rods vertically into metal fingers on the racking platform. A *rod basket and hanger* is located further up on the mast and is used to rack three sucker rods at a time. The *tubing platform*, used to rack tubing, is located lower on the mast.

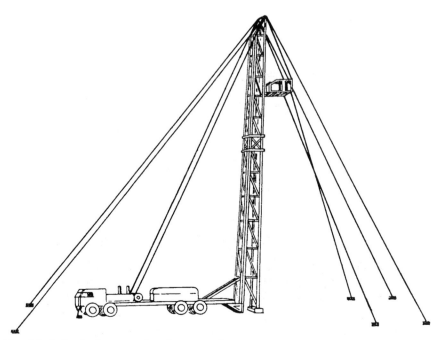

Fig. 22–1. Radiating guy wires from a well-servicing unit mast

A *structural mast* is made of angular steel. It can also telescope up and also has a single racking platform or both a rod basket and a tubing platform (plate 22–2). Masts and poles are stabilized with radiating guy wire attached to anchors in the ground (fig. 22–1).

A well-servicing unit has a two-, three- or four-person crew consisting of a service unit operator, derrick operator, and one or two floor persons. During transit to the next workover, the mast is folded down (plate 22–3). Unlike a drilling rig, work is done on a well-servicing unit only during daylight, except offshore.

Plate 22–1. A single pole well-servicing unit with mast telescoped up

Plate 22–2. Well service unit with a structural mast that is pulling tubing. Note the tubing platform (lower) and rod basket (higher).

Plate 22–3. A well-servicing unit with the mast folded down for travel

A *carrier unit* is a highway vehicle designed for workover with an engine-hoist-mast system. The hoist is driven by one or two diesel engines on the unit. Unlike a well-servicing unit, the power to drive the carrier unit on the highway comes from the hoist engines. Hydraulic pumps pivot and raise the mast. Some carrier units have the mast on the back (*back-in unit*), and others have the mast and cab on the same end (*front-in unit*) for better driver visibility. Hydraulic leveling jacks adjust the unit on uneven ground.

A *coiled tubing unit* (fig. 22–2) is a well-servicing unit that uses tubing wound around a reel on the truck to raise and lower equipment in the well. Coiled tubing is a continuous length of flexible, steel tubing up to 19,000 ft (5,791 m) long with a diameter that is often 1¼ in. (3.2 cm). It is unwound from a reel, gripped by friction blocks, and fed into an *injector* as it goes into the well.

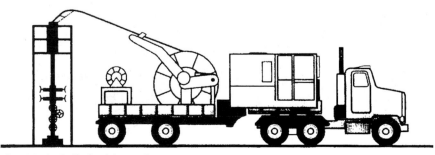

Fig. 22–2. Coiled tubing unit

A coiled tubing unit reduces trip time compared to using straight tubing joints that have to be screwed on or off the workstring each 30 ft (9.1 m) on a common well-servicing unit. Because the tubing is injected into the well through a control head that maintains a pressure seal, coiled tubing can be used on high-pressure wells. Drilling can also be done with a coiled tubing unit using larger diameter (2 in.–5 cm) tubing. Coiled tubing, however, cannot be rotated, and a turbine mud motor is used to drive the bit.

A *concentric tubing workover* uses smaller diameter and lighter equipment than normal that is run down through the tubing string. A lightweight *macaroni rig* is designed to run ¾- and 1-in. (1.9- and 2.5-cm) diameter tubing into the well.

Snubbing units are designed for workovers in wells under high pressure. They use unidirectional slips to grip and force tubulars into or out of a well as one of the two preventers on the well are alternately engaged or while using a solid rubber stripper head. The tubulars are forced into the well using blocks and wireline or hydraulic power.

Well Intervention on Offshore Wells

Well intervention is work on a producing well. It can include repairing, replacing, or installing equipment, workover, well stimulation, and production logging. Well intervention on offshore wells can be very expensive, and an interruption in production can be very costly if the well is shut in. Offshore well intervention is designed to be very efficient and, if possible, done without shutting down production.

For offshore well servicing, a service company is often contracted. On a small production platform, a wireline unit and small hoist can be used. A socket with a *sinker bar* (a weight) is attached to the end of the wireline to run it down the well. On larger platforms, satellite wells, and subsea wells, well intervention is done with a barge, jackup, or semisubmersible that has workover equipment and living quarters for the workover crews. The mast is located on two steel beams that project over the side of the vessel to position the mast over the well to be serviced.

Deep-water production platforms have one or two derricks that are left on the platform for workovers, and the barge, jackup, or semi is not needed. For highly deviated and subsea wells, *through-the-flowline* (*TFL*) or *pumpdown* equipment has been developed to workover wells. The equipment is circulated down the flowline into the production tubing, activated by pressure, and then recirculated out the tubing. For well stimulation, a specially designed ship with hydraulic fracturing equipment can be used.

Preparing the Well

The well is *killed* (flow is stopped) by filling the well with a *kill fluid* such as brine, drilling mud, oil, or a special liquid before most workovers. Often the kill fluid is pumped down the casing-tubing annulus and back up the tubing string. After the well has been killed, blowout preventers are usually installed, and the tubing string and other downhole equipment are removed. Blowout preventers may not be necessary on a low-pressure well.

Well Problems

Sand cleanout

Loose sand from unconsolidated sandstone reservoirs can clog the bottom of the well, causing a *sand control problem*. Without removing the production tubing, either a coiled tubing or macaroni rig can be used to run a small-diameter workstring of tubing down the production tubing. Saltwater pumped down the workstring tubing picks up the sand and circulates back up the workstring tubing–production tubing annulus.

If the production tubing is removed, a workstring of larger diameter tubing is run down the well. The saltwater is pumped down the casing-tubing annulus, and the sand and water are circulated up the tubing. A *bailer* or *sand pump* on a wireline can also be run down the tubing string to remove the sand.

Loose sand grains in a reservoir can be stabilized by pumping an epoxy resin into the well to glue the sand grains together adjacent to the well. The Wild Mary Sudik Well in the Oklahoma City field (1930) is a famous example of a sand control problem. It blew out for 11 days because of the loose sand mixed in with the oil.

Well cleanout

Because of the rapid drop in temperature and pressure between the reservoir and the bottom of the well, calcium carbonate, barium sulfate, calcium sulfate, and magnesium sulfate can precipitate out of oilfield brine to form *scale* (a salt coating) in the tubing. Chemicals, called *scale inhibitors*, can be pumped down the well to dissolve and remove the scale.

Tubing can be clogged with waxes from a waxy crude. A *paraffin knife* (fig. 22–3a) or *paraffin scraper* (fig. 22–3b) run on a wireline through the tubing can be used to remove the wax. A *hot oil treatment* uses heated crude oil, usually from the separators, that is pumped down the well by a service company (a *hot oiler*) to dissolve the wax. The oil is then pumped back out of the well. Chemicals (*paraffin solvents*) can be pumped into the well and flowlines to remove the wax.

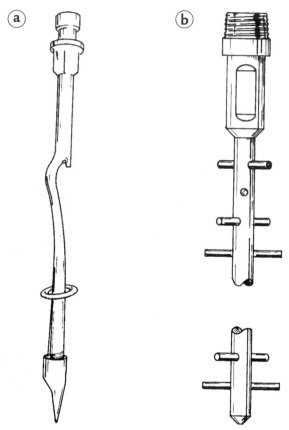

Fig. 22–3. (a) Paraffin knife and (b) paraffin scratcher

Pulling rods

The sucker-rod string on a beam-pumping unit can break due to corrosion or wear. The intact, upper portion of the sucker-rod string is pulled (*pulling rods*) and unscrewed into joints with a *power rod tong* or manually with a metal circle called a *back-off wheel* or *circle wrench*. A fishing tool, either a *sucker-rod overshoot* or *mousetrap*, is used to remove the lower, broken part of the sucker-rod string.

The rods also have to be pulled when repairing the tubing or the downhole pump. When they are pulled, they are either laid on the ground when using a single-mast unit or are racked vertically, three at a time, in the rod hanger when using a double-pole or structural mast unit. If the well is under pressure, a *rod blowout preventer* is attached to the top of the well for safety when pulling rods.

Pulling and repairing tubing

When production in a well falls, it could be due to a leak in the tubing string caused by corrosion, stress, or abrasion on the tubing string from the sucker-rod string. The tubing is pulled and inspected. As the tubing is being run back into the well, it can be pressure-tested with a portable hydraulic pressure rig. Collapsed tubing can be opened with a *tubing swage* (fig. 22–4) run on a wireline several times through the tubing string.

Fig. 22–4. Tubing swage

Downhole pump

Falling production in a well can also be due to a malfunctioning downhole pump. The sucker-rod string is pulled to retrieve a rod or insert pump. Both the sucker-rod and tubing string have to be pulled to retrieve a tubing pump. The pump seals and parts are then inspected and repaired if necessary.

Casing repair

Collapsed casing in the well can be opened with a *casing roller* (fig. 22–5a) that uses a series of rollers on the sides. It can also be reamed out with a *tapered mill* (fig. 22–5b) that is run on a workstring and rotated. If the collapsed casing cannot be opened, the well will have to be *sidetracked* (drilled out around the collapsed casing). Leaks in the casing can be located with pressure tests in the well. Casing holes are repaired with a casing patch, scab liner, or expandable casing patch. The *casing patch* is a steel patch glued in place with epoxy resin. A *scab* or *scab off liner* is a liner run into the well to the casing leak level. The liner is then sealed in place with packers or cement. An *expandable casing patch* is a steel tubular with sealing elements that is expanded across the damaged area with an expander mandrel on a workstring. Expandable casing and liners can also be used to close perforations.

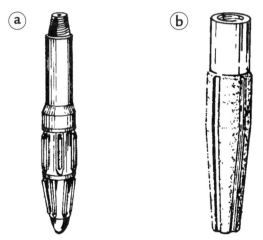

Fig. 22–5. (a) Casing roller and (b) tapered mill

If the upper part of the casing string is damaged but not cemented into the well, it can be cut with a *chemical cutter* (see chap. 16, fig. 16–6) and retrieved. The chemical cutter is run into the well on a wireline. A chemical propellant in the tool is activated by an electrical signal, and high-temperature, corrosive fluid jets out of the cutter ports under high pressure to slice through the casing. The casing is then pulled from the well. A casing overshot or patch tool and new casing is then run in the well.

Secondary cementing

Primary cementing is the cement job done on casing when it is originally run. *Secondary cementing* is done later on a well during a workover. A *cement bond log*, a type of sonic log, can be run in a cased well to determine where and how well the cement has set behind the casing. Gaps in the cement behind the casing are called *holidays* and can be filled by *squeeze cementing* (fig. 22–6). The casing adjacent to the holiday is perforated, and the zone is isolated with packers. Cement is then pumped under pressure down the well, through the perforations, and into the holiday. Cement squeezing can also be used to repair casing leaks.

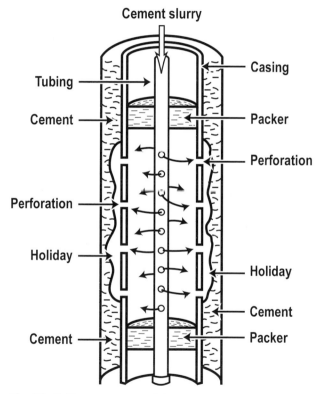

Fig. 22–6. Cement squeeze job

A cement squeeze job is either a bradenhead squeeze or packer squeeze. A *bradenhead squeeze* is a relatively low-pressure cement squeeze job in which the cement is pumped down a tubing string or drillstring (workstring). The workstring is positioned just above the zone to be squeezed. The workstring-casinghead annulus (*bradenhead*) is closed. Pressure is applied through the workstring to squeeze cement through the perforations. A *packer squeeze* is a relatively high-pressure cement squeeze job. A packer is used to seal the workstring-casing annulus above the zone to be squeezed. The cement is pumped down the workstring, and pressure is applied.

Swabbing

Swabbing is the removal of water or drilling mud from a well (*unloading*) so that the oil and gas can flow into a well. A *swab job* is done both after the well is completed to remove the last of the drilling mud or completion fluid and to restore production in a producing well.

A truck-mounted *swabbing unit* with a short mast is used to run a swab tool down the tubing string on a wireline (fig. 22–7). A *swab tool* is a hollow steel rod with rubber swab cups. When the swab tool is raised, the swab cups seal against the tubing to act as a piston and lift (*swab out*) the liquid out of the well. A *lubricator* (a length of casing or tubing) is temporarily attached above the valve on the tubinghead or casinghead to provide a pressure seal. The swab tool can be run into a well under pressure through the lubricator so the well does not have to be killed during swabbing. An *oil saver* is used on top of the lubricator to retain any oil coming up on the wireline.

Sometimes a gas well will not flow because of water filling the well. Soap sticks can be dropped into the tubing (*soaping the well*) to form gas bubbles in the water to help lift the water out of the well.

Replacing gas lift valves

Gas lift valves can stick in an open or closed position. If the gas lift valve was installed in a *gas lift mandrel*, a side pocket in the tubing, it can be retrieved and run back in on a wireline. If not, the tubing string must be pulled to retrieve the valve.

Replacing packers

Packers are designed to seal the casing-tubing annulus. A tubing or completion packer (fig. 19–10) is commonly used in well completion and sometimes has to be repaired or replaced. Packers are retrievable or nonretrievable. The *retrievable packer* is easily removed by pulling the tubing string. The *nonretrievable packer* is made of drillable material such as soft metal and has to be milled out.

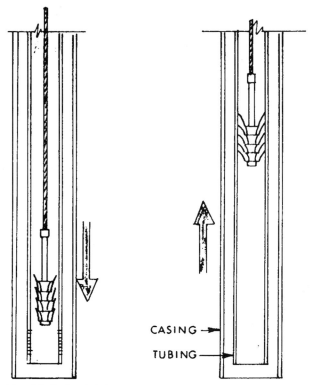

Fig. 22–7. Swab job

CASING →

TUBING →

Recompletion

A well is *recompleted* by abandoning the original producing zone and completing in another zone. The well can be either *drilled deeper* to complete in a deeper zone or *plugged back* to complete by perforating the casing in a higher zone. The original producing zone must be sealed with cement in one of three methods. A cement squeeze job can be used to plug perforations in the depleted zone. Cement can also be pumped down a workstring until it fills the well to the desired level to form a *cement plug*. A *bridge plug* can be used to mechanically seal that level of the well, and then a *dump bailer*, a long cylinder filled with cement slurry, can be run into the well to place cement on top of the bridge plug.

Reservoir Mechanics

Reservoir Drives

Pressure on the fluids in a reservoir rock causes the fluids to flow through the pores into the well. This energy that produces the gas, oil, and water is called the *reservoir drive* or *reservoir energy* and comes from fluid expansion, rock expansion, and/or gravity. The type of reservoir drive controls the production characteristics of that reservoir.

There are four different types of reservoir drives for oil reservoirs. Every oil reservoir has at least one and sometimes two of these reservoir drives. The relative importance of each reservoir drive can change with production. Gas reservoirs have only one of two types of reservoir drives.

Oil reservoir drives

A *dissolved-gas*, *solution-gas*, or *depletion drive* oil reservoir is driven by gas dissolved in the oil. In the subsurface, the oil is under high pressure and has a considerable amount of natural gas dissolved in it. When a well is drilled into the reservoir and production is initiated, pressure on the oil in the reservoir decreases, and gas can bubble out of the oil. Expanding gas bubbles in the pores of the reservoir force the oil through the rock into the well. The expanding volume of oil and rock as the pressure drops also helps the drive.

A dissolved-gas drive reservoir has a very rapid decline in both reservoir pressure and oil production rate as the oil is produced (fig. 23–1). Because of the rapid reservoir pressure drop, any flowing wells have to be put on pumps early. Little or no water is produced during production from this type of reservoir. There is a very rapid gas/oil ratio increase near the end of production. A dissolved-gas drive is very inefficient and will produce relatively little of the original oil in place from the reservoir. A *secondary gas cap* located on the subsurface oil reservoir can be formed by gas bubbling out of the oil.

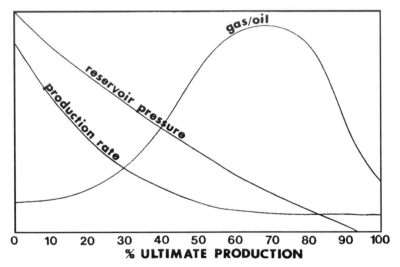

Fig. 23–1. Characteristics of a dissolved-gas drive oil field. (Modified from Murphy, 1952.)

A *free gas cap expansion drive* oil reservoir is driven by gas pressure in the free gas cap above the oil. The expanding free gas cap pushes the oil into the wells. Any solution gas bubbling out of the oil adds additional energy. A free gas cap expansion drive reservoir has a moderate decline in both reservoir fluid pressure and production rate as the oil is produced (fig. 23–2). A sharp rise in the gas/oil ratio as the oil is produced from a well shows that the expanding free gas cap has reached the well, and further oil production will be very limited from that well. This type of reservoir is best developed with wells producing only from the oil portion of the reservoir, leaving the gas in the free gas cap to supply the energy. Usually little or no water is produced. The recovery of oil in place from this type of reservoir is moderate.

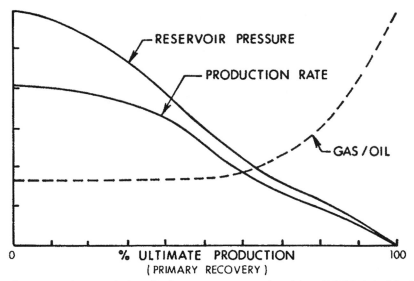

Fig. 23-2. Characteristics of a free gas cap expansion drive oil field. (Modified from Murphy, 1952.)

Water drive reservoirs are driven by the expansion of water adjacent to or below the oil reservoir. The produced oil is replaced in the reservoir pores by water. The water can either come from below the oil reservoir in a *bottomwater drive* or from the sides in an *edgewater drive.* An active water drive maintains an almost constant reservoir pressure and oil production through the life of the wells (fig. 23-3). The amount of water produced from a well sharply increases when the expanding water reaches the well and the well *goes to water.* The recovery of oil in place from a water drive reservoir is relatively high.

Gravity is also a drive mechanism. It is present in all reservoirs, as the weight of the oil column causes oil to flow down into the well. It is most effective in a very permeable reservoir with a thick oil column or a steep dip. Gravity drive is common in old fields that have depleted their original reservoir drive. Downdip wells will have higher production rates than those of updip wells. In a *gravity drainage pool*, the rate of oil production is usually low compared to other drives, but oil recovery can be very high over a long period of time.

Many oil reservoirs have several reservoir drives and are called *combination* or *mixed-drive reservoirs.* The relative importance of the reservoir drive will change with time during production. In the later stages of oil production from a dissolved-gas drive reservoir, gravity drainage becomes significant. The most efficient reservoir drive system is a combination of a free gas cap expansion and a water drive sweeping the oil from both above and below into the wells.

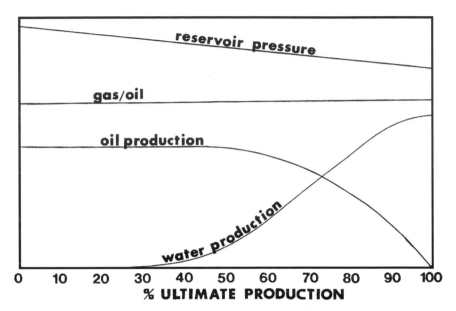

Fig. 23-3. Characteristics of a water drive oil field. (Modified from Murphy, 1952.)

The East Texas oil field has an active water drive. The oil reservoir is in contact with a very extensive aquifer below it in the Woodbine Sandstone. The expanding water forces the oil up into the wells. The original oil-water contact in the East Texas field, when the field was discovered in 1930, was at 3,320 ft (1,012 m) below sea level (fig. 23-4). As oil was produced from 30,400 wells, the bottomwater drive caused the oil-water contact to rise to a level of 3,245 ft (989 m) below sea level by 1965.

Wells on the west side of the field went to water first. Wells on the east side will have the longest production history and go to water last (fig. 25-5). Water invading a reservoir is called *water encroachment*. By 1993, the field had produced 5,135 million bbl (816 million m³) of oil. Only 4% of the field's original reserves, about 210 million bbl (33 million m³) of oil, was left to be produced in the next 10 to 15 years. The recovery will be more than 82% of the oil in place. This extremely high oil recovery is due to (1) the strong water drive, (2) the high porosity (about 30%) and permeability (thousands of millidarcys) of the Woodbine Sandstone reservoir, and (3) the low-viscosity oil.

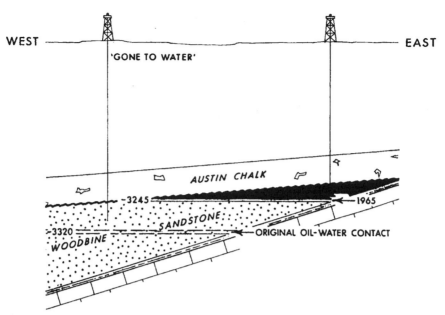

Fig. 23-4. Cross section of the East Texas oil field showing the movement of the oil-water contact with production from 1930 to 1965. (Modified from Landes, 1970.)

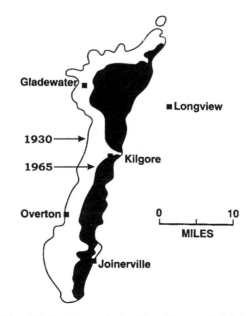

Fig. 23-5. Map of the East Texas oil field showing water encroachment from 1930 to 1965. (Modified from Landes, 1970.)

The Turner Valley field, located to the southwest of Calgary, Alberta, is formed by a large drag fold on a thrust fault along the disturbed belt (fig. 23–6a). The reservoir rock is primarily Mississippian age limestone. The field was discovered in 1913 by drilling with a cable tool drilling rig next to a gas seep. Because the oil reservoir was 2,000 ft (610 m) deeper and located to the west of the free gas cap, it was first believed that it was just a gas field. The gas was wet and was produced in large quantities. The condensate was removed in a gas processing plant at Turner Valley and sold for gasoline. It was called skunk gas because it contained sulfur and stunk like a skunk when burned in an automobile. There was little use for the natural gas. Almost all the dry gas from the gas processing plant was flared in a location called Hell's Half Acre.

In 1930, oil was discovered below the gas cap (fig. 23–6b). Unfortunately, although Turner Valley holds about 1 billion bbl (160 million m^3) of oil, the free gas cap, which was the oil reservoir drive, was depleted. Less than 12% of the oil will ultimately be produced. The reservoir is too deep to economically repressure it. The gas in a free gas cap should never be produced before or during oil production because it supplies the energy to produce the oil.

The reservoir drive of an oil field can be determined from both the nature of the reservoir and from production characteristics. Isolated reservoirs that are encased in shales such as shoestring sandstones and reefs or those cut by sealing faults often have dissolved-gas drives. If the reservoir has a large free gas cap, it has a free gas cap expansion drive; if not, it probably has a dissolved-gas drive. Extensive sandstones and other reservoirs that connect to large water aquifers often have water drives. Abnormally high pressure suggests that the reservoir is isolated and does not have a water drive.

Reservoir pressures and oil production will also indicate the type of reservoir drive. A rapid decrease in both reservoir pressure and oil production is characteristic of a dissolved-gas drive. Shutting in wells will not cause the reservoir pressure to build up. An active water drive has almost constant reservoir pressure and oil production. If the reservoir pressure does decrease, shutting in the wells allows the reservoir pressure to increase to almost its original pressure.

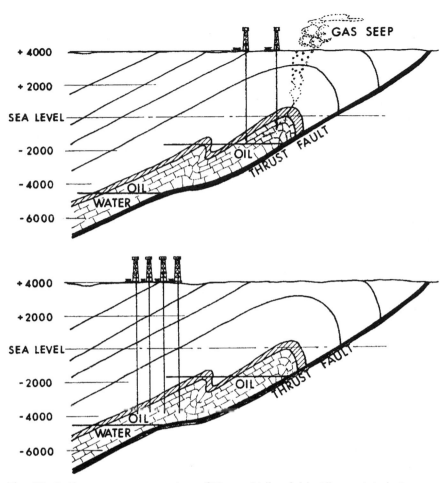

Fig. 23–6. East-west cross section of Turner Valley field, Alberta (a) during gas production from free gas cap (1913–1930) and (b) during oil production after 1930. (Modified from Gallup, 1982.)

Gas reservoir drives

Gas reservoirs have either an expansion-gas or a water drive. An *expansion-gas* or *volumetric drive* is due to the pressure on the gas in the reservoir. When a well is drilled into the high-pressure reservoir, the well has relatively low pressure. The high-pressure gas in the pores of the reservoir expands out into the well. This drive recovers a relatively large amount of original gas in place in the reservoir.

A *water-drive* gas reservoir is similar to a water-drive oil reservoir and is due to expanding water adjacent to or below the reservoir. It is not as effective as an expansion-gas drive because the water flows around and traps pockets of gas in the reservoir. It has a moderate recovery of gas in place.

Maximum Efficient Rate

The *maximum efficient rate* (*MER*) is the maximum rate at which a well or field can be produced without wasting reservoir energy or leaving bypassed oil in the reservoir. It generally ranges from 3 to 8% of the recoverable oil reserves per year. This rate provides for an even rise in the oil-water contact along the bottom of the reservoir (water encroachment) and holds the gas/oil ratio to a minimum. The MER of a field can be accurately determined only after reservoir drives have been identified and productivity tests have been run.

Petroleum Production

World Oil reported an average global oil production rate of 73.6 million bbl/d (11.7 m³/d) in 2010. Of that, Saudi Arabia averaged 8.2 million bbl/d (1.3 million m³/d) (11%) and the United States averaged 5.5 million bbl/d (0.9 million m³/d) (7.4%).

The API reported that in 2008 there were a total of 1,004,606 producing wells in the United States. Of those, 48% were gas and condensate wells and 52% were oil wells. Texas had the most wells. More than half (54%) of all the world's producing oil wells were located in the United States. Russia was second, followed by China and Canada. The average oil production per oil well in the United States was 10.2 bbl/d (1.6 m³/d).

A *petroleum engineer* is an engineer who is trained in drilling, testing, and completing a well and producing oil and gas. A *reservoir petroleum engineer* is in charge of maximizing the production from a field to obtain the best economic return.

Well and Reservoir Pressures

Tubing pressure is measured on the fluid in the tubing, whereas *casing pressure* is measured on the fluid in the tubing-casing annulus. The pressure gauge at the top of a Christmas tree measures tubing pressure. *Bottomhole pressure* is measured at the bottom of the well. It is measured either as *flowing*, with the well producing, or *shut-in* or *static*, after the well has been shut in and stabilized for a period of time such as 24 hours (fig. 24–1). *Downdraw* is the difference between shut-in and flowing pressure in a well.

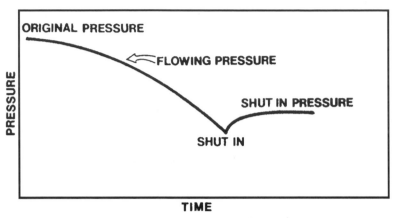

Fig. 24–1. Flowing and shut-in pressure in a well

The original pressure in a reservoir before any production has occurred is called *virgin, initial,* or *original pressure*. During production, reservoir pressure usually decreases. Reservoir pressure can be measured at any time during production by shut-in bottomhole pressure in a well. A *pressure bomb*, an instrument that measures bottomhole pressure, can be run into the well on a wireline. A common pressure bomb consists of a pressure sensor, recorder, and a clock-driven mechanism for the recorder. It is contained in a metal tube about 6 ft (1.8 m) long. The chart records pressure with time as the test is being conducted. Temperature can also be recorded on a similar instrument. Another instrument, an *electronic pressure recorder*, can be run on a conductor wire.

Well Testing

Tests on a well are run by the well operator, a specialized well tester, or a service company to determine the optimum production rate. They can use equipment available on the site or portable test equipment. After the well has been completed, a potential test can be run. The *potential test* determines the maximum gas and oil that the well can produce in a 24-hour period. It uses the separator and tank battery on the site to hold the produced fluids. Potential tests can also run periodically during production and may be required by some government regulatory agencies.

A *productivity test* is run to determine the effect of different production rates on the reservoir. It is made with portable well test equipment (fig. 24–2) that measures the fluid pressure at the bottom of the well when it is shut-in and then during several different stabilized rates of production. The measurements are used to calculate the absolute open flow and the maximum production rate that the well can produce without damaging the reservoir.

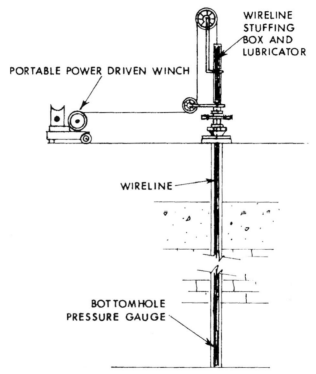

Fig. 24–2. Productivity test equipment

For wells that have a central processing unit, periodic *production tests* can be made to determine how much each well is producing. These tests are run manually or automatically. Oil well test data typically include oil production, water production, gas rate, gas/oil ratio, and flowing tubing pressure. Gas well test data typically include gas rate, condensate production, water production, flowing tubing pressure, and condensate/gas ratio.

Pressure transient testing on a well involves measuring pressures and their flow rates. One type, a *drawdown test*, measures the shut-in bottomhole pressure and then the pressure change as the well is put on production and the pressure drops to a stable, flowing pressure. A *buildup* test measures the flowing bottomhole pressure and then the pressure change as the well is shut in and the pressure rises to a stable, shut-in pressure. A *multirate test*, such as a four-point test, measures the flowing bottomhole pressure at different, stabilized flow rates.

Deliverability is the ability of the reservoir at a given flowing bottomhole pressure to move fluids into the well. *Maximum potential flow* or *absolute open flow* (*AOF*) is the maximum flow rate into a well when the bottomhole pressure is zero. It is a theoretical flow rate that is calculated from a multivariate test. The *production index* (*PI*) of a well is the downhole pressure drawdown in pounds per square inch (psi) divided by the production in barrels per day (bbl/d). Wells on land usually have a PI of greater than 0.1 psi/bbl/day, whereas offshore wells have a PI greater than 0.5. *Inflow performance relationship* (*IPR*) is similar to PI because it plots drawdown against production but is more accurate in that it also accounts for reservoir drive, increasing gas/oil ratios, and relative permeability changes with production.

Gas wells are tested with routine production tests that measure the amount of gas, condensate, and water produced. A *backpressure test* measures the shut-in pressure and the pressures at different stabilized flow rates to determine the well deliverability.

Cased-Hole Logs

After a reservoir in a well has been depleted, a decision must be made to either plug and abandon or recomplete the well. To recomplete, a new oil or gas reservoir must be identified behind the casing. Only the natural gamma ray and neutron porosity logs can be run in a cased-hole. A *pulsed neutron log* is a type of neutron log that emits pulses of neutrons into the formation and measures returning gamma rays. It can distinguish gas and oil from water in the reservoir and is used to find gas and oil reservoirs located behind the casing.

Production Logs

Production logs are run in producing wells to evaluate a problem. They are run either on a wireline through the tubing or on a tubing string. There are several types of production logs.

Tracer logs are used to detect fluid movement in a well. A radioactive tracer is injected into the well at a specific location, and its movement is tracked by recording gamma rays. A *continuous flowmeter* uses propellers on a vertical shaft to measure fluid flow up a well to make a continuous record of flow versus depth in the well. A *packer flowmeter* uses a packer to seal the well at that depth to ensure that all the fluids flow up through the flowmeter in the packer and are measured.

A *noise log* uses a microphone to detect and amplify any sounds in a well. The log can locate where fluids are flowing into the well, and the frequency of the sound can be used to distinguish between liquid and gas. A *temperature log* measures the temperature of fluid filling a well. Before the temperature log is run, the well is shut in for a period of time to allow the temperatures to come to equilibrium. Because expanding gas cools when entering a well, it can be located by a temperature log.

A *manometer* measures pressure in the well at a specific depth, and a *gradiometer* measures a continuous profile of the pressure gradient. A *water-cut meter* measures the amount of water in the fluid filling the well. A *collar log* has a casing-collar locator that uses either a magnetic detector or scratcher to locate the casing collars in a well. It is used to accurately find locations in the well. A collar log is used with a natural gamma ray log to locate where to perforate the casing.

Decline Curves

A *decline curve* is a plot of oil or gas production rate with time made for a single well or an entire field (fig. 24–3). Production rate will decline with time as the reservoir pressure decreases. The *initial production (IP)* of a well is the first 24 hours of production and is usually the highest. As the production rate declines, the well eventually becomes a *stripper well* that is barely profitable. Stripper wells are defined in the United States as producing less than 10 bbl (1.6 m³) of oil per day over a 12-month period or 60 Mcf (2,000 m³) of gas per day at maximum flow rate, and they receive special tax advantages. The API reported that in 2007 there were a total of 396,537 stripper oil wells in the United States that accounted for about

77% of the total US oil wells. They produced an average of 2 bbl/d (0.32 m³/d) and accounted for 15.7% of total US oil production.

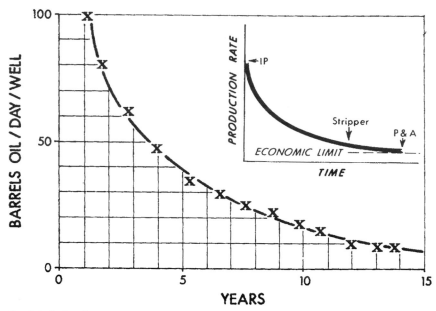

Fig. 24–3. Decline curve

The *economic limit* of a well is when production costs equal net production revenue. It depends on how deep the well is, how much water it produces, where the well is located, and several other factors. When the economic limit of a well is reached, it is plugged and abandoned, or improved oil recovery is initiated. Most wells are designed with a 15- to 20-year life.

The shape of the oil decline curve depends on the reservoir drive (fig. 24–4). Solution-gas reservoirs have a very sharp decline, whereas water-drive reservoirs have almost constant production for the life of the wells. The shape of a free gas cap expansion drive curve is between the curves for solution-gas and water-drive reservoirs. The decline curves for wells producing from a fractured reservoir in a tight sandstone or dense limestone such as the Spraberry field of Texas, or a gas shale such as the Barnett Shale, are very distinctive (fig. 24–5). The well can have a high IP as oil drains rapidly through the very permeable fractures. As the fractures drain, the production rapidly drops. Within a short period, the production settles to a long and steady rate as the oil drains slowly from the relatively impermeable rock into the fractures.

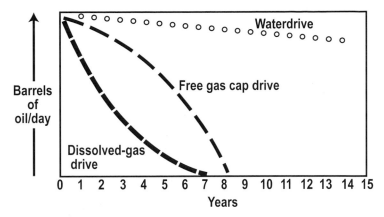

Fig. 24–4. Reservoir drive decline curves

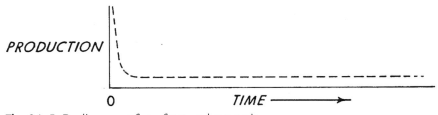

Fig. 24–5. Decline curve for a fractured reservoir

Bypassing and Coning

Drilling and completing a well is an economic investment. The best return on that investment is to produce the gas and oil as fast as possible to recover costs and make a profit as soon as possible.

Many reservoirs, however, are not homogeneous, and there are pockets of oil or gas in less permeable areas. In a water-drive reservoir, the water flows in to replace the oil or gas as it is being produced. If the oil and gas are produced too fast, the water can flow around pockets of oil and gas in less permeable areas in a process called *bypassing* (fig. 24–6). Bypassing seals the oil (*bypassed oil*) and gas (*bypassed gas*) in that area and prevents it from being produced from existing wells. To prevent significant bypassing and have maximum ultimate production, the oil and gas should be produced at a slower rate to allow less permeable zones time to drain.

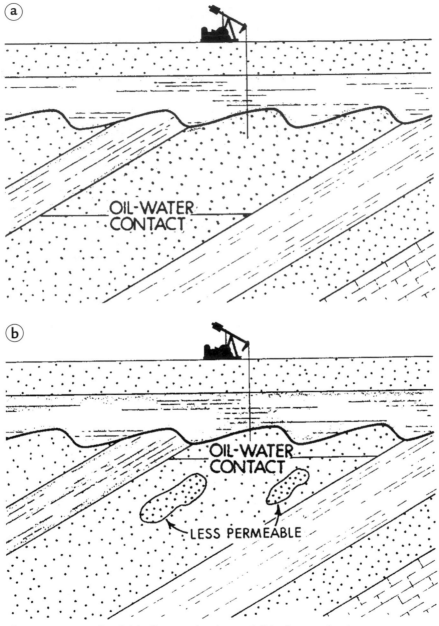

Fig. 24–6. Bypassing (a) before production and (b) after production

Coning is caused by oil being produced too fast. The oil-water contact is sucked up in a bottomwater drive reservoir (fig. 24–7), or the gas-oil contact is sucked down in a free gas cap expansion drive reservoir. This can cause permanent damage to the well. Horizontal wells can be used to prevent coning. If coning does occur on a horizontal well, it is called *cresting*.

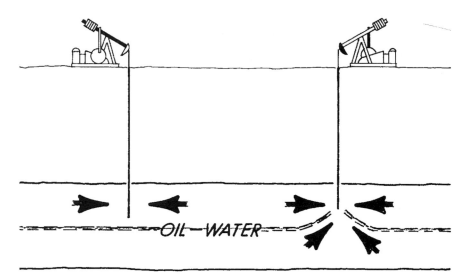

Fig. 24–7. Coning

Cycling

As reservoir pressure drops during gas production from a retrograde gas reservoir, condensate separates out of the gas in the reservoir. The liquid coats the subsurface grain surfaces and is very difficult, if not impossible, to recover. To prevent condensate from separating in the subsurface reservoir, *cycling* is used. Produced gas is stripped of natural gas liquids on the surface. The dry gas is then reinjected through injection wells into the reservoir to maintain the reservoir pressure.

Well Stimulation

Several well stimulation methods can be used to increase the well production rate. These include acidizing, explosive fracturing, and hydraulic fracturing.

Acidizing

A well can be *acidized* or given an *acid job* by pumping acid down into the well to dissolve limestone, dolomite, or any calcite cement between sediment grains. HCl (*regular acid*), HCl mixed with HF (*mud acid*), and HF (*hydrofluoric acid*) are commonly used. HCl is effective on limestones and dolomites, and HF is used for sandstones. For formations with high temperatures, acetic and formic acids are used. To prevent the acid from corroding the steel casing and tubing in the well, an additive called an *inhibitor* is used. A *sequestering agent* is an additive used to prevent the formation of gels or precipitates of iron that would clog the pores of the reservoir during an acid job.

Two types of acid treatment are matrix and fracture acidizing. During *matrix acidizing*, the acid is pumped down the well to enlarge the natural pores of the reservoir. During *fracture acidizing*, the acid is pumped down the well under higher pressure to fracture and dissolve the reservoir rock. After an acid job, the spent acid, dissolved rock, and sediments are pumped back out of the well during the *backflush*. An acid job is also used to remedy skin damage on a wellbore and is called a *wash job*.

Explosive fracturing

From the 1860s until the late 1940s, explosives were commonly detonated in wells to increase production. *Well shooting* or *explosive fracturing* was done with liquid nitroglycerin in a tin cylinder called a *torpedo*. It was run down the well and detonated on the bottom. The explosion created a large cavity that was then cleaned out, and the well was completed open hole. The person in charge of the nitro was called the *shooter*. The technique was both effective and dangerous.

Hydraulic fracturing

Hydraulic fracturing was developed in 1948 and has effectively replaced explosive fracturing. In 2010, over 60% of all wells completed in the United States were fraced. During a *frac job* or *hydraulic fracturing* (fig. 24–8), a service company injects large volumes of frac fluids under high pressure into the well to fracture the reservoir rock (plate 24–1). Frac jobs are done either in an open-hole or a cased well with perforations. A common *frac fluid* is a gel formed by water and *polymers*, long organic molecules that form a thick liquid when mixed with water. Oil-based frac fluid and foam-based frac fluids using bubbles of nitrogen or carbon dioxide can be used to minimize formation damage. Typically, about 0.5% of the frac fluid is composed of additives similar to those used in drilling fluids. The frac fluid is transported out to the frac job in large trailers.

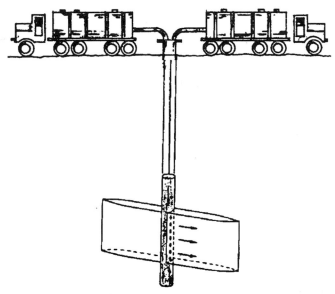

Fig. 24–8. Hydraulic fracturing

Plate 24–1. Hydraulic fracturing an oil well

A frac job is done in three steps. First, a pad of frac fluid is injected into the well by several large pumping units mounted on trucks to initiate fracturing the reservoir. Next, a slurry of frac fluid and propping agents are pumped down the well to extend the fractures and fill them with propping agents. *Propping agents* or *proppants* are small spheres that hold open the fractures after pumping has stopped. The propping agents are commonly well-sorted quartz sand grains. In high-pressure wells, ceramic or aluminum oxide microspheres are used. The well is then *backflushed* in the third stage to remove about 10 to 30% of the original frac fluid.

Crosslinked frac fluids that have a high viscosity necessary to carry the propping agents down the well can be used. A *breaker fluid* is then injected into the well to make the crosslinked frac fluid more fluid and easier to remove during backflush.

Medium and hard (brittle) formations are best for fracturing, because loose formations (unconsolidated) do not permit the propping agents to hold open the fractures. All the equipment used during the frac job is driven onto the site. The frac fluid is mixed and stored in *frac tanks*. The frac fluid is mixed with proppants in a *blender*. Pump trucks are connected to a manifold to pressurize the pad and the slurry and pump them down the well. A *wellhead isolation tool* can be connected to the top of the well to protect the wellhead from the high pressures and abrasive propping agents. The frac job is monitored and regulated from the *frac van*.

Frac jobs are described by the amount of frac fluid and proppants used. The average modern frac job uses 60,000 gallons (227,000 liters) of frac fluid and 100,000 pounds (45,000 kg) of sand. A *massive frac job* is a very large frac job (plate 24–2). There is no exact definition of a massive frac job, but it typically uses more than 1 million gallons (3.8 million liters) of frac fluid and 5 million lb (2.3 million kg) of sand.

A *frac pack* or *frac/pack* uses a viscous gel and a relatively high concentration of sand proppants. It forms relatively short but wide fractures. Frac packs are common in offshore wells.

Hydraulic fracturing is a very common well-stimulation technique that increases both the rate of production and ultimate production. It increases the production rate from 1½ to 30 times the initial rate with the highest increases in tight reservoirs. Ultimate production is increased from 5 to 15%. It is used in all tight gas sand reservoirs and as a common remedy for skin damage in a wellbore.

A well can be fraced several times during its life. In some instances, however, hydraulic fracturing can harm a well by *fracing into water*. The hydraulically induced fractures extend vertically into a water reservoir that floods the well with water.

Plate 24–2. Aerial photograph of a massive frac job. The well is in the center with lines of pumping units and frac fluid trailers on either side. (Courtesy of Halliburton.)

Disposal of Oilfield Brine and Solution Gas

The natural gas produced with oil often creates a disposal problem. It comes from the separators at very low (atmospheric) pressure, and there is usually no market for it. Even with a market, the gas would have to be compressed to pipeline pressure. In the past, it was often burned (*flared*) in the oil fields. This is against the law today in most countries. Flaring still occurs in some situations when any other gas disposal method is not practical or during well testing. The produced gas can be used to increase the ultimate oil production from the reservoir by reinjecting it into the subsurface reservoir in a *pressure maintenance system* (fig. 24–9). Produced wet gas is first gathered and is usually stripped of valuable natural gas liquids. It is then compressed and pumped into an *injection well*. In a saturated oil field, the gas is injected into the free gas cap. In an undersaturated oil field, the gas is injected into the oil reservoir.

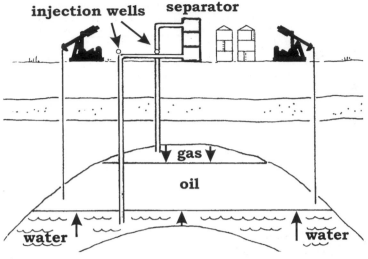

Fig. 24–9. Pressure maintenance system

Gas from the separators can also be given to the landowner to heat his or her home and operate irrigation pumps. This *farmer's gas* can be part of the lease agreement before any wells are drilled. Gas from the separator could also be used to operate equipment in the field such as the engine used to drive a beam-pumping unit.

Oilfield brine from the separators can also be pumped down another injection well into the subsurface reservoir below the oil-water contact as part of the pressure maintenance system. If there is no injection well system available for the well or field, the oilfield brine or the water removed from natural gas is stored in a metal (see chap. 20, plate 20–4) or fiberglass *saltwater tank*. The fiberglass tank is more resistant to corrosion.

A *saltwater disposal well* is used to pump the brine or water into a subsurface reservoir rock (fig. 24–10). The disposal well has to be permitted by a government agency and must meet specific criteria. The oilfield brine cannot be injected into a subsurface freshwater reservoir. The reservoir must already contain naturally saline waters that cannot be used for drinking or irrigation. The reservoir must also be able to sustain the increased pressure of the injected water without leaking into another freshwater reservoir.

If there is no disposal well, the brine is stored in an open fiberglass or metal tank to evaporate and reduce the volume. When the tank is eventually filled, a service company (a *water hauler*) is used to transport the brine to a commercial saltwater disposal well.

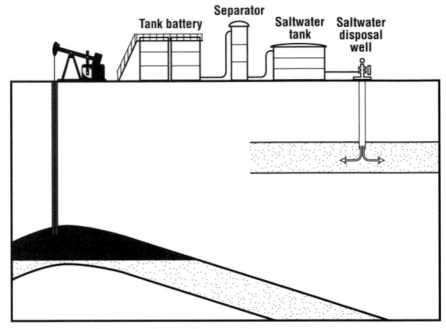

Fig. 24–10. A saltwater tank and disposal well

Surface Subsidence

During production, reservoir pressure decreases, and water usually flows in from the sides and bottom to replace the produced fluids. If water does not replace the produced fluids, the subsurface reservoir rock can compact and the surface of the ground subsides (fig. 24–11). This has happened in the Wilmington oil field in Long Beach, California, which has been producing since the 1930s. Beginning in the 1940s, surface subsidence in the shape of a bowl was noted. The center of the bowl has now subsided a total of 29 ft (8.8 m) leaving much of the city below sea level. A massive water injection program has stopped the subsidence, and the city is now protected from seawater flooding by a dike.

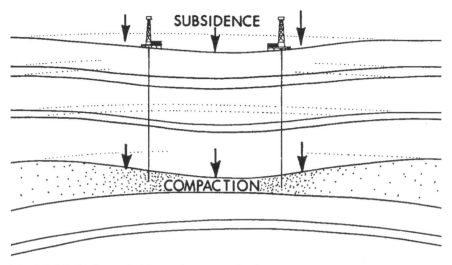

Fig. 24–11. Surface subsidence due to production

The bottom of the North Sea above the Ekofisk oil field has subsided several tens of feet because of compaction of the Ekofisk Chalk reservoir. The subsidence was first noticed in 1984 after the casing in several wells had collapsed, and the level of the boat dock on the platform became submerged. The elevation of the Ekofisk production platform had dropped to a dangerous level. In 1987, the legs of the platform were cut, the deck jacked up, and extensions spliced into the legs.

Corrosion

Corrosion is the chemical degradation of metal. It can be a problem during both drilling and production. Corrosion occurs when metal is exposed to air, moisture, or seawater, or by chemicals such as oxygen, carbon dioxide (*sweet corrosion*), or hydrogen sulfide (*sour corrosion*) in the produced fluids. *Total acid number* is a measure of the acidity and corrosiveness of a crude oil. It is a number expressed in mg KOH/g. Higher numbers are more corrosive.

Exposed metal surfaces on equipment are painted for protection. *Inhibitors* are chemicals that are injected to coat steel in the well and the production facilities with a thin film. The inhibitor can be injected either automatically or manually in periodic batches into the casing-tubing annulus of the well. A concrete coating can be used to protect the insides of flowlines. The tubing in injection and disposal wells is often lined with plastic. Large metal structures such as pipelines and offshore production structures can be shielded by *cathodic protection*. It involves charging the structure with an electrical charge to prevent corrosion.

Production Maps

A *well status map* is used to analyze production from a field and identify problem wells. The map shows the location of all wells in a field. Producing wells have the well number, barrels of oil and water production per day, and the gas/oil ratio next to them. Injection wells have the well number, barrels of water injected per day, pressure, and cumulative injection in thousands of barrels. A *cumulative production map* lists the total amount of water, gas, and oil that each well has produced up to a specific date. *Bubble maps* are used to show how much each well has produced (fig. 24–12). A circle is drawn around each well with the radius of the circle (*bubble*) proportional to either the well's cumulative production (CUM) or its initial production (IP) of gas, oil, or water.

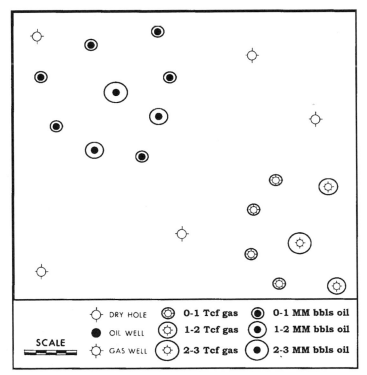

Fig. 24–12. Bubble map

Stranded Gas

Stranded gas is natural gas that has no market. Large reservoirs of stranded gas occur in western Siberia, northwestern Canada, Alaska, and the Middle East. Natural gas can be converted into a liquid to decrease its volume and transport it to a market either as liquefied natural gas or synthetic crude oil. When methane is compressed and cooled to –269°F (–167°C), it becomes a liquid called *liquefied natural gas* (*LNG*). LNG occupies 1/645 the volume of natural gas. Special tankers can then be used to transport the LNG across the sea to markets. The largest conventional gas field in the world is shared by the countries of Qatar (North Dome field) and Iran (South Pars field) (fig. 24–13). It will produce 1,200 trillion ft^3 (36 trillion m^3) of natural gas. It also contains 19 billion bbl (3 billion m^3) of recoverable condensate. The trap is a broad anticline, the reservoir rock is the Khuff Formation with dolomite and limestone, the seal is overlying salt, and the source rock is a Silurian age black shale. The reservoir rock in the South Pars field averages 9% porosity and 26 md permeability. Qatar has extensive LNG facilities and is the world's largest LNG exporter.

Gas-to-liquid involves mixing natural gas and air in a reactor to form *synthesis gas* (CO and H). The synthesis gas is then put in another reactor to form synthetic crude oil. Qatar also has the world's largest gas-to-liquid facility.

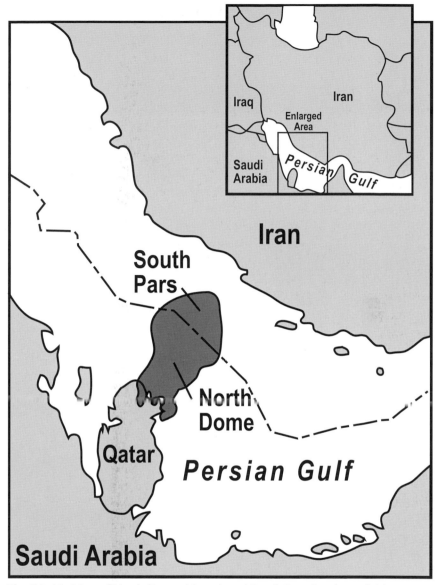

Fig. 24–13. Map of North Dome (Qatar) and South Pars (Iran) gas fields

Reserves

According to several sources, including the API, there was an estimated total of about 1.34 trillion bbl (213 million m^3) of proved oil reserves in the world in 2008. Of that, 56% was in the Middle East, 13% in Canada, and 1.4% in the United States. The Energy Information Agency estimated that in the United States in 2009, there were 22.3 billion bbl (3.5 billion m^3) of proved oil reserves, with 25% in Texas, 19.5% offshore, 16% in Alaska, and 12.7% in California.

Oil & Gas Journal estimated that 6,609 Tcf (187 billion m^3) of proved gas reserves existed in the world in 2010. The Energy Information Agency estimated that in the United States in 2009, there were 284 Tcf (7.9 trillion m^3) of proved dry gas reserves, with 30% in Texas, 12.7% in Wyoming, and 8.5% in both Oklahoma and Colorado.

Recovery Factor

The amount of oil or gas in the subsurface reservoir is called *oil in place* (*OIP*) and *gas in place* (*GIP*). *Recovery factor* is the percentage of OIP or GIP that the reservoir will produce. The recovery factor for an oil reservoir depends on (1) the viscosity of the oil, (2) the permeability of the reservoir, and (3) the reservoir drive.

Typical values for oil and gas reservoir drives are shown in table 25–1.

Table 25–1. Recovery factors

oil reservoir drives	recovery factors
solution gas	5 to 30%
free gas cap expansion	20 to 40%
water	35 to 75%
gravity	50 to 70%
gas reservoir drives	
expansion gas	75 to 85%
water	60%

(Modified from Sills, 1992.)

Shrinkage Factor and Formation Volume Factor

The amount of natural gas dissolved in crude oil in the subsurface reservoir is called the *dissolved, reservoir,* or *solution gas/oil ratio* and is expressed in scf/bbl. It depends upon the temperature and pressure of the reservoir and the chemistry of the oil. In general, deeper reservoirs have higher dissolved gas/oil ratios. If the oil has dissolved all the gas it can hold under those conditions, it is *saturated*. An oil reservoir with a free gas cap is saturated. If there is no free gas cap, the oil reservoir is *undersaturated* and can hold more gas.

When the oil is produced, the reservoir pressure decreases to surface pressure, and gas bubbles out of the oil. The amount of gas and oil produced on the surface is called the *producing gas/oil ratio*. It is similar to the dissolved gas/oil ratio but can be higher if some gas is being produced from the free gas cap.

As gas bubbles out of the oil on the surface, the volume of the oil decreases. The amount to which one barrel of oil decreases in volume on the surface is called the *shrinkage factor* (fig. 25–1). It is expressed as a decimal that ranges from 1.0 to 0.6 and depends on the amount of gas that bubbles out of the oil. A stabilized barrel of oil under surface conditions (60°F temperature and 14.7 psi pressure or 15°C and 101.325 kPa) is called a *stock-tank barrel of oil*. Crude oil reserves are always reported in stock-tank barrels of oil. The number of barrels of oil under reservoir conditions that need to be produced to shrink to a stock-tank barrel of oil is called the *formation volume factor* (*FVF*) (fig. 25–1). It usually varies from 1.0 to 1.7 and is the inverse of the shrinkage factor. A FVF of 1.4 is characteristic of high-shrinkage crude oil, and 1.2 is low-shrinkage crude oil. The FVF can be estimated from the producing gas/oil ratio. The higher the producing gas/oil ratio, the larger the FVF.

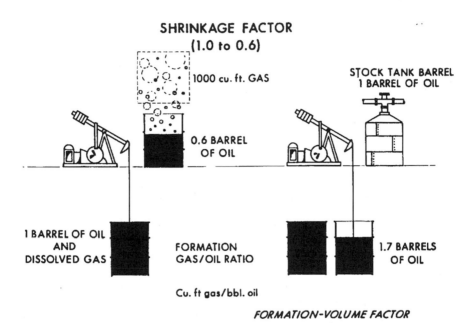

Fig. 25–1. Shrinkage factor and formation volume factor (FVF)

Gas in a subsurface reservoir is under high pressure and temperature. When the gas is produced, the pressure and temperature decrease to surface conditions, and the gas expands in volume. The volume of natural gas in the subsurface reservoir that expands to 1 standard cubic foot on the surface is called the *gas formation volume factor* (B_g). It depends on reservoir temperature and pressure and gas composition. Natural gas is measured in *standard cubic feet* (*scf*), the cubic feet of natural gas under surface conditions defined by law. Natural gas reserves are always reported in standard cubic feet (scf).

Reserve Calculations

Reserves are the amount of oil and gas that can be produced from a well or field in the future under current economic conditions using current technology.

Oil reserves

Oil reserves can be computed both volumetrically and by decline curves. The *volumetric* or *engineering formula* for oil reserves for a single well or an entire oil reservoir is:

$$\text{stock-tank bbl of oil} = \frac{V \times 7758 \times \varnothing \times S_o \times R}{FVF}$$

V is the volume of the oil pay zone drained by a well or wells expressed in units of acre-feet. One *acre-foot* is the volume generated by 1 acre of surface area and 1 foot deep (fig. 25–2). An acre-foot of volume can hold 7,758 barrels of oil. The porosity ($\varnothing$) of the reservoir is expressed as a decimal and is usually determined from well logs or cores. S_o is oil saturation expressed as a decimal and is determined from electrical resistivity well logs or cores. *R* is the recovery factor express as a decimal that is estimated from the reservoir drive, reservoir permeability, and oil viscosity. The formation volume factor (*FVF*) can be estimated from the producing gas/oil ratio or determined from laboratory analysis of produced fluids.

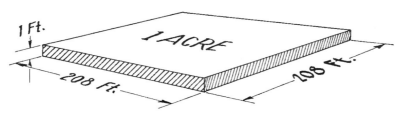

Fig. 25–2. Acre-foot

The *decline curve method* uses production data to fit a decline curve to project estimated future oil production. It is assumed that the production will decline on a reasonably smooth. The curve can be expressed by mathematics or plotted on graph paper with an arithmetic production scale to estimate future production (fig. 25–3a). If the well produces with a dissolved gas or free gas cap expansion drive, plotting the curve on a logarithmic production scale will yield a straight line for the decline curve (fig. 25–3b).

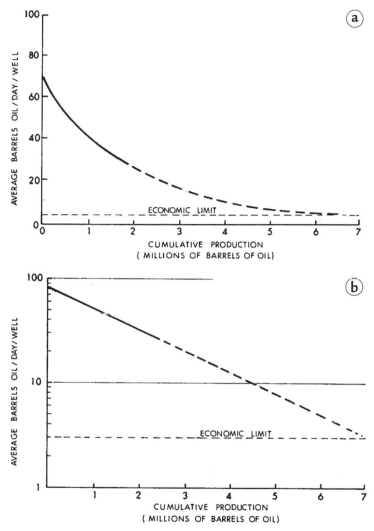

Fig. 25–3. Decline curve method of estimating reserves: (a) arithmetic scale and (b) logarithmetic scale

Gas reserves

Gas reserves are computed volumetrically and by P/Z plots. For a gas reservoir without any oil, the *engineering* or *volumetric formula* is:

$$\text{standard cubic feet of gas} = \frac{V \times 43{,}560 \times \text{Ø} \times S_g \times R}{Bg}$$

V is the volume of the gas reservoir in acre-feet (1 acre-foot can hold 43,560 standard cubic feet). Porosity (Ø) is reservoir porosity expressed as a decimal and is usually determined from well logs or cores. S_g is gas saturation that is determined from electrical resistivity well logs or cores. R is the recovery factor and is determined from the reservoir drive and permeability. B_g is the gas formation volume factor that is determined from tables of reservoir temperature and pressure and gas composition.

Gas from an associated gas reservoir with oil is more difficult to calculate because the gas comes from both the free gas cap and solution gas that bubbles out of the oil as the reservoir pressure drops.

For a single gas well, reserves are estimated from a P/Z *plot* (fig. 25–4). P is the reservoir pressure measured in the well. It will decrease as gas is produced from the well. Z is the *compressibility factor* that compensates for natural gas not behaving as an ideal gas under high pressure and temperature conditions of the subsurface reservoir. It varies between 1.2 and 0.7 and is determined from tables of temperature, pressure, and gas composition. P/Z plotted against cumulative production is a straight line. Where it intersects the abandonment pressure is the ultimate gas production from that well. *Abandonment pressure* is the lowest gas reservoir pressure before the well is plugged and abandoned. It is usually the lowest pressure that the gas pipeline will accept. This is usually between 700 to 1,000 psi (49 to 70 kg/cm^2). When economics permits, the life of a gas well can be extended with a compressor to increase the produced gas pressure to exceed pipeline pressure.

Materials balance method

The *materials balance method* for a gas or oil reservoir uses an equation that relates the volume of oil, water, and gas that has been produced from that reservoir and the change in reservoir pressure to calculate the remaining oil and gas. It assumes that as fluids from the reservoir are produced, there will be a corresponding change in the reservoir pressure that depends on the remaining volume of oil and gas (fig. 25–5).

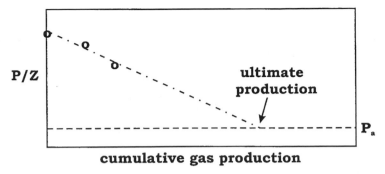

Fig. 25–4. *P/Z* method of estimating gas reserves

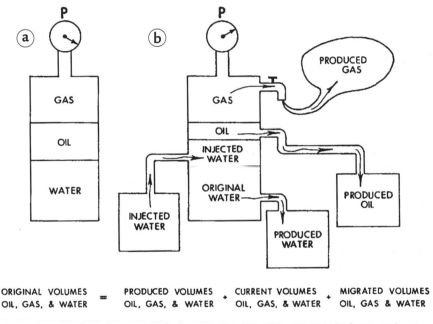

ORIGINAL VOLUMES OIL, GAS, & WATER	=	PRODUCED VOLUMES OIL, GAS, & WATER	+	CURRENT VOLUMES OIL, GAS, & WATER	+	MIGRATED VOLUMES OIL, GAS & WATER

Fig. 25–5. Model of materials balance equation: (a) reservoir before production and (b) reservoir during production

Types of Reserves

The U.S. Securities and Exchange Commission (SEC) defines reserves based on the probability that they exist (table 25–2).

Table 25–2. Reserve types

reserves name	probability of existence
proved or proven reserves	at least 90%
probable reserves	at least 50%
possible reserves	at least 10%

Developed reserves are presently being produced from wells, whereas *undeveloped reserves* are known to exist but are not currently being produced.

Improved Oil Recovery

Primary production is the oil produced by the reservoir drive energy. It depends on the type of reservoir drive, oil viscosity, and reservoir permeability, but averages 30 to 35% of the oil in place and can be as low as 5%. This leaves a considerable amount of oil in the reservoir after the pressure has been depleted. Because of this, *improved oil recovery*, the application of engineering techniques to produce more oil after primary production, is commonly used. *Ultimate oil recovery* is the total production from a well or field by primary production and improved oil recovery that is justified by economic conditions.

A typical gas reservoir will produce 80% of the gas by primary production. Because so little gas is left in the depleted reservoir, gas fields are plugged and abandoned after primary production.

Waterflood

A *waterflood* involves injecting water through injection wells (fig. 26–1) into the depleted oil reservoir. It can be initiated either before or after the original reservoir drive has been fully depleted. The water sweeps some of the remaining oil through the reservoir to producing wells. A waterflood can recover 5 to 50% of the remaining oil in place.

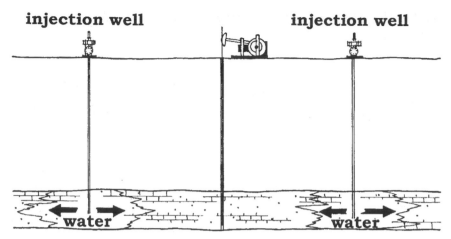

Fig. 26–1. Waterflood

The injected water is often oilfield brine from the separators but can also be water from other sources that has been treated. The injected water must be compatible with the producing formations and not cause reactions that decrease the permeability of the formation being flooded. Suspended solids that can plug the pores are removed from the injection water by filtration. Organic matter and bacteria that produce slimes are neutralized by biocides. Oxygen is removed from the water to prevent corrosion. The water is either pumped under pressure down the well or is fed by gravity from storage tanks on a higher elevation such as a hill.

The injection wells can be either drilled or converted from producing wells (plate 26–1). Waterfloods are described by the aerial pattern of the wells and are either spot or line drives. The common *five-spot pattern* has four water-injecting wells located at the corners of a square with a producing well at the center (fig. 26–2a). The pattern is repeated in the field so that four injection wells surround each producing well and four producing wells surround each injection well. A *line drive* has alternating lines of producers and injectors and can be either *direct* (fig. 26–2b) or *staggered* (fig. 26–2c). An *edge waterflood* uses injection wells along the margin of the field. The injected water drives oil up and toward the producing wells in the center.

The waterflood usually becomes uneconomical and is abandoned when the produced water cut becomes too high. This can be greater than 99% water cut in some waterfloods. Waterfloods are most effective in solution gas drive reservoirs where there is relatively little primary production. Some waterfloods may take up to two years of water injection before any increase in production occurs. This *fill-up time* is caused by the time it takes gas bubbles in the pores of the reservoir to be compressed and redissolved in the remaining oil.

Plate 26–1. Water injection well, Burbank oil field waterflood, Oklahoma

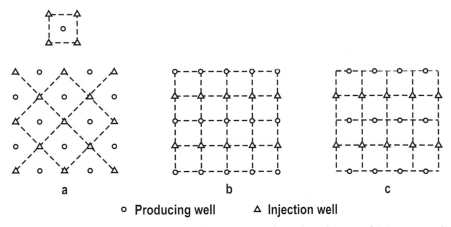

○ **Producing well**   △ **Injection well**

Fig. 26–2. Waterflood patterns: (a) five-spot, (b) direct line drive, and (c) staggered line drive

In many oil fields, the reservoir is not homogeneous, and the waterflood is not as efficient. Fluids such as water will always flow along the path of least resistance. A reservoir rock might have a zone of high permeability, such as a well-sorted bed of sandstone or a vugular or fractured zone in limestone. As the water sweeps through the reservoir, the injected water flows fastest through the most permeable zone (a *thief zone*) and reaches the producing well to cause a *breakthrough*. Once a breakthrough occurs, the rest of the water will tend to flow through that permeable zone *bypassing* oil in the less permeable portions of the reservoir. The sooner the water breaks through, the less efficient the waterflood.

Gravity also affects waterflooding. Because water is heavier than oil, water tends to flow furthest along the bottom of the reservoir because of *gravity segregation*. This leaves the oil relatively untouched in the top of the reservoir.

In one variation of a waterflood, heated water is injected to make the oil more fluid. The water can also be treated with *polymers* (long, chain-like, high-weight molecules) that increase the viscosity of the water. *Alkaline or caustic flooding* uses an alkaline chemical such as sodium hydroxide mixed with the injected water. The chemicals react with the oil in the reservoir to improve the amount of recovery.

Enhanced Oil Recovery

During *enhanced oil recovery* (*EOR*), substances that are not naturally found in the reservoir are injected into the reservoir. EOR includes thermal, chemical, and gas-miscible processes. It can be initiated after primary production or waterflooding.

Miscible gas drive

A *gas-miscible process* involves injecting a gas into the reservoir that dissolves in the oil. *Inert gas injection* uses either carbon dioxide (CO_2), nitrogen (N), or liquefied petroleum gas (LPG). The injected gas should not corrode metal equipment in the well, should not mix with natural gas in the reservoir to form an explosive combination, and should be relatively inexpensive.

During a *carbon dioxide flood*, carbon dioxide gas is usually brought to the project by pipeline from carbon dioxide wells or an industrial source. Large natural reservoirs of carbon dioxide gas occur in many areas such as the McElmo dome in Colorado and the Jackson dome in Mississippi. It is also available as a by-product of natural gas plants and from coal gasification.

When carbon dioxide is injected into the reservoir, it is usually *miscible* with the oil (dissolves in the oil), making the oil more fluid. The carbon dioxide gas then pushes the fluid oil through the reservoir toward producing wells. It can often recover about 35% of the remaining oil.

Because of the very low viscosity of the carbon dioxide, it tends to finger and break through to producing wells, leaving unswept areas in the reservoir. To prevent this, alternating volumes of water and gas can be injected into the reservoir in a *water-alternating-gas* (*WAG*) process. Carbon dioxide floods have produced 1.5 billion bbl (0.36 billion m³) of oil in the United States and it is predicted that they will produce an estimated additional 87.1 billion bbl (13.8 m³) of oil.

Nitrogen can also be injected into the oil reservoir in a process similar to a carbon dioxide flood. The largest oil field in Mexico, the Cantarell field in the Gulf of Mexico continental shelf, is under nitrogen injection. The field is named after a fisherman, Rodesindo Cantarell, who reported a natural oil seep in the Gulf to PEMEX, the Mexican national oil company, leading to the field's discovery. It produces 20 to 24 °API gravity oil primarily from carbonate breccia in anticline traps located in four fault blocks. The field was discovered in 1976 and produced an average of 1.16 million bbl of oil per day (180 thousand m³/d) during its first year in 1981. After the production declined, it was put on nitrogen flood in 2000 and reached peak production of 2.1 million bbl of oil per day (330,000 m³/d) in 2003. It has since been on a steep production decline. The nitrogen used in the flood comes from the largest nitrogen production plant in the world, which separates it out of air. Liquefied petroleum gas is also miscible with oil and is used in a *LPG drive*. The source of the LPG (propane or a propane-butane mixture) is usually wet gas.

Chemical flood

A *chemical flood* is a process in which different fluids are injected into the depleted reservoir in separate batches (*slugs*). The fluids, each serving a different purpose, move as separate *fronts* from the injection wells through the reservoir rock toward the producing wells (fig. 26–3). In a *micellar-polymer* flood, a slug of reservoir water is first injected to condition the reservoir as it moves ahead of other slugs of injected chemicals. Next, a slug of surfactant solution is injected into the reservoir. The *surfactant* acts as a detergent, reducing the surface tension of the oil and washing the oil out of the reservoir pore spaces. The oil forms small droplets suspended in the water called a *microemulsion*. The next slug is water-thickened by polymers. Pressure on the polymer water from the injection wells drives the surfactant and oil microemulsion front ahead through the reservoir rock toward producing wells (fig. 26–3). A chemical flood can be used only for sandstone reservoirs because carbonates absorb the surfactants. It can recover about 40% of the remaining oil but is an expensive process.

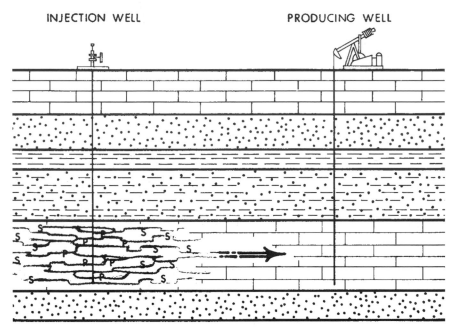

Fig. 26–3. Chemical flood: S is surfactant and P is polymer.

Thermal recovery

Thermal recovery techniques utilize heat to make heavy oil (<20 °API gravity) more fluid for recovery. *Cyclic steam injection* or the *huff 'n' puff* method uses single wells to inject steam into the heavy oil reservoir for a period of time such as two weeks during the *injection period* (fig. 26–4a). During the following *soak period*, the well is shut in for several days to allow the steam to heat the heavy oil and make it more fluid. The same well is then used to produce the heated heavy oil with a sucker-rod pump during the *production period* for a similar period of time to the injection period (fig. 26–4b). Steam injection and pumping are alternated for up to 20 cycles until it becomes ineffective.

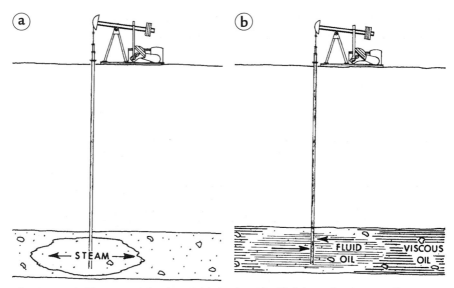

Fig. 26–4. Cyclic steam injection: (a) injection—huff, (b) production—puff

A *steamflood* or *steamdrive* uses both injection and production wells. Superheated steam is pumped down injection wells into a heavy oil reservoir. The steam heats the heavy oil to greatly reduce its viscosity. As the steam gives up its heat, it condenses into hot water that drives the oil toward producing wells. The pattern of injection and producing wells in a steamflood is similar to that of a waterflood, but the wells are very closely spaced. The recovery will vary between 25 to 65% of the oil in place.

A steamflood is being used in the Kern River field, Bakersfield, California (plate 26-2). The field was discovered in 1899 by digging a pit on the banks of the Kern River next to an oil seep. The oil is 12 to 16 °API gravity. The reservoir is 500 to 1,300 ft. (152 to 396 m) deep and consists of unconsolidated sands with 28 to 30% porosity and 1 to 5 darcys (D) permeability. Primary recovery was 15%, but with the steamflood, begun in the mid-1950s, the recovery will be 55%. Steamflooding is also being used on several Bolivar Coastal fields in Venezuela and in Alberta, Canada.

Steam-assisted gravity drainage (SAGD) is also used for heavy oil production. Two horizontal laterals are used (fig. 26-5). The horizontal laterals are parallel, with one lateral about 16 ft (5 m) directly above the other. Steam is continuously injected into the upper lateral. The heated heavy oil drains by gravity into the lower lateral. SAGD is commonly used in Alberta.

Plate 26–2. Kern River field steamflood, California

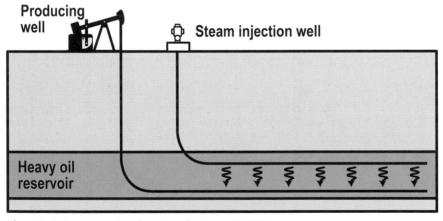

Fig. 26–5. Steam-assisted gravity drainage (SAGD)

A *fireflood* or *in-situ combustion* involves setting the subsurface oil on fire. If the well is shallow, the fire can be started with either a phosphorus bomb or a gas burner run into the well. Pumping air into the reservoir to start the fire by spontaneous combustion works in deeper reservoirs. Once the oil is burning, large volumes of air must be injected into the reservoir to sustain the fire. Pumping air is a large expense in a fireflood and increases with depth of the producing formation as more and bigger air compressors are required.

The fire generates heat, causing the oil to become more fluid. The large volume of hot gases generated by the fire drives the heated oil toward producing wells (fig. 26–6). A fireflood will fail if there is not enough oil in place to sustain the fire. The most common fireflood is *forward combustion*, in which the fire and injected air originate at the injection well. The oil flows toward the producing wells. In *dry combustion*, only air is injected. In *wet combustion* or *combination of forward combustion and waterflooding* (*COFCAW*), water and air are injected either together or alternately. The generated steam from water helps drive the oil.

The recovery from a fireflood can be 30 to 40% of the oil in place. Corrosion of equipment is a problem because of the high temperatures and corrosive gases that are generated. Time-lapse seismic methods can be used to trace the movement of the subsurface fire front.

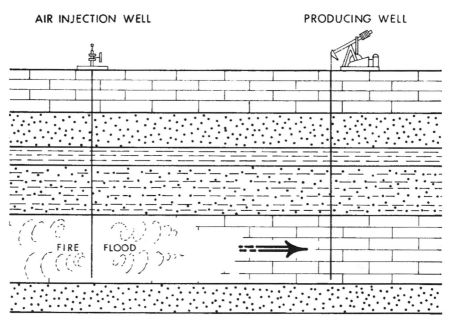

Fig. 26–6. Fireflood

Efficiency

The effectiveness of a waterflood or EOR project is described by sweep and displacement efficiencies. *Sweep efficiency* is a ratio of the pore volumes that are contacted by the injected fluid to the total reservoir pore volume. Both horizontal and vertical sweep efficiencies are computed. Sweep efficiency is strongly influenced by the *mobility ratio*, the ratio of the driving fluid viscosity to the viscosity of the oil being displaced. Mobility ratios close to one are most efficient. *Displacement efficiency* is the ratio of oil volume that is swept by the process to the oil volume in place before the process.

Unitization

When more than one company is operating in a field, the field can be *unitized* to coordinate a fieldwide effort to increase ultimate production. A *unit operator* is appointed to direct a pressure maintenance, waterflood, or EOR project in that field. The costs and production are shared, proportional to each member's acreage or reserve position in the field. Unitization can be either voluntary or forced by government decree.

The Prudhoe Bay oil field, Alaska, is unitized for pressure maintenance and waterflood. Produced gas and water, along with treated seawater, are being injected into the reservoir. The original reserve estimate for Prudhoe Bay was 9.6 billion bbl (1.5 billion m^3) of oil, but because of the pressure maintenance and waterflood, along with reservoir management, the reserve estimate is now 13 billion bbl (2.1 billion m^3) of oil.

Plug and Abandon

Both dry holes and producing wells, either onshore or offshore, that have been depleted must be properly *plugged and abandoned* (P & A). The procedure is required by law to prevent saltwater from polluting fresh groundwater reservoirs and involves cementing the borehole. First, if possible, casing is cut and pulled for salvage. All depleted producing formations are sealed by placing cement plugs at those levels (fig. 26–7). Near-surface freshwater reservoirs are also protected by cement. A mechanical plug is used to bridge the wellbore at a specific depth to control the cement level. The upper portion of the well adjacent to the freshwater reservoirs is then cemented.

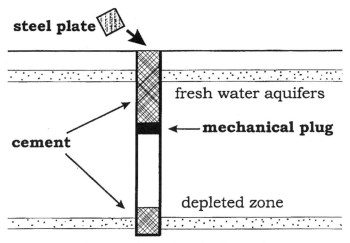

Fig. 26–7. A plugged and abandoned well (P & A)

The job can be as simple as cementing the upper 100 ft (30.5 m) of the well. A more complex job also might involve cementing above and below all high-pressure and all permeable zones in the well. The casing is cut 6 ft (2 m) below the surface, and a steel plate is welded to the top of the casing. The surface hole is then filled with dirt, and a marker is installed.

Offshore wells are plugged and abandoned just like land wells. All the subsea equipment is retrieved. Any part of the well that sticks above the mudline, such as casing, is cut off to not leave a navigational hazard. On an abandoned offshore production platform, usually only the deck equipment, such as the modules, is salvaged. The jacket can be either tipped over to lie on the ocean bottom or cut off below sea level to prevent an obstruction to navigation. The abandoned structure on the ocean bottom makes an excellent fish reef.

Unconventional Oil and Gas

Unconventional crude oil can be immature or degraded crude oil, or it can be conventional oil that occurs in a reservoir of very low permeability that needs to be stimulated for the oil to flow. Unconventional natural gas can occur in a reservoir that is not a typical porous and permeable reservoir rock. It can occur in coal, very-low-permeability rocks, and ice. These deposits need specialized methods for production and can be more expensive to produce than conventional deposits.

The number of oil and gas discoveries had been increasing for several decades until the mid-1980s, when it started to decrease along with the decreasing amount of reserves found each year. Unconventional oil and gas will become more important with time to replace conventional oil and gas.

Tight Formations

Tight formations are very-low-permeability reservoirs. They can sometimes be produced by horizontal wells that intersect vertical natural fractures (see chap. 17, fig. 17–11) and hydraulic fracturing (fracing) that open existing fractures and create new fractures.

Gas shale

A *gas shale* is a black shale source rock that has generated gas and oil, but not all the gas (*shale gas*) has been expelled from the rock. It is a combination source rock, reservoir rock, and seal. The gas can range from methane to wet gas with significant amounts of condensate. The gas is contained in natural vertical fractures (joints), in pores spaces, and adsorbed onto organic matter. Gas shale is produced with horizontal wells that intercept naturally occurring vertical joints and hydraulic fracturing to open and enlarge the joints. Although a massive frac job is effective on a gas shale, it is too expensive to make the well economic.

A special *slickwater frac* is used that costs about 30% of a massive frac job. Two of the biggest expenses on a massive frac job are pumping the thick frac fluid and the proppants. A slickwater frac pumps water with a friction reducer additive as the frac fluid. This greatly reduces the necessary pumps. The frac fluid also uses a very low concentration of special proppants that have a low density. Slickwater fracs form relatively long but narrow fractures. A horizontal gas shale well will typically have a significantly high initial production rate than a comparable vertical well. Typical initial production rates (IPs) in most gas shales such as the Barnett, Eagle Ford, and Marcellus are 2 to 3 MMcf/d, whereas the Haynesville averages 5 to 6 MMcf/d. The ultimate recovery for a horizontal well, however, will be only slightly larger than that for a vertical gas shale well.

Because the horizontal sections (laterals) are 3,000 ft (914 m) to more than 10,000 ft (3,048 m) long, hydraulic fracturing must be done in sections called *stages* (fig. 27–1). Each stage is typically 500 ft (152 m) or 1,000 ft (305 m) long. Twelve frac stages are common. The first stage is at the toe end of the lateral. A plug is used to isolate that stage and it is fraced, and another plug is then used to isolate the next stage closer to the heal and it is fraced. A gas shale well can be fraced several times during the life of the well. The recovery factor is estimated to be about 15 to 35% of the gas in place.

There are very few dry holes drilled into gas shale. The infrequent dry hole is commonly caused by mechanical failure during drilling. Some gas shale wells, however, produce at a significantly higher rate than others primarily due to concentrations of natural fractures. Hydraulic fracturing opens up the existing fracture network and fills them with proppants. The most economic gas shale wells produce gas rich in natural gas liquids. Areas near a natural fault are not good for shale gas production because some of the gas has escaped along the fault, and during hydraulic fracturing the fault takes most of the frac fluid.

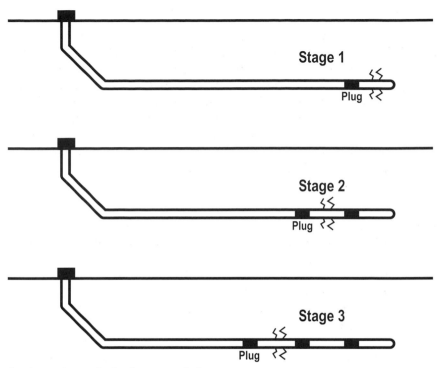

Fig. 27–1. Stages in fracing a gas shale

An environmental concern is the water used during hydraulic fracturing. Millions of gallons of water can be used during one frac job. The *flow back* is frac water and possible formation water that flow to the surface after the frac job. This water has to be treated and then reinjected or disposed of. There is also concern that hydraulic fracturing can penetrate and pollute freshwater reservoirs.

There are a very large number of gas shales in the United States (fig. 27–2), Canada, and throughout the world. The first gas shale well in the United States was dug in 1821 in Fredonia, New York, to 27 ft (8 m). The gas was used for street lights. Those and all other shale gas wells were marginally economic until the early years of this century, when the Barnett Shale of the Fort Worth basin in Texas was exploited using modern technology. Gas reserves in the United States were rapidly declining until gas shales were recently developed. The Energy Information Service estimated that the United States has at least 827 Tcf (765 Bm³) of technically recoverable shale gas, more than enough for 100 years at the present rate of consumption.

Fig. 27-2. Location of U.S. gas shales. (Modified from the Energy Information Administration, http://www.eia.gov/oil_gas/rpd/shale_gas.jpg, 2011.)

The best gas shales (1) are relatively shallow so drilling costs are low, (2) have a high organic carbon content, (3) have a high concentration of natural fractures, (4) have normal or high pressures, (5) are rich in natural gas liquids, (6) are thick, and (7) are brittle. A substance that is *brittle* means that it will break instead bending (elastic). A brittle shale responds to hydraulic fracturing better than an elastic shale. Shales with some silica or calcium carbonate are more brittle than shales with more clay minerals. The brittleness in the Barnett Shale comes from its relatively high content of siliceous microfossils such as diatoms. It has a silica content of 40 to 60% and a clay content of 20 to 30%.

The Mississippian age Barnett Shale of the Fort Worth basin in Texas ranges from 6,000 to 10,000 ft (1,830 to 3,050 m) deep. The shale contains from 50 to 300 Bcf of gas in place per section (one square mile). The basin will produce an estimated 32 to 45 Tcf (0.9 to 1.3 Tm^3) of shale gas. Several wells in the Barnett Shale produce so much condensate that they are classified by the state as oil wells.

The Cretaceous age Eagle Ford Shale of south Texas is a very productive gas shale that averages 250 ft (76 m) thick. The shale is brittle because of its calcite content. It underlies the Austin Chalk and outcrops in the San Antonio area. It dips to the southeast where it produces at depths of 4,000 to 14,000 ft (1,220 to 4,270 m). In the deeper part to the southeast, it produces dry gas (fig. 27-3). To the northwest, it produces wet gas and condensate. The natural gas liquids content of the wet gas production greatly enhances the gas shale economics. In the shallow updip portion, it produces crude oil that occurs primarily in natural fractures. The oil recovery is expected to be only 5%.

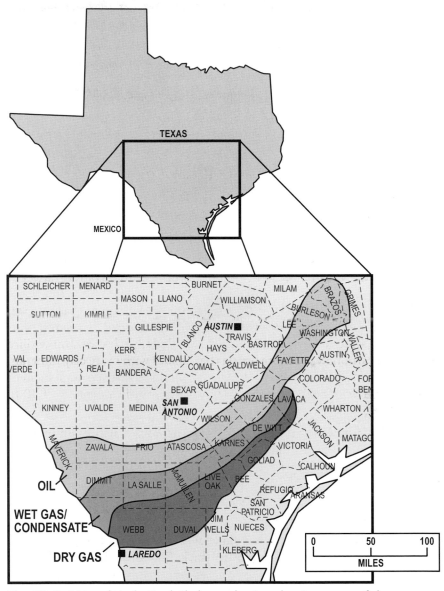

Fig. 27–3. Map of Eagle Ford Shale production showing areas of dry gas, wet gas/condensate, and oil production. (Modified from the Energy Information Administration, http://www.eia.gov/oil_gas/rpd/shaleusa9.pdf, 2010.)

The Devonian age Marcellus formation gas shale underlies a large area of Pennsylvania and New York and is very close to large, East Coast natural gas markets. Tens of thousands of wells had been drilled through the Marcellus Shale to deeper targets until a slickwater frac was first performed on it in 2004, proving it to be an economical gas shale. The shale produces primarily dry gas in north-central Pennsylvania where it is about 250 ft (76 m) thick. The gas is rich in natural gas liquids and is overpressured in southwest Pennsylvania, where it is about 100 ft (30 m) thick. It contains 20 to 100 Bcf of gas in place per section (640 acres), less than that of the Barnett Shale. The Marcellus Shale is rich in total organic matter where a significant amount of the porosity is located.

Tight gas sands

Tight gas sands are very-low-permeability (<0.1 md) sandstone gas reservoirs. They occur in large, continuous deposits without any conventional trapping mechanism such as anticlines. A tight sand gas well is best developed with a horizontal well and hydraulic fracturing. It is estimated that the United States has about 300 Tcf of technically recoverable gas in tight gas sands (fig. 27–4).

Fig. 27–4. United States tight gas sand map. (Modified from the Energy Information Administration, http://www.eia.doe.gov/oil_gas/rpd/tight_gas.pdf, 2010.)

The granite wash reservoirs of the Northern Texas panhandle and western Oklahoma were formed when the Wichita-Amarillo uplift exposed granite to weathering during the Pennsylvanian period. They cover 2,000 square miles (5,180 km^2) at a depth of 11,000 to 17,000 ft (3,350 to 5,180 m). The granite wash occurs in 50 to 70 different layers that range in composition from sandstone to conglomerate with boulders. Although tight, the granite wash has higher porosities and permeabilities than gas shales. The advantages of granite wash wells are that they have a very high natural gas liquids content, high initial production rates, and a relatively lower decline rate per year (50 to 60%) compared to shale gas wells. Some wells produce over 1,000 bbl (159 m^3) of natural gas liquids per day. Although the granite wash was first drilled in 1954, it was not until horizontal well and multiple-stage slickwater fracing was applied that it became economic.

Bakken Formation

The late Devonian to Early Mississippian age Bakken Formation underlies the middle and central portions of the Williston basin in North Dakota and Montana and the provinces of Saskatchewan and Manitoba (fig. 27–5). It consists of (1) an upper organic-rich black shale; (2) a middle sandstone, dolomite, and limestone reservoir rock; and (3) a lower organic-rich black shale. Both the upper and lower shales are proven source rocks. The reservoir rock is estimated to contain 3 to 4 billion bbl of light (41 °API gravity), sweet recoverable oil along with 1.8 Tcf (51 B m^3) of gas and 148 MM bbl (23.5 MM m^3) of natural gas liquids. It averages 40 ft (13 m) in thickness, has an average porosity of 5%, and has a very low permeability that averages 0.04 md. Depth to the Bakken Formation varies from a few thousand to over 10,000 ft (3,050 m) deep in the center of the basin.

Fig. 27–5. Map of Bakken Formation

The key to Bakken production is the vertical and subvertical fractures in the reservoir rock. These are exploited with horizontal wells and hydraulic fracturing. The first field in the Bakken Formation was the Elm Coulee oil field, discovered in 1997. It will eventually produce 200 MM bbl (32 MM m³) of oil.

Coal Bed or Coal Seam Gas

Wells are often drilled into coal to produce *coal bed* or *coal seam gas*, which is pure methane gas. The methane gas was produced during the natural conversion of woody organic matter into coal. The gas occurs adsorbed to the surface of the coal along open natural fractures called *cleats*. There are two sets of cleats, face and butt, that are perpendicular to each other and are at right angles to the coal bedding (fig. 27–6). A face cleat is continuous, and butt cleat is not. Because of the large internal surface area of the coal along the cleats, the coal bed contains six to seven times more gas than an equivalent volume of a conventional gas reservoir. Coal itself has very low permeability, but cleats give the coal permeability.

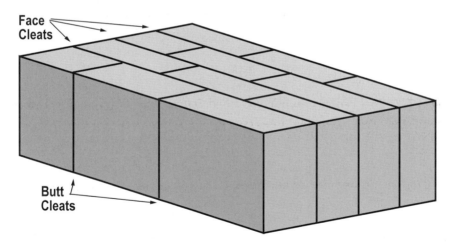

Fig. 27–6. Coal cleat terminology

Methane gas will not be released from the coal face because of the pressure on it. A coal seam gas well first produces water for a period of time, up to a year, as the fractures drain in a process called *dewatering*. As the pressure in the coal cleat decreases, methane gas is released (*desorbed*) from the surface of the coal and flows into the well. The United States is estimated to have 270 Tcf of recoverable coal bed gas. Most of the present coal bed gas production in the United States is located in New Mexico, Colorado, and Alabama (fig. 27–7).

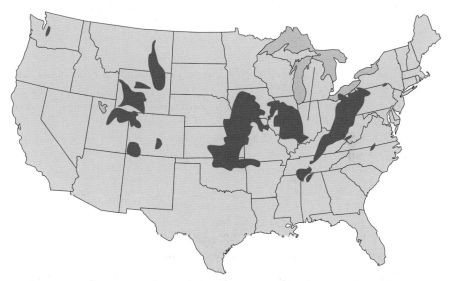

Fig. 27–7. Coal bed gas deposits in United States. (Modified from The Energy Information Administration, http://en.openei.org/wiki/File:EIA-coalbed-gas.pdf, 2009.)

The Cretaceous age Fruitland Formation coal of the San Juan basin in New Mexico and Colorado has 50 Tcf (1.4 Tm^3) of coal bed gas in place, with recoverable reserves of 7.8 Tcf (223 Bcf m^3). It is up to 70 ft (21 m) thick. The produced gas also contains 3 to 13% CO_2. The wells are typically completed open-hole with a cavity that has been enlarged by an underreamer tool (*open-hole cavity* or *cavity completion*) (fig. 27–8). After dewatering, the gas production typically peaks at 1 to 6 million cf/d (28 to 170 million m^3/d) per well and then declines to less than 5% of that a year.

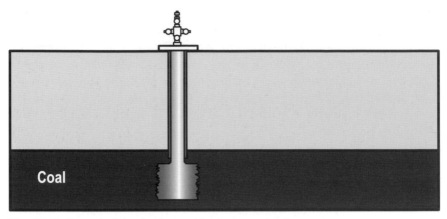

Fig. 27–8. Open-hole cavity for coal bed gas production

Oil Shales

Oil shales are sedimentary rocks rich in a type of organic matter called *kerogen*. When the oil shale is heated to about 660°F (350°C), kerogen is transformed into a type of crude oil called *shale oil*. Oil shales are organic-rich source rocks that are old enough but have never been buried deeply enough for heat to transform the organic matter into oil. They are immature source rocks. The inorganic sediments are commonly clay, fine-grained quartz and calcite, and salts. Oil shales occur in 100 major deposits in 27 countries including Australia, Brazil, and Russia and are estimated to contain over 10 trillion barrels (1.59 trillion m^3) of oil in place. The largest and richest deposits are in the western United States. These high-grade oil shales yield more than 25 bbl (4 m^3) of oil per ton of shale.

The oil shales in the Eocene age Green River Formation of the Rocky Mountains (fig. 27-9) are actually calcareous muds and salts that were deposited in a lake. Algae are a major source of the organic matter. In the Piceance basin of northwestern Colorado, the Green River Formation is estimated to hold about 600 billion bbl (95 billion m^3) of oil in high-grade oil shale. Large oil shale deposits are also located in the Green River and Washakie basins of Wyoming and the Uinta basin of Utah. Because the oil shale crops out, it can be surface-mined.

Another production method is to dig a large cavern in the oil shale. The oil shale on the sides of the cavern is set on fire for heat to generate oil in the subsurface. High production cost is a major factor preventing the commercial development of oil shales.

Fig. 27–9. United States oil shales. (Modified from J. R. Dyni, 2006.)

Tar Sands

Tar sands are composed of very heavy oil (*bitumen*) that is 8 to 14 °API gravity and is mixed water and sediments such as sand. The oil is too viscous to be produced by conventional methods. Very large tar sand deposits occur in northern Alberta and Venezuela. They are estimated to contain a total of 4 trillion barrels (636 billion m³) of oil in place, significantly more than the world's conventional oil reserves. It is thought that the tar originated as good quality crude oil that was degraded on the surface during the geological past into tar.

The largest of three tar sand deposits in Alberta is found in the Athabaska tar sands (fig. 27–10), which have been estimated to contain 173 billion bbl (27.5 billion m³) of recoverable bitumen from the shallowest 10% of the entire deposit. The tar sands occur along an angular unconformity (fig. 27–11) and are thought to have formed from the same high-quality oil located in the Devonian reef fields of Alberta (fig. 10–28). The oil, however, seeped on the surface during the Cretaceous time and was degraded into tar.

The surface deposits are typically 130 to 200 ft (40 to 60 m) thick and are composed of 10% bitumen, 4% water, and 86% solid sediments. They are being mined with large steam shovels and treated with hot water and caustic soda. Between 90 to 100% of the bitumen floats to the top and is separated. The bitumen is then treated with refining processes to produce *synthetic crude oil* that is about 32 °API gravity with very low sulfur (<0.2%). Two tons of oil sands are needed to make one barrel of synthetic crude oil.

Steam-assisted gravity drainage (SAGD) and cyclic steam injection are also used in Athabaska tar sand production.

Fig. 27–10. Location map of tar sands in Alberta

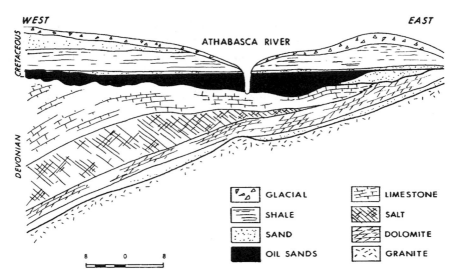

Fig. 27–11. Cross section of Athabaska tar sands, Alberta. (Modified from Jardine, 1974.)

Gas Hydrates

Gas or *clathrate hydrate* is ice with gas, primarily methane, physically trapped in the cagelike ice crystals. It has the appearance of snow or ice. The gas is very densely packed in the ice crystal. One volume of gas hydrate can be melted to yield 168 volumes of methane gas measured at standard temperature and pressure. Gas hydrates occur both in deep (>1,000 ft or 305 m) ocean sediments and in permafrost. Permafrost is permanently frozen soil that is common in Arctic areas. It is difficult to accurately calculate the amount of gas trapped in gas clathrates, but it is in the order of hundreds of thousands of Tcf of gas, far more than conventional natural gas reservoir reserves.

When drilling for conventional oil and gas offshore, the well sometimes drills through gas hydrates. The heat caused by drilling can cause the gas hydrates to break down and the well can sluff in. Gas hydrate zones are cased as soon as they are drilled to prevent this.

Similar gas hydrates sometimes form during conventional natural gas production as the gas expands and cools. These hydrates can restrict or block flowlines.

The greatest challenge is how to economically produce the gas from gas hydrates. The gas hydrates in permafrost, located on land in Arctic areas, have the most promise because they will be the least expensive to exploit. Wells drilled into the gas hydrate can cause a decrease in pressure that can cause the gas hydrate to break up and the gas to be liberated.

Glossary

A. (1) ampere and (2) area.

AA. (1) after acidizing and (2) as above.

A/. acidized with.

a. acres.

aa. as above.

AADE. American Association of Drilling Engineers (www.aade.org).

AAPG. American Association of Petroleum Geologists (www.aapg.org).

AAPL. (1) American Association of Petroleum Landmen (www.aapl.org). (2) American Association of Professional Landmen (www.landman.org).

ab. above.

abandonment pressure. The gas pressure at which a gas well must be abandoned as the produced gas pressure decreases. It is often pipeline pressure at 1,000 psi.

abd. (1) abandon, (2) abandoned, and (3) abundant.

abd-gw. abandoned gas well.

abd-loc. abandoned location.

abdogw. abandoned oil and gas well.

abd-ow. abandoned oil well.

abdt or **abnd.** abundant.

abnormal high pressure. Pressure in a subsurface reservoir that is higher than expected from normal hydrostatic pressure at that depth; cf. normal pressure. (geopressure and overpressure)

absolute open flow. The maximum rate that a gas well can produce at zero bottom-hole pressure. (maximum potential flow) (AOF)

absorption tower. A vertical, steel vessel where natural gas bubbles up through a light-hydrocarbon liquid that removes natural gas liquids.

abst or **abstr.** abstract.

abstract or **abstract of title.** A record of the ownership and all the transfers of ownership of a tract of land. It is used in the title examination for a parcel of land and is made by a landman (abstractor) or abstract company to establish a clear title to that tract of land. (abst or abstr)

abt. about.

abun. abundant.

abv. above.

AC. alternating current.

ac. (1) acid, (2) acidizing, (3) acre, and (4) acreage.

accelerator. An additive that increases the rate of a process such as cement setting; cf. retarder.

accum. accumulative.

accumulator. Skid-mounted steel cylinders that contain hydraulic fluid under pressure. They are located next to a drilling rig and are used to operate the rams on the blowout preventers.

acd or **acdz.** (1) acidize and (2) acidized.

ac-ft. acre-feet.

acid gas. A gas such as hydrogen sulfide or carbon dioxide that forms an acid with water. It can cause corrosion of metal equipment.

acidizing or **acid job.** A well stimulation technique used primarily on limestone reservoirs. Acid is poured or pumped down the well to dissolve the limestone and increase fluid flow. Hydrochloric acid or mud acid, a mixture of hydrochloric and hydrofluoric acids, is commonly used. During matrix acidizing, the acid dissolves the reservoir. During acid fracturing, acid is pumped down the well under pressure to fracture the reservoir.

acoustic impedance. Sound velocity times density of a rock layer.

acoustic velocity log. A wireline well log that measures sound velocity through the rocks in microseconds per foot (µsec/ft). The porosity of the rock can be calculated from the sound velocity of the rock. (sonic log) (AVL)

acre. An area of 43,560 square ft. There are 640 acres in a square mile (section). A square that is 209 ft on a side is an acre. (a or ac)

acreage. Leased land.

acreage contribution agreement. A type of well support agreement in which a nondrilling party will transfer a lease or leases to a drilling party that drills a well at a specific location. Information from that well will be shared with the nondrilling party; cf. bottom-hole and dry-hole contribution agreements.

acre-foot. The volume formed by a surface area of 1 acre that is 1 foot deep. It can hold 7,758 barrels of oil. Acre-feet are used to describe oil reservoir volume and to calculate reserves. (ac-ft)

acrg. acreage.

ACS. American Chemical Society (www.acs.org).

ACT. automatic custody transfer.

AD. authorized depth.

add. additive.

ADDC. Association of Desk and Derrick Clubs (www.addc.org).

additive. A substance that is added to a fluid to cause an effect. Drilling mud additives include (a) thinners and thickeners, (b) weighting material, (c) friction reducers, and (d) clay stabilizers. A common cement additive is an accelerator. (add)

addn. addition.

adj. adjustable.

adpt. adapter.

adspn. adsorption.

AE. asphalt emulsion.

aeromagnetic. An exploration method that uses a magnetometer mounted in a stinger that protrudes out the back of an airplane to measure the strength of the earth's magnetic field. The plane is flown at a constant height. Over rugged terrain, a helicopter can be used.

AESC. Association of Energy Service Companies (www.aesc.net).

AF. after fracturing (hydraulic).

AFE. authority or authorization for expenditure.

AFP. average flowing pressure.

AGA. American Gas Association (www.aga.org).

age. A time subdivision of epochs (e.g., Maastrichtian age); cf. era and period.

AIChE. American Institute of Chemical Engineers (www.aiche.org).

AIME. American Institute of Mining, Metallurgical and Petroleum Engineers (www.aimeny.org).

AIR. average injection rate.

air-balanced beam-pumping unit. An oil well beam-pumping unit that uses a piston that moves up and down in a compressed air cylinder to offset the weight of the sucker-rod string.

air drilling. Rotary drilling with air pumped down the drillstring instead of circulating drilling mud. (pneumatic drilling)

air gun. A common seismic source used in the ocean. It is a metal cylinder that is towed in the water behind a ship. The air gun is continuously filled with high-pressure air from the ship. A high-pressure air bubble is periodically released into the water from the air gun for the energy impulse. Several air guns of different sizes are often used at the same time in an air gun array.

Albian. An age of geological time from 112 to 99.6 million years ago. It is part of the Cretaceous period.

algorithm. The precise procedure for a numerical or algebraic procedure. Algorithms are used to program computers for seismic data processing.

alk. alkalinity.

alkaline flood. An improved oil recovery method that uses alkaline chemicals in the injection water to increase oil recovery. (caustic flood)

allow. Allowable.

allowable. The amount of gas or oil that a regulatory agency permits a well, lease, or field to produce during a period of time such as a month. (allow)

alt. Alternate.

alternating current. An electric current that periodically reverses its direction. In the United States, alternating current makes 60 complete cycles per second (Hertz). Alternating current is made by an electrical generator and is used in homes and offices for lighting and appliances; cf. direct current. (AC)

amb. Ambient.

ambient. The temperature and pressure of the surrounding environment. (amb)

American Petroleum Institute. A national trade organization for the US oil and natural gas industry located in Washington, DC. Its mission is to influence public policy for a strong, viable US oil and natural gas industry. The API conducts and sponsors petroleum industry research and maintains and publishes over 500 equipment, operating, and safety standards. It collects, maintains, and furnishes statistical information on drilling activity, drilling results, and well completions (www.api.org). (API)

amine unit. Natural gas processing equipment that uses organic bases (amines) to absorb and remove hydrogen sulfide and carbon dioxide.

amor. amorphous.

amorphous. Structureless, without crystals. (amor)

amplitude anomaly. A brightening (bright spot) or dimming (dim spot) of a seismic reflector over a local area.

amp. ampere.

amplitude-versus-offset analysis. A seismic method in which the amplitude of a reflector is compared at different offsets (source to detector distances). It is used to locate gas reservoirs and identify carbonate rocks. (AVO analysis)

amt. amount.

AN. anhydrite.

anchor hole. The lowest portion of a well. It is located below the pay zone and is used to accommodate equipment such as a logging tool. (rat hole)

ang. angular.

angular unconformity. An ancient erosional surface with the sedimentary rock layers below the unconformity tilted at an angle to those above.

anhy. anhydrite.

anhydrite. A salt mineral composed of $CaSO_4$; cf. gypsum. (anhy) (AN)

ann. annulus.

annular preventer. A cylinder at the top of a blowout-preventer stack containing rubber with steel ribs. Pistons compress the rubber to close around any size or shape of pipe in the well.

annulus. The space between two concentric cylinders such as between the tubing and casing strings. (ann)

anoxic basin. A basin in which the bottom waters lack oxygen, and organic matter can be preserved.

Ant. antonym. (opposite)

anticline. A large, long, upward fold of sedimentary rocks. A circular uplift is a dome. An anticline can trap petroleum; cf. syncline.

antifoam. An additive used to reduce foam.

anti-whirl drilling bit. *See* walk to the right.

AOF. absolute open flow. (gas well)

AOFP. absolute open flow potential.

API. American Petroleum Institute. (www.api.org)

API gravity. *See* °API gravity.

API unit. A unit of radioactivity used with natural gamma ray logs.

API well number. A 10- or 12-digit number assigned to every well drilled in the United States. Digits 1 and 2 are state codes, digits 3 through 5 are for county, parish, or offshore, digits 6 through 10 identify the well, and digits 11 and 12 record a well property such as sidetracking.

app. appears.

appraisal well. A well drilled out from a discovery well to determine the extent of a new field. (step out or delineation well)

appx. (1) approximate and (2) approximately.

Aptian. An age of geological time from 121 to 112 million years ago. It is part of the Cretaceous period.

aquifer. (1) A water-bearing rock and (2) a permeable rock.

aquitard. A rock through which fluids cannot pass. (impermeable rock); cf. reservoir rock.

arch. A long uplift in rocks.

aren. arenaceous.

arenaceous. sandy; cf. argillaceous. (aren)

arg. argillaceous.

argillaceous. shaly; cf. arenaceous. (arg)

Ark or **ark.** (1) arkose and (2) arkosic.

arkose and **arkosic sandstone.** Sandstone derived from the weathering of granite. (granite wash) (Ark and ark)

arnd. around.

ARO. at rate of.

array. Several geophones connected to a single channel to record as a unit. Arrays are described by their geometry such as in-line, perpendicular, cross, and diamond. Several arrays make a spread. A source array is several seismic sources fired at the same time.

artesian well. A well in which water flows to the surface under its own pressure.

artificial lift. A system used to lift crude oil and water up the tubing string in a well that will not flow by itself (e.g., sucker-rod pump, electric submergible pump, gas lift, and hydraulic pump).

AS. after shot.

ASAP. as soon as possible.

ASD. abandoned well, salvage deferred.

ASGMT. assignment.

ASP. abandoned well, salvaged and plugged.

Asph or **asph.** asphalt.

asphalt. A brown to black, solid composed of high-molecular weight, hydrocarbon molecules. (Asph, asph, and aspt)

asphalt-based crude oil. A refiner's term for crude oil that contains little paraffin wax and has a residue of asphalt. It will yield a relatively high percentage of high-grade gasoline and asphalt when refined; cf. mixed-based crude oil and paraffin-based crude oil.

aspt. asphalt.

assgd. assigned.

assgmt. assignment.

assoc. associated.

associated gas. Natural gas that is either dissolved in crude oil or is in contact with crude oil in the reservoir. It is produced during crude oil production; cf. nonassociated gas.

assy. assembly.

ASTM. American Society for Testing and Materials. (www.astm.org)

astn. asphaltic stain.

AT. after treatment.

Athabaska oil or **tar sands.** A very large deposit of very heavy oil, water, and sand that crops out on the surface of northern Alberta in Canada. It is mined by large steam shovels and heated to separate the heavy oil, sand, and water. The heavy oil is treated by a refinery process called cracking to produce synthetic crude oil that is similar to diesel oil.

at wt. atomic weight.

atm. (1) atmosphere and (2) atmospheric.

atoll. A circular or elliptical reef with a central lagoon.

att. (1) attempt and (2) attempted.

aulacogen. A long, narrow rift in a continent, often filled with thick sediments.

auth. authorized.

authigenic. A mineral that was formed by a chemical reaction in the subsurface; cf. detrital. (AUTHG)

authority or **authorization for expenditure.** A cost estimate of doing something. The AFE of drilling a well is estimated both as a dry hole and as a completed well. (AFE)

auto. automatic.

autotracking. A computer process that traces an individual seismic reflector through a cube of 3-D seismic data to make a horizon slice of that seismic horizon.

av and **avg.** average.

AVC. automatic volume control.

AVL. acoustic velocity log.

AVO. amplitude versus offset.

AVT. apparent vertical thickness.

AW. acid water.

awtg. awaiting.

aux. auxiliary.

avail. available.

avg. average.

axis. The center of a fold.

az. azimuth.

azimuth. The direction of a horizontal line usually measured in degrees clockwise from true or magnetic north; e.g., North 70° East; cf. dip. (az)

B and **b.** barrel.

B. formation volume factor.

B/. (1) base and (2) base of.

BA. barrels of acid.

backflush. To pump an injected fluid back out of a well. The frac fluid is backflushed after a frac job.

back-in interest. A working interest in a well that becomes effective at a future time.

back-in unit. A service unit used for well workover that must be backed up to the well.

back off. To unscrew threaded pipe.

back off operation. A method used to remove stuck pipe from a well. A string shot located above the stuck point on the drillstring is exploded as torque is applied to unscrew the pipe. A fishing tool is then used to retrieve the pipe left in the well.

backpressure test. A gas well test that measures pressures at different flow rates to calculate well deliverability.

bacterial degradation. The removal of lighter, shorter, hydrocarbon molecules from crude oil by bacteria. It leaves tar and asphalt.

bail. To remove liquid from a well. (blg)

bald-headed anticline or **structure.** A buried anticline with production along the flanks but no production on the crest. Erosion removed all the potential reservoir rocks from the crest. (scalped anticline)

barefoot completion. A well with casing run and cemented down to the top of the reservoir rock. The reservoir rock is left uncased; cf. set-through completion. (open-hole and top-set completion)

bar finger sand. A long, narrow sand body deposited as a distributary mouth bar on a prograding delta.

barite. A mineral composed of $BaSO_4$. It is used as an additive to make drilling mud heavier.

barrel. The English system measure of crude oil volume. A barrel contains 42 U.S. gallons and is equivalent to 0.159 cubic meter; 7.5 barrels of average weight oil weighs about one metric ton. (B, b, and bbl)

barrels-of-oil equivalent. The amount of natural gas that has the same heat content as an average barrel of oil. It is about 6,000; cf of gas. (oil equivalent gas or energy equivalent barrels) (BOE)

Bas and **bas.** basalt.

basal conglomerate. A soil zone located on an unconformity.

basalt. The most common volcanic rock. It is very fine-grained and dark in color. Basalt is called scoria when it contains gas bubbles. (Bas and bas)

base map. A map that shows the location of (1) wells that have been drilled or (2) seismic lines and shot points.

basement rock. Unproductive rocks for petroleum that underlie sedimentary rocks. Basement rock is usually igneous and/or metamorphic rock. Granite is a common basement rock. (Bm and bsmt)

base plate. *See* pad.

basic sediment and water. The solid and water impurities in crude oil. (BS&W)

basin. A large area with a relatively thick accumulation of sedimentary rocks (10,000 to 50,000 ft). The deep part of the basin where the crude oil and natural gas forms is called the kitchen. The shallow area surrounding the deep basin is called the shelf. A basin can be partially filled with sediments, such as the Gulf of Mexico, or completely filled with sediments, such as the Anadarko basin of Oklahoma.

BAT. before acid treatment.

bat. battery.

batch. A single treatment in contrast to continuous.

batholith. A large irregular subsurface intrusion of igneous rock.

battery. *See* tank battery.

BAW. barrels of acid water.

BB. bridged back.

bbl. barrel.

bbl/D. barrels per day.

bbl/min. barrels per minute.

BC. (1) barrels of condensate and (2) bottom choke.

Bcf. billion cubic feet.

Bcf/D. billion cubic feet per day.

BCPMM. barrels of condensate per million cubic feet of gas.

BD or **B/D.** barrels per day.

bd. bed.

bdd. bedded.

BDA. breakdown acid.

BDF. below derrick floor.

bdg. bedding.

BDO. barrels of diesel oil.

BDP. breakdown pressure.

Bdst. boundstone.

BDT. blow-down test.

beam-pumping unit. An oil well rod pumping unit with a walking beam that pivots on a Samson post. A sucker-rod string connects the walking beam to a downhole pump on the bottom of the tubing string. The prime mover is commonly an electrical motor. *See also* rod pumping system.

bedding. layers in sedimentary rocks. (bdg)

behind the pipe. Crude oil or natural gas located in the rock behind the casing in a well. It is not currently being produced.

benchmark crude oil. A crude oil used as a standard for comparing the properties of other crude oils and to set prices. West Texas Intermediate is the benchmark for the United States. Brent is used internationally. Dubai is used in the Middle East.

bent. (1) bentonite and (2) bentonitic.

bentonite. The commercial name for clay used to make common water-based drilling mud. Bentonite swells and forms a gel when exposed to freshwater. It comes in dry sacks. (bent)

bent housing a metal housing for a downhole mud or turbine motor. It has a fixed or adjustable bend of 1° to 3° behind the motor that serves as a bent sub.

bent sub. A short section of pipe with a small angle machined into it. The angle on some bent subs can be adjusted from the surface. It is used on a steerable downhole assembly.

BF. barrels of fluid.

BFH. barrels of fluid per hour.

BFMW. barrels of formation water.

BFO. barrels of frac oil.

BFPD. barrels of fluid per day.

BFW. barrels of formation water.

BFWTR. barrels of freshwater.

BG. background gas.

BGC. barrels of gas condensate.

BHA. bottom-hole assembly.

BHC. bottom-hole choke.

BHCP. bottom-hole circulating pressure.

BHCS. borehole compensated.

BHCT. bottom-hole circulating temperature.

BHFP. bottom-hole flowing pressure.

BHFT. bottom-hole flowing temperature.

BHL. bottom-hole location.

BHP. bottom-hole pressure.

bhp. brake horsepower.

BHSIP. bottom-hole shut-in pressure.

BHSP. bottom-hole static pressure.

BHST. bottom-hole static temperature.

BHT. bottom-hole temperature.

bid round. *See* license round.

bicenter and **bi-center bit.** A drilling bit with both a pilot bit on the bottom and a reamer on one side. It is designed to drill and ream a larger diameter hole than the inner diameter of the casing through which it passes.

billing interest. *See* working interest.

bin. The area (square or rectangular) in which all seismic reflection midpoints that fall within it are used to make a common midpoint gather for a 3-D seismic survey.

bio. biotite.

bioclastic. composed of shell fragments. (biocl)

biofacies. *See* facies.

biogenic gas. Methane gas produced by bacterial on organic matter at relatively shallow depths; cf. thermogenic gas. (microbial gas)

biostratigraphy. The use of microfossils to study and identify sedimentary rocks.

Biot. biotite.

biotite. A common mineral formed of thin black sheets; cf. muscovite. (black mica) (bio and biot)

bioturbation. The disturbance and mixing of sediments by burrowing animals and plant roots.

birdfoot delta. A delta with several lobes protruding out into a basin.

Bit and **bit.** bitumen.

bit. The cutting tool used on the end of a drillstring during drilling. A tricone bit is commonly used in rotary drilling. *See also* tricone bit, diamond bit, and polycrystalline diamond compact bit.

bit breaker. A plate placed in the rotary table to grip the bit. It enables the rotary table to screw or unscrew the bit from the drillstring as the bit is held stationary.

bitumen. Solid hydrocarbons such as tar in sedimentary rocks. It is soluble in organic solvents and is less than 10 °API gravity; cf. kerogen. (Bit and bit)

BL. barrels of load.

bl. black.

black mica. *See* biotite.

black oil. An oil that contains a relatively high percentage of long heavy nonvolatile hydrocarbon molecules; cf. volatile oil. (low-shrinkage oil)

Bld. boulder.

bld. bailed.

bldg. building.

bledg. bleeding.

bleed. (1) To slowly drain off a gas or liquid. (2) Oil bubbling out from a fresh core.

blg. bailing.

blind rams. Two large metal blocks with flat surfaces that are closed across the top of well to shut it in. They are used in a blowout-preventer stack; cf. pipe rams.

blk. (1) black and (2) block.

blkt. blanket.

blnd. blend.

BLO. barrels of load oil.

blo. blow.

block. A pulley on a drilling rig. (blk)

BLOTBR. Barrels of load oil to be recovered.

blowout. An uncontrolled flow of fluid from a well.

blowout preventer and **blowout-preventer stack.** A series of rams and spools mounted vertically on top of a well below the drill floor. They are bolted to the top of the well and are designed to close the well when drilling. The rams are operated by pneumatic pressure. (BOP stack)

BLPD. barrels of liquid per day.

blr. bailer.

BLS. below land surface.

BLW. barrels of load water.

Bm. basement.

bn. brown.

bnd. banded.

BNO. barrels of new oil.

BNW. barrels of new water.

BO. (1) barrels of oil and (2) backed out.

BOE. (1) barrels of oil equivalent and (2) blowout equipment.

BOEMRE. Bureau of Ocean Energy Management, Regulation and Enforcement (www.boemre.gov).

BOL. barrels of oil load.

boll weevil. An inexperienced oilfield worker.

bonus. money paid to a mineral rights owner for signing a lease.

book reserves. To calculate reserves that have been located by drilling and add them to a company's assets. Reserves are often confirmed by several levels of management and an independent consultant before they are booked. *See also* reserves.

boot sub. A fishing tool. Drilling mud is circulated down the inside of the tool to flow out the bottom and pick up small pieces of junk on the bottom of the well. The mud then circulates up along the outside of the tool where the junk falls into a basket; cf. junk basket.

BOP. blowout preventer.

BOPD. barrels of oil per day.

BOPH. barrels of oil per hour.

BOS. brown oil stain.

bot. bottom.

bottom-hole assembly. The drill collars, subs, and bit on the bottom of the drillstring. (BHA)

bottom-hole contribution agreement. A type of well support agreement in which a nondrilling party will pay monies to a drilling party that drills a well at a specific location. Information from that well will be shared with the nondrilling party; cf. dry hole contribution and acreage contribution agreements.

bottom-hole pressure. Fluid pressure on the bottom of a well. It can be either static or flowing.

bottomwater. Water located in the reservoir below the crude oil; cf. edgewater.

boundstone. A type of limestone formed by organisms still in their original positions such as reef rock. (Bdst)

box. A female-threaded connection that mates with a pin (a male connection); cf. pin.

BP. (1) bridge plug, (2) back pressure, and (3) bull plug.

BPD. barrels per day.

BPH. barrels per hour.

BPV. backpressure valve.

BR. (1) building rig and (2) building road.

br. brown.

brackish water. A mixture of freshwater and brine. (brak or brksh)

braided stream. A stream with numerous intertwining channels separated by gravel bars.

brak. brackish.

break. To separate an emulsion such as oil-in-water.

break out. To unscrew tubulars such as drillpipe; cf: make up.

breakthrough. To have injected water flow through a reservoir and reach a producing well.

breccia. A conglomerate with angular particles. (Brec or brec)

Brent or **Brent Blend.** An internationally recognized benchmark crude oil. It has 38 °API gravity and 0.3% sulfur. Brent is a mixture of oils from 15 oil fields in the North Sea.

brg. bearings.

bridge plug. An expandable tool that is run in a cased well to seal the well at that depth. Bridge plugs are (1) permanent, (2) drillable, or (3) retrievable. (BP)

bridging material. *See* lost circulation material.

bright spot. A relatively intense seismic reflection. It can be off the top of a gas-filled sandstone reservoir.

brine. Water that has more salt than seawater that has 35 parts per thousand salinity; cf. freshwater.

brit. brittle.

British thermal unit. The English system unit used to measure the heat content of natural gas. Natural gas has about 1,000 British thermal units of heat per cubic foot. It is equal to about one kilojoule in the metric system. (Btu)

brk. (1) break and (2) broke.

brkn. broken.

brksh. brackish.

brn. brown.

brownfield. An oil or gas field that is know and has been developed to some stage; cf. greenfield.

BRPG. bridge plug.

BRT. below rotary table.

brt. bright.

BS. (1) basic sediment and (2) bit size.

BS&W. basic sediment and water.

bsg. bushing.

bsmt. basement.

BSW. barrels of saltwater.

BTM or **btm.** bottom.

btm chk. bottom choke.

btmd. bottomed.

btry. battery.

Btu. British thermal unit.

btw. between.

BU. build up.

bu. buff.

bug. A microfossil.

bug picker. *See* micropaleontologist.

build angle. To increase the inclination of a deviated well; cf. drop angle and maintain angle.

buildup test. A type of pressure transient test that measures the flowing bottom-hole pressure; then the pressure change as a well is shut in, and the pressure rises to a stable shut-in pressure.

bullheading. (1) A method used to pump treating fluids into a formation. The formation is isolated with a packer and the treating fluids are pumped down the tubing string. If no packer is used, it is called bradenheading. (2) a method used to kill a well by pumping kill mud into the kill line and well annulus.

bull wheel. A spool of drilling line on a cable tool rig.

Bur or **bur.** burrow.

BUT. butane.

butane. A hydrocarbon composed of C_4H_{10}. It is a gas under surface conditions and is found in natural gas. Butane has two isomers: isobutane and n-butane. (BUT and C_4)

button tricone bit or **insert tricone bit.** A tricone bit in which holes have been drilled into the steel cones and buttons of hard tungsten-steel carbide have been inserted. It is used to drill hard rocks; cf. milled-teeth tricone bit.

buttress sand. Sand deposited on top of an unconformity.

BW. barrels of water.

BWL. barrels of water load.

BWPD. barrels of water per day.

bypassing. Water flow around relatively impermeable rocks containing oil and gas in the reservoir during production or a waterflood.

C. (1) concentration, (2) coal, (3) Celsius, and (4) centigrade.

c. core.

C/. contractor.

cable tool rig. An older type of drilling rig, often constructed of wood, that pounds a hole in the ground by raising and lowering a bit on a cable. It was replaced by rotary drilling rigs during 1900–1930. (standard tools) (CTR) cf. rotary drilling rig

CaCO₃. calcium carbonate. It is the chemical composition of the mineral calcite and the rock limestone.

cal. (1) calories, (2) calcite, (3) calcitic, and (4) caliper survey.

calcareous. A rock containing calcium carbonate ($CaCO_3$). (calc)

calcite. A common mineral composed of $CaCO_3$. Limestones and most seashells are composed of calcite. (Ca, cal, Calc, and calc)

caliper log. A wireline well log that measures the diameter of the wellbore. (CL, CAP, CAL, cal, and CALP)

calorific value. The heat content per unit volume of natural gas. It is measured in Btu per cubic feet.

Cambrian. A period of geological time from 542 to 488 million years ago. It is part of the Paleozoic era. (Camb)

Campanian. An age of geological time from 83.5 to 71.3 million years ago. It is part of the Cretaceous period.

C&A. compression and adsorption.

C&C. circulation and conditioning.

cantilevered jackup. A jackup with the derrick mounted on steel beams over one side of the deck. This is in contrast to a derrick located over a drilling slot (keyway) indented into the deck.

cantilevered mast. A mast that is assembled horizontally and then pivoted vertically into place by the drawworks and traveling block on the drilling rig.

CAOF. calculated absolute open flow.

CAP and **cap.** capacity.

caprock. (1) an impermeable rock layer that forms the seal on top of an oil or gas reservoir. (seal) (2) the insoluble rock on the top of a salt plug. (CR)

carb. carbonaceous.

carbonaceous. containing carbon (C). (carb)

carbon dioxide. A colorless, odorless gas composed of CO_2. It can occur as an impurity (inert) in natural gas and is used for enhanced oil recovery.

carbon dioxide flood. An enhanced oil recovery method. Carbon dioxide (CO_2) is injected into a depleted oil reservoir to mix with the remaining oil and push it toward producing wells.

Carboniferous. A time period from 359 to 299 million years ago that is used in Europe. In North America, the Mississippian period is equivalent to the Lower Carboniferous, and the Pennsylvanian period is equivalent to the Upper Carboniferous. It is part of the Paleozoic era. (Carb)

carried interest. A partial ownership in a well that does not bear any expenses up to a point such as to the casing or through the tanks; cf. working interest.

carrier bed. A permeable rock layer along which fluids can flow.

carrier unit. A self-propelled, well-servicing rig.

carve out. To take out of (e.g., an overriding royalty is carved out of a working interest).

CASD. cased off.

cased-hole log. A wireline well log that can be run in a well that has been cased. e.g., natural gamma ray log and neutron porosity log; cf. open-hole log.

casing. Relatively thin-walled, large-diameter (commonly 5½ to 13⅜ in.), steel pipe that has threaded connections on each end. Different API classes of casing range from 16 to over 42 ft long. Each section is called a joint. Joints of casing are screwed together to form a casing string. (CSG and csg) *See also* casing string.

casing drilling. To drill a well by rotating a casing string and bit. Drillpipe is not used. The casing is left in the well, and the drilling assembly is retrieved up the casing on a wireline.

casing-free pump. A hydraulic pump that uses only one tubing string. The power oil goes down the tubing string, and the produced fluids come up the tubing-casing annulus; cf. parallel-free pump. *See also* hydraulic pump.

casinghead. A forged or cast steel fitting on the lower part of the wellhead. It seals the annulus between two casing strings. The casing hanger that suspends a casing string is located in the casinghead. Each casing string has a casinghead; cf. tubinghead. (CH and csg hd)

casinghead gas. Natural gas that bubbles out of oil on the surface at the well. (CHG)

casinghead gasoline. *See* condensate.

casing point. (1) The depth to which a casing string has been set in a well. (CP or csg pt) (2) The time after drilling and testing a well that a decision has to be made to either complete (case) or plug and abandon the well.

casing pressure. Pressure on the fluid in the casing-tubing annulus. It can be either flowing or static; cf. tubing pressure. (CP and csg prss)

casing program. The lengths, diameters, and other specifications of different casing strings that are to be used in a well.

casing pump. A large sucker-rod pump held in position on the bottom of a well by a packer.

casing roller. A long tapered tool with rollers on the sides that is run on a drillstring to roll out collapsed casing in a well.

casing string. A long length of many casing joints screwed together. A casing string is run in a well and cemented to the sides of the well during a cement job. Some types of casing strings are surface, intermediate, and production. All casing strings run back up to the casinghead on the surface; cf. liner string.

catch a log. To receive a hard copy or electronic copy of a well log that has just been recorded.

cathead. A hub on a shaft (catshaft) on the drawworks of a drilling rig that is used to pull a line (catline) to lift or pull equipment.

cathodic protection. A method that applies an electrical charge to a metal structure such as a pipeline or offshore platform to prevent corrosion.

catwalk. A flat, steel walkway that is elevated and connects stock tanks or instillations.

caustic flood. *See* alkaline flood.

Cav and **cav.** cavern.

cave. The collapse of well walls into the hole. (sluff)

cavings. Small rock particles that have fallen off the well walls and down the well. (cvsg)

CB. (1) core barrel, (2) core bit, (3) changing bit, and (4) changed bit.

CC. (1) casing collar, (2) casing cemented (depth), and (3) cumulative cost.

CCL. casing collar locator.

CCM. circulate and condition mud.

CD. contract depth.

CDP. common-depth-point.

cdsr. condenser.

CEC. cation exchange capacity.

cell. cellar.

cellar. A rectangular pit dug below the floor of a large drilling rig to hold the blowout preventers. It is usually lined with boards or cement. (cell)

Celsius. The temperature scale that has replaced centigrade. It is based on a 0° freezing point and a boiling point of 100° for freshwater. To convert Celsius temperature to Fahrenheit temperature, multiply by 9/5 and add 32°. (C)

cem. cement.

cement. (1) Minerals that naturally grow between clastic grains and solidify a sedimentary rock. (2) Portland cement used to bind the casing strings to the well walls. (Cmt, cmt, and cem)

cement bond log. A type of sonic log that determines where and how well cement has set behind casing. (CBL) *See also* holiday.

cementing head. An L-shaped fitting that is attached to a wellhead for a cement job. It conducts the wet cement (slurry) from the cement pumps down the well.

cement job. To cement casing into a well. Primary cementing is done when the casing is originally run in the well. Secondary cementing is done later during a workover on a well.

Cenomanian. An age of geological time from 99.6 to 93.5 million years ago. It is part of the Cretaceous period.

Cenozoic. An era of time from 65.5 million years ago until today. The Cenozoic is divided into the Tertiary and Quaternary periods. It is known as the age of mammals.

centi. 1/100.

centigrade. The metric temperature scale that has been replaced by Celsius. Degrees centigrade is the same as degrees Celsius. *See also* Celsius.

centimeter. A metric unit of measurement equal to 1/100 meter and 0.3937 inch. (cm)

centipoise. A unit of viscosity equal to 1/100 poise. *See also* poise. (cp)

centralizer. An attachment to the outside of a casing string that uses bowed steel bands to keep the string centered in the well.

central processing unit. (1) a common separator and tank battery for several oil wells. (2) gas-conditioning equipment that services several gas wells. (CPU)

CF. (1) casing flange and (2) cubic feet.

cf. compare.

cf. cubic feet.

CFG. cubic feet of gas.

CFGPD. cubic feet of gas per day.

CFGPH. cubic feet of gas per hour.

C$_5$. pentane.

C$_4$. butane.

CG. (1) corrected gravity and (2) connection gas.

cg. (1) coring and (2) coarse grained.

Cgl, **cgl**, or **cglt.** conglomerate.

c-gr. coarse-grained.

CH. choke.

CH. (1) casinghead and (2) chat.

ch. (1) chert, (2) choke, and (3) chain.

C/H. cased-hole.

chain tongs. A hand tool used to tighten or loosen pipe.

chalk. An extremely fine-grained limestone composed of calcareous microfossils such as coccoliths. It naturally has high porosity but very low permeability. (Chk, chk, and CK)

chance of success. *See* success rate.

channel. (1) A single seismic-recording unit. The geophones in an array are recorded together on one channel. (2) A cavity in the cement behind casing in a well.

charcoal test. A test used to measure the amount of condensate in natural gas. Activated charcoal is used to absorb the condensate from a volume of natural gas.

charthouse. A gas meter shelter. (meterhouse)

chat. A driller's term for conglomerate. (CH)

check shot. A method used to determine the seismic velocities of rock layers in a well. The seismic source is located on the surface next to the well. A geophone is raised in the well to measure seismic velocities at various depths. It is similar to a vertical seismic survey, but the geophone stations are located further apart.

chemical cutter. A tool used to cut pipe in a well with jets of hot caustic chemicals under high pressure.

chemical flood. An enhanced oil recovery method in which batches of chemicals are injected into a depleted oil reservoir (e.g., micellar-polymer flood).

chert. A sedimentary rock composed of amorphous quartz. (flint) (ch and cht)

CHG. casinghead gas.

chiller. A heat-exchanger vessel that uses cooling to remove natural gas liquids from natural gas.

Chk and **chk.** chalk.

chk. choke.

chkbd. checkerboard.

chkd. checked.

chky. chalky.

chl. channel.

chng. (1) change and (2) changed.

choke. A constriction or orifice that restricts flow in a line. It is described by its diameter in $1/64$ inch. (ch, CK, and chk)

choke manifold. A series of pipes and valves located next to a drilling rig. It is designed to guide fluids from the well to the mud tanks, reserve pit and other areas, and direct kill mud to the well.

CHP. casinghead pressure.

Christmas tree. The vertical structure of pipes, fittings, values, chokes, and gauges that are bolted to the wellhead of a gas well or flowing oil well to control the flow from the well. (production tree and tree) (Xtree)

cht. chert.

chty. cherty.

CI and **C.I.** contour interval.

CIBP. cast iron bridge plug.

CIP. cement in place.

CIRC and **circ.** (1) circulate and (2) circulating.

circulate. To pump drilling fluid down the drillstring and back up along the outside of the drillstring. (CIRC or circ)

circulate out. To pump drilling fluid down a well while not drilling to remove well cuttings or gas.

circulation. The movement of drilling mud down through the drillstring and back up through the drillstring-casing annulus.

CJPF. casing jets per foot.

CK. (1) choke, (2) chalk, and (3) filter cake.

ck. check.

CKF check for flow.

CL. center line.

Cl and **cl.** clay.

Cl. (1) chlorides and (2) salinity.

cl. chlorine.

clastic. A sedimentary grain that has been transported and deposited as a whole particle such as a sand grain. (detrital) (clas)

clastic ratio map. A map that uses contours to show the ratio of (1) conglomerates, sandstones, and shales to (2) limestones, dolomites, and salts in a formation.

clay. A fine-grained particle less than $1/256$ mm in diameter; cf. silt and sand. (Cl and cl)

clay mineral. A very-fine-grained mineral formed by a layered molecular structure of aluminum, silicon, and oxygen atoms. Bentonite is a clay mineral used to make drilling mud.

clean oil. Crude oil that is below a maximum basic sediment and water content and meets pipeline specifications. (pipeline oil)

clean sands. Well-sorted sands; cf. dirty sands.

cleat. An open, natural fracture in a coal bed. Usually there are two sets of clints that are perpendicular to the coal bed and each other. Face cleats are continuous, whereas butt cleats are not.

cleavage. A flat surface formed by a mineral breaking along a crystal plane. It is described by the quality and number of cleavage planes (i.e., mica has one excellent cleavage).

cln. clean.

closure. The vertical distance in a petroleum trap between the top of a reservoir rock down to the spill point.

clsd. closed.

cm. centimeter.

CMP. common-mid-point.

Cmt and **cmt.** (1) cement, (2) cemented, and (3) cementing.

cmtd. cemented.

cmtg. cementing.

CN. conglomerate.

CNL. compensated neutron log.

cntr. (1) centered and (2) container.

cntrt. contorted.

CO. (1) clean out, (2) cleaned out, (3) cleaning out, (4) circulate out, (5) crude oil, and (6) coal.

co. change out.

coal. A sedimentary rock composed primarily of carbonaceous material formed by woody plant remains transformed by heat and time. Lignite, bituminous, and anthracite are types of coals. (c and CO)

coal bed and **coal seam gas.** Methane gas generated during coal formation. It is adsorbed to the surface of the coal in pores and along natural fracture (clint) surfaces. The coal must be dewatered to decrease the pressure before the gas can be produced.

coastal plain. A plain with underlying thick sediments deposited along an ocean margin.

coccolith. A calcium carbonate plate from a small, single-cell animal (coccolithophore) that lives floating in the ocean. It is so small that it can only be identified by a scanning electron microscope. A relatively pure deposit of coccoliths is called chalk.

COF. calculated open flow.

COFCAW. combination of forward combustion and waterflooding.

COH. coming out of hole.

coiled tubing. High-strength flexible steel tubing that is usually 1¼ inches in diameter. It comes in a continuous length wrapped around a reel. (CT)

coiled tubing unit. A well service unit with a reel of continuous coiled tubing on the back. It is used for running equipment down a well during a workover or for drilling a well. Before the coiled tubing goes down or comes out of the well, it goes through a pipe straightener on top of the well.

collar. A short steel cylinder with female threads. It is used to join pipe joints with male threads. (coupling and tool joint) (colr)

collar log. A production log that records the depth of each casing collar in a well.

colr. collar.

com. common.

combination of forward combustion and waterflooding. A fireflood in which air and water are alternately injected into the reservoir. The steam generated from the water helps drive the oil toward producing wells. (wet combustion) (COFCAW)

comm. (1) communicate and (2) communicated.

commingle. To mix production from (1) two or more zones in a well (subsurface commingling) or from (2) two or more wells (surface commingling).

common-depth-point stacking or **common-mid-point stacking.** A seismic acquisition and processing method in which numerous different reflections off the same subsurface point within a bin are assigned a common depth or midpoint and combined to reduce noise and reinforce the reflector. (CDP and CMP stacking)

comp. (1) complete and (2) completion.

compaction. The volume decrease of sediments by pressure during burial.

compaction anticline. An anticline formed by compaction of softer sedimentary rocks over and along a harder reef or bedrock hill.

company person or **representative.** An employee of the operator who works with the tool pusher to make sure a well is being drilled to specifications. (coordinator)

compensated log. A wireline well log that has been adjusted to irregularities in wellbore size and roughness.

compensated neutron log. *See* neutron log. (CNL)

compl. (1) complete and (2) completed.

completion. To install equipment to prepare for producing gas and/or oil from a well.

completion card. A form published by a commercial firm containing information on the drilling and testing history and the geological and producing characteristics of a specific well.

completion fluid. An inert fluid, usually treated water or diesel oil. It is used (1) in the casing-tubing annulus of a producing well to prevent casing corrosion or (2) in a well to replace drilling mud during well completion to prevent formation damage and still maintain pressure control.

completion packer. A packer run on the bottom of the tubing string to seal the space (annulus) between the tubing and casing. (tubing packer)

completion report. *See* well completion report.

completion rig or **unit.** A derrick and hoisting unit used after the drilling rig has been released to run the final string of casing in a well. It is smaller and less expensive than the drilling rig.

compliant platform or **tower.** An offshore production facility that is anchored on the bottom. The upper part is free to move within a restricted radius (e.g., spar and tension-leg platform)

comp nat. completed natural.

compounder. A system of pulleys, belts, shafts, chains, and gears that transmit power from the prime movers to the drilling rig.

compr. compressor.

compressibility factor. *See* Z factor.

compression. forces that push together; cf. tension.

compression and **compressional fault.** A fault formed by compressional stress (e.g., reverse and thrust fault); cf. tensional fault.

compressional wave. A wave that causes particles to move back and forth as the wave passes through. It is similar to sound waves in air. Compressional waves are recorded during normal seismic and along with shear waves during multicomponent seismic; cf. shear wave (primary and pressure wave) (P wave)

compression ratio. A ratio of the volume of uncompressed gas divided by the volume of the same gas compressed by a compressor (e.g., 10.1).

compression test. A test used to determine the condensate content of natural gas. A gas sample is compressed and then allowed to expand to cool it. Condensate separates from the cooling gas.

compressor. A device that increases the pressure of gas and can cause the gas to flow. Compressors use pistons in cylinders, rotating vanes on a shaft, and other methods. (compr) *See also* compression ratio.

compr sta. compressor station.

compt. completed.

computer-generated log. A log made by a computer from two or more logging measurements. It can be generated at the wellsite (quick-look log) or in a computing center. (computer center log)

conc. concentrate.

concentric tubing workover. A workover that uses smaller than normal equipment that can be run down a tubing string.

concession. (1) The contract between a host country and a company to explore and drill. (2) The land or ocean bottom covered by that contact.

concession agreement. *See* tax royalty participation contract.

concrete gravity production platform. *See* gravity storage production platform.

COND or **cond.** condensate.

cond. (1) condition, (2) conditioned, (3) conditioning, and (4) conduct.

condensate. A light-hydrocarbon mixture that is a liquid under surface conditions but is a gas mixed with natural gas under subsurface reservoir conditions of high temperature. It is composed of pentane (C_5), hexane (C_6), heptane (C_7), octane (C_8), nonane (C_9), and decane (C_{10}) and is almost pure gasoline in composition. Condensate is very light in density and is transparent to yellowish in color. It is produced during the production and transportation of natural gas. Wet gas contains condensate. It is classified as crude oil by government regulatory agencies. (casinghead gasoline, drip, drip gasoline, drips, natural gasoline, distillate, and white oil) (COND, cond, and CONDS)

conditioning. (1) Preparing and altering drilling mud. (2) Circulating drilling mud in a well for a period of time to prepare the well for logging or another process.

condr. conductor (pipe).

CONDS. condensate.

condt. conductivity.

conductivity. The ability of a material to conduct electrical current. It is the inverse of resistivity. Conductivity is recorded on well logs in units of mho per meter. (condt)

conductor casing. A length of large-diameter steel pipe that is pile-driven or drilled into the land surface or ocean floor before drilling. It is used to stabilize the soil as a well is being drilled and to attach the blowout preventers. (conductor pipe and structural casing)

conductor hole. A large-diameter shallow hole drilled to hold the conductor pipe.

conductor pipe. *See* conductor casing.

C_1. methane.

conf. (1) confirm, (2) confirmed, and (3) confirming.

confirmation well. A well drilled after the discovery well to prove the extent of a new petroleum reservoir.

cong. conglomerate.

conglomerate. A poorly sorted sedimentary rock with rounded, pebble- to clay-sized grains; cf. breccia. (Cgl, cgl, cglt, and cong) (CG)

Coniacian. An age of geological time from 89 to 85.8 million years ago. It is part of the Cretaceous period.

coning. The drawing of (1) the oil-water contact up or (2) the gas-oil contact down into an oil reservoir. The contact has the shape of a cone around the well and is caused by too rapid oil production.

conn. connection.

connate water. Saline subsurface water.

cons. considerable.

consol. consolidated.

consolidated sediments. Sediments bound together into a relatively hard sedimentary rock. (indurated sediments) (consol) *See also* unconsolidated sediments.

cont. continue.

contact. The boundary between two rock layers. (Ctc or ctc)

contactor. A metal vessel that causes a gas passing through it to come in contact with a chemical on a bed. A contactor can use a solid desiccant to remove water from natural gas.

Contam and **contam.** (1) contamination and (2) contaminated.

contribution agreement. *See* support agreement.

continental drift. A relatively old theory that the continents were all joined in one supercontinent (Pangaea) that broke up during the Mesozoic era with the fragments drifting across the earth. *See also* seafloor spreading and plate tectonics.

continental rise. A large wedge of sediments at the base of a continent slope in water depths of 5,000 to 13,000 ft.

continental shelf. A shallow platform that surrounds the continents. It extends from the beach and slopes out at less than 1° to an ocean depth of about 450 feet where the shelf break (an abrupt change of slope) is located. Offshore drilling and production occurs on the continental shelf; cf. continental slope.

continental slope. The slope (3° to 4°) leading down from the shelf break on the continental shelf to the deep ocean bottom. Deep-water drilling and production occurs on the continental slope; cf. continental shelf.

continuous flowmeter. An instrument that measures fluid flow versus depth in a well.

contour. A line of equal value such as elevation or thickness on a map.

contour interval. The difference in value between two adjacent contours. (CI or C.I.)

contract depth. The depth to which a well is to be drilled in a drilling contract.

converted wave. A seismic wave that has reflected off a seismic reflector and been transformed into another type of wave. A P wave that was converted to an S wave (SP wave) is an example.

coord. coordinates.

coordinator. *See* company person and representative.

COPAS. Council of Petroleum Accountants Societies (www.copas.org).

coquina. A sedimentary rock composed of broken shells. (coq)

Cor. coral.

cor. corner.

core. (1) A cylinder of rock drilled from a well that can be either (a) a full-diameter core (3½ to 5 in. in diameter) or (b) a sidewall core. (1 in. in diameter) (c or cr). (2) To drill and obtain a core.

core barrel. A tubular run above a rotary coring bit. It has both (1) an outer core barrel to rotate and cut the core and (2) an inner barrel to remain stationary and receive the core. The barrels are separated by ball bearings. (CB)

corr. (1) correct, (2) correction, and (3) corrosion.

correlation. The matching of rock layers.

correlation log. A wireline well log with a 1-in. to 100-ft scale; cf. detail log.

corro. corrosion.

corrosion. The chemical and biological degradation or abrasive wearing away of metal. Sweet corrosion is due to CO_2. Sour corrosion is due to H_2S. (corr and corro)

corrosion inhibitor. A chemical applied either in batches or continuously to prevent corrosion in a well. It usually forms a coating on metal.

cost oil. Produced oil that is available to a multinational company for sale to reimburse previous expenditures. *See also* production sharing contract.

COTD. cleaned out to total depth.

CO_2. carbon dioxide.

coupling. A short steel cylinder with female threads. It is used to connect pipe joints with male threads. (collar and tool joint) (cplg)

CP. (1) casing point and (2) casing pressure.

cp. centipoises.

cplg. coupling.

CPM. cycles per minute.

CPU. central processing unit.

CPS. (1) counts per second and (2) cycles per second.

CR. cap rock.

cr. (1) core and (2) creek.

crack a valve. To slightly open a valve to start flow.

cracking. A refining process that breaks long-chained hydrocarbons into more valuable, short-chained hydrocarbons.

craton. Land covered by sedimentary rocks that surrounds a shield.

crd. cored.

crest. The top of a structure such as an anticline.

Cretaceous. A period of geological time from 145.5 to 65.5 million years ago. It is part of the Mesozoic era. (Cret)

crevasse splay. Sediments deposited to the side of a delta through a break in the levee.

crg. coring.

crit. critical.

crk. creek.

crn blk. crown block.

crooked hole. A well with a large deviation along the wellbore that was not made on purpose; cf. deviated well and straight hole.

crooked hole country. An area with dipping hard rock layers that cause crooked holes when drilling wells.

crossbedded. A sedimentary rock that displays crossbeds. (x-bd, x-bdd, X-bdd, and XBD)

crossbeds. Sedimentary rock layers, usually in sandstone, deposited at an angle up to 36° from horizontal in dunes or ripples by air or water currents.

crossover sub. A short section of pipe used in a downhole assembly to connect pipes of different sizes or thread types.

crown block. A fixed steel frame with steel wheels on a horizontal shaft. It is located at the top of a derrick or mast. The hoisting line goes through the crown block; cf. traveling block. (crn blk)

crown land. Land owned by the federal or provincial government in Canada.

crude oil. A naturally occurring liquid composed of thousands of different chemical compounds, primarily hydrocarbons. The molecules range from 5 to more than 60 carbon atoms in length. Crude oil color is commonly black to greenish-black but can also be yellowish to transparent. Crude oil is described by density (°API gravity) and percent sulfur content (sweet and sour). It is measured in barrels (bbl) in the English system and metric tons or cubic meters in the metric system. (CO) *See also* petroleum.

crude stream. Crude oil from a single field or a mixture from fields that is offered for sale by an exporting country.

cryogenic plant. An installation that uses a natural gas–driven turbine to cool the gas to a very low temperature and remove natural gas liquids. (expander plant)

CSA. casing set at.

cse gr. coarse grained.

C_7. heptane.

CSG and **csg.** casing.

csg hd. casinghead.

csg press. casing pressure.

csg pt. casing point.

C_6. hexane.

CSPG. casing packer.

CT. carbide test.

Ctc or **ctc.** contact.

CTD. corrected total depth.

ctd. coated.

Ctgs. cuttings.

C_3. propane.

CTR. cable tool rig.

ctr. center.

C_2. ethane.

CU. clean up.

cubic foot. The English system unit of natural gas volume measurement. It is the volume of a cube, one foot on a side. One thousand cubic feet (Mcf) is commonly used. (cf or c^3)

cubic meter. The metric system unit of natural gas and crude oil volume measurement. It is the volume of a cube, 1 meter on a side; 1 cubic meter is equal to 6.29 barrels of oil or 35.3 cubic feet of gas. (m^3)

cu ft/bbl. cubic feet per barrel.

CUM. cumulative.

cumulative. The total gas, oil, and/or water production from a well or wells. (CUM)

CUSH and **cush.** cushion.

cushion. Water or oil filling a tubular or well. It is used to control pressure. (CUSH and cush)

cut. To dilute something.

cuttings. Rock flakes made by the drill bit. (well cuttings) (Ctgs)

CV. control valve.

Cvg. caving.

cvgs. cavings.

C/W. completed with.

c-wave. mode-conversion wave.

cx. coarse.

cycle time. The time elapsed from the beginning to the end of a project.

cyclic steam injection. An enhanced oil recovery method used for heavy, viscous oil. A well is used to first pump steam into the subsurface reservoir to heat the oil and make it more fluid. The same well is then used to pump the heated, heavy oil. (huff 'n puff)

cycling. The injection of produced gas back into a retrograde condensate reservoir to slow the drop of reservoir pressure and the separation of condensate in the reservoir.

cyclothem. Alternating marine and nonmarine sedimentary rocks.

cyl. cylinder.

D. (1) depth and (2) dome.

d. (1) darcy, (2) diameter, and (3) day.

DA. daily allowable.

daily drilling report. A report made by the tool pusher or company person covering the last 24 hours of activities on a drilling rig. It usually includes total depth at report time, footage drilled during last 24 hours, activities such as tripping and repairs, formations drilled, mud measurements, and supplies used. The report is made from the tour reports. The International Association of Drilling Contractors has published a standard report. (morning report) (DDR)

daily mud cost. The amount of money spent on drilling fluids during the past 24 hours of rig operation. (DMC)

daily well cost. The amount of money on drilling operations during the past 24 hours of rig operation. (DWC)

D&A. (1) dry and abandoned and (2) drilled and abandoned.

D&C. drill and complete.

D&D. (Association of) Desk and Derrick (Clubs) (www.addc.org).

Danian. An age of geological time from 65 to 61 million years ago, part of the Paleocene epoch.

darcy. The unit of permeability. 1 darcy permeability in a porous medium allows 1 cubic centimeter of fluid with 1 centipoise viscosity to flow in 1 second through a cross section of 1 square centimeter along a length of 1 centimeter with a pressure differential of 1 atmosphere. The plural is darcys. A millidarcy (md) is $\frac{1}{1,000}$ darcy. (d)

dat. datum.

datum. A level surface to which contours are referred, such as sea level. (dat and DM)

days from spud of well. How many days a rig has been drilling.

days left on location. How many more days the rig will be drilling down to total depth.

daywork drilling contract. A drilling contract based on a cost per day during drilling to contract depth. It is commonly used offshore; cf. footage contract and turnkey drilling contract.

DB. (1) diamond bit and (2) drill break.

DBOS. dark brown oil stains.

DC. (1) direct current, (2) drill collar, (3) depth correction, (4) dual completion, and (5) daily cost.

DD. (1) drilling deeper and (2) drilled deeper.

dd. dead.

DDR. daily drilling report.

dead oil. Crude oil that (1) will not flow from a rock or (2) has no dissolved gas.

dec. (1) decide and (2) decision.

decatherm. A unit of heat equal to 1 million British thermal units (Btu).

decline curve. A plot of oil or gas production rate from a well versus time.

deconvolution. A computer process that restores the seismic echoes to their original seismic-source form to make subsurface reflections sharper and reduce noise.

decr. decrease.

deep water. Water depth greater than 1,000 ft. Ultra-deep water is greater than 5,000 ft. deep.

deg. degree.

°API gravity (degrees API gravity). A measure of the density of a liquid or a gas. Freshwater is 10. Average weight oil is 25 to 35, heavy oil is below 25, light oil is 35 to 45, and condensate is above 45 °API gravity. (API gravity) (gr API)

dehydrator. A vessel that uses either a solid or liquid desiccant to remove water from natural gas.

delay rental lease. A type of lease in which a delay rental payment must be made each year during the primary term to the lessor if drilling has not commenced; cf. paid-up lease.

delineation well. A well drilled to the side of a discovery well to determine the extent of the new field. (step out and appraisal well)

deliverability. (1) The normal flow rate of a gas well. (2) The ability of a reservoir to move fluids into a well at a given flowing bottom-hole pressure. (delv)

delr. deliver.

delta. Sediments deposited by a river emptying into an ocean.

delta switching. A process in which a river abandons an old delta for a shorter route to the ocean and builds a new delta.

Δt (delta T). The sound velocity through a rock measured by a sonic log. The units are in microseconds per foot. (interval transit time)

delv. deliverability.

demulsifier. A chemical used to break an emulsion.

DENL. density log.

dense limestone. Limestone with no permeability.

density. (1) Weight of a substance per unit volume such as gm/cm^3 or lbm/gal. (r) (2) perforations per foot.

density log. *See* formation density log. (DENL, D/L, and DL)

depl. depletion.

depletion drive. *See* dissolved gas drive.

depth of investigation. A measure of how far back in the rock from the wellbore a well log measurement is made. It is usually the distance that causes 50% of the log measurement response.

derrick. A steel tower with four legs that sits on the drill floor of a drilling rig. It must be raised vertically in sections in contrast to a mast.

Derricks are used for hoisting on offshore drilling rigs. The term *derrick* is sometimes incorrectly used for mast; cf. mast. (drk)

derrick floor. *See* drill floor. (DF)

derrick operator. A member of the drilling crew who is second in charge. The derrick operator stands on the monkeyboard when making a trip to rack drillpipe. The derrick operator is also in charge of the circulating system. (monkeyman)

desander and **desilter.** Metal cones used on a drilling rig to centrifuge drilling mud from a well to remove fine-grained well cuttings. Drilling mud flows through the desanders and desilters after flowing across the shale shaker.

designer well. A high-angle deviated well or horizontal well with more than one intended target.

det. detector.

detail log. A wireline well log hard copy with a 5-in. to 100-ft scale; cf. correlation log.

deterministic. A process or method in which there is an exact relationship between the variables and the outcome can be predicted with certainty; cf. stochastic.

detr. detrital.

detrital. A sediment grain that has been transported and deposited as a whole particle such as a sand grain; cf. authigenic. (clastic) (detr)

dev. (1) deviate, (2) deviated, and (3) deviation.

developmental geologist. A geologist who specializes in the exploitation of discovered petroleum fields; cf. exploration geologist.

developmental well. A well drilled in the known extent of a field; cf. wildcat well.

deviated well. A well drilled out at an angle from a straight hole; cf. crooked hole and straight hole.

deviation. The angle of a wellbore from vertical. (drift angle)

deviation drilling. *See* directional drilling.

Devonian. A period of geological time between 416 and 359 million years ago. It is part of the Paleozoic era and is known as the age of fish. (Dev)

dew point. The temperature at which a liquid starts to separate out of a gas as it is being cooled. (DP)

DF. (1) derrick floor, (2) drill floor, and (3) drilling fluid.

DFE. derrick floor elevation.

DFP. date of first production.

DG. (1) dry gas and (2) developmental gas well.

DHC. dry hole contribution.

dia. diameter.

Diag and **diag.** diagenesis.

diagenesis. The processes that form sedimentary rock from loose sediments (e.g., cementation and compaction). (Diag or diag)

diamond bit. *See* natural diamond bit.

diatom. A tiny single-cell plant that floats in water. It has a round, pill-box-shaped shell composed of silicon dioxide.

diatomaceous earth and **diatomite.** A sedimentary rock composed primarily of siliceous, diatom shells. It is a reservoir rock for heavy oil in California.

diesel-electric rig. A modern rotary drilling rig that uses diesel engines to drive electrical generators that make alternating current. The alternating current is changed into direct current by a silicon controlled rectifier (SCF). Equipment on the rig floor, such as the drawworks, is driven by AC or DC electric motors that are more efficient; cf. mechanical rig. (electric rig, electric-drive rig, and motor-generator rig)

diesel engine. An engine that uses diesel oil as a fuel. Unlike a gasoline engine that uses spark plugs, a diesel engine uses the heat generated by the compression of the diesel fuel-air mixture in the cylinder for ignition. Diesel engines are used as prime movers on drilling rigs.

diff. difference.

differential wall pipe sticking. The adherence of a drillstring to the sides of a well due to suction.

digital. Data that have been converted into numbers. Seismic data are digitized in a binary number system in which the seismic reflections are represented either by the number 0 or 1. This allows the data to be recorded and stored on magnetic tapes and processed by computers. Seismic data used to be recorded by analog on a sheet of paper; cf. analog.

dike. A layer of igneous intrusion that cuts preexisting layers; cf. sill.

DIL. dual induction log.

dilut. diluted.

dim spot. A portion of a seismic reflector with a less intense reflection amplitude. A porous or gas-saturated reef overlain by shale can cause a dim spot; cf. bright spot.

dip. The angle and direction in which a plane such as a sedimentary rock layer or fault goes down in the ground. It is measured at right angles to the strike; cf. strike.

dipmeter and **dip log.** A wireline well log that measures the dip of each rock layer in a well. The data are plotted on a tadpole or stick plot. (DM and DIP)

dip-slip fault. A fault with predominately vertical displacement. It can be either a normal or reverse dip-slip fault; cf. strike-slip fault.

dir and **direc.** direction.

direct current. Electrical current that flows only in one direction. It is made by batteries and used in flash lights and to start engines; cf. alternating current. (DC)

direct hydrocarbon indicator. A bright spot, flat spot, or other evidence of a petroleum deposit on a seismic record. (DHI)

directional drilling. Drilling a well (deviated well) out at an angle on purpose. The well is drilled with either a steerable downhole assembly or a rotary steerable motor. (deviation drilling)

directional survey. A well survey that measures the angle and orientation of the wellbore using a magnetic compass or gyroscope. (dir sur and DS)

dir sur. directional survey.

dirty sands. Poorly sorted sands; cf. clean sands.

disc. discovery.

disch. discharge.

disconformity. An ancient, erosional channel. The sedimentary rock layers above and below the disconformity are parallel. It is a type of unconformity.

discovery well. A well that locates a new petroleum deposit, either (1) a new field or (2) a new reservoir.

disman. dismantled.

displ. (1) displace and (2) displaced.

displacement efficiency. The ratio of the volume of oil sweep divided by the volume of oil in place in a reservoir during waterflood or enhanced oil recovery.

disposal well. A well used to inject an unwanted fluid, usually oilfield brine, into the subsurface; cf. injection well.

dissolved-gas drive. A reservoir drive in which the drop in reservoir pressure during production causes dissolved gas to bubble out of the oil and force the oil through the reservoir rock. It has a relatively low oil recovery efficiency. (solution gas and depletion drive)

dissolved-gas/oil ratio. The standard cubic feet of natural gas dissolved in one barrel of oil in the reservoir. (formation and solution gas/oil ratio)

DIST. distillate.

distillate. *See* condensate.

distributary. A river channel outlet on a delta.

distributary mouth bar. A sand bar deposited in front of a distributary on a delta.

disturbed belt. A zone of thrust faults that moved during the formation of a mountain range. (overthrust belt)

division order. A form that establishes the distribution of production revenues and the assessment of costs to working interest owners on a well or lease on fee land. (D.O.)

division order analyst. A person responsible for the distribution of oil and gas production revenues and maintenance of division orders.

dk. dark.

D/L. density log.

DM. (1) drilling mud and (2) datum.

DMC. daily mud cost.

dn. down.

dns. dense.

DO. (1) drill out, (2) drilled out, (3) dolomite, (4) drilling obligation, and (5) developmental oil well.

D.O. division orders.

DOC. drilled out cement.

DOD. drilled out depth.

DOE. Department of Energy (www.doe.gov).

doghouse. The enclosure that houses the seismic recording equipment on a recording truck or seismic ship.

dogleg. A relatively sharp turn (>3° per 100 ft) in a well.

Dol, dol, and **dolo.** dolomite.

dolomite. A mineral composed of $CaMg(CO_3)_2$. It is formed by the natural alteration of calcite. A rock composed of dolomite is called dolostone. (DO, Dol, dol, and dolo)

dolostone. A sedimentary rock composed primarily of dolomite mineral grains. It forms from the natural chemical alteration of limestone and is often a reservoir rock.

dom. dominate.

dome. A circular or elliptical uplift in sedimentary rocks. It can form a petroleum trap. (D)

DOP. drilled out plug.

double. Two tubular joints; cf. single, treble, and fourble.

double-barrel separator. A separator with two horizontal steel cylinders mounted vertically. The upper cylinder receives the produced fluids and makes an initial gas-liquid separation. The lower cylinder completes the oil-water separation. (double-tube separator)

double-pole unit. A well-servicing unit with two steel, telescoping tubes that are braced together for a mast. It is more efficient that a single-pole unit because the rods can be hung in the mast in doubles and the tubing in singles. The unit can have either one or two drums and is operated by a three- or four-person crew; cf. single-pole unit.

double section. The same section of rock encountered twice by drilling through a reverse fault; cf. lost section. (repeated section)

double-tube separator. See double-barrel separator.

downdip. In a direction located down the angle of a plane such as a sedimentary rock layer or fault; cf. updip.

downdraw. The difference between static and flowing pressure in a well.

downhole mud and **turbine motor.** A motor that is driven by drilling mud pumped down the drillstring. It rotates the bit located below it.

downthrown. The side of a dip-slip fault that moved down; cf. upthrown.

down-to-the-basin fault. A dip-slip fault that moves down on the basin side. *See also* growth fault.

DP. (1) drillpipe and (2) dew point.

D/P. drilled plug.

dpg. deepening.

dpn. deepen.

DPT. deep pool test.

dpt. depth.

DPU. drillpipe unloaded.

dr. (1) drive and (2) drum.

drag fold. A fold formed along a fault plane. It is caused by friction of one side of the fault against the other side when the fault moved.

drawdown test. A type of pressure transient test that measures the shut-in bottom-hole pressure; then the pressure change as a well is put on production, and the pressure drops to a stable, flowing pressure.

drawworks. A drum in a steel frame used on the floor of a drilling rig to raise and lower equipment in a well. It is driven by the prime movers. Hoisting line is wound around the reel. (drwks)

dress. (1) To mill a fish to prepare the surface for a fishing tool. (2) To sharpen a drag bit.

drift angle. The angle of a wellbore from vertical. (deviation)

drillable. An object made of a substance that is designed to be run into a well and then removed by drilling it into particles and retrieving the particles.

drill bit. The tool on the end of a drillstring that cuts the well. Two types of drill bits are rotary-cone and fixed cutter. The most common of the rotary-cone bits is the tricone bit, and the most common fixed cutter bit is the polycrystalline diamond compact (PDC) bit.

drill break. A change in drill penetration rate recorded on the rate of penetration curve on a mud log or drilling time log. It is usually the result of drilling into a different rock such as from shale into limestone. (drilling break) (DB)

drill collar. A relatively heavy, thick-walled, large-diameter pipe run on the bottom of a drillstring above the bit. It comes in 31-ft sections with both a male-threaded and female-threaded end. (DC)

driller. The person in charge of the drilling rig crew on that tour. The driller operates the machinery. (drlr)

driller's cabin. A climate-controlled enclosure on the drill floor where the driller sits in a special chair and overlooks the drill floor through glass. Computer and touch screens and joy sticks allow the driller to monitor and control drilling and the iron roughnecks from the cabin.

driller's depth. *See* driller's total depth.

driller's method. A technique used to control a well that has a kick. After the blowout preventers have been thrown, the kick-diluted mud is replaced with original mud under pressure during the first circulation. The original mud under pressure is then replaced with kill mud during the second circulation; cf. wait-and-weigh method.

driller's total depth. The depth of a well to the bottom measured from the rotary table by driller's counting the drillpipe joints in the well. (rotary total depth)

driller's report. *See* tour report.

drill floor. The elevated, flat, steel surface on which the derrick or mast sits and most of the drilling activity occurs. It is supported by the substructure. (derrick or rig floor) (DF)

drill-in fluid. A liquid that replaces drilling mud when the well is being drilled into the gas or oil pay zone. It prevents formation damage but retains pressure control.

drilling and spacing unit. The area, such as 40 acres, upon which only one producing well can be located. It is declared by a government regulatory agency. (DSU)

drilling barge. A barge with a drilling rig mounted on it. It is used for exploratory drilling in shallow, protected waters.

drilling break. *See* drill break.

drilling console. A panel on the drill floor. It contains the weight indicator, mud pump pressure, rotary table torque, pump strokes, rate of penetration, and other indicators.

drilling contract. The legal agreement between the operator and the drilling contractor to drill a well. Three types are footage, day rate, and turnkey contracts. The International Association of Drilling Contractors has published standard contracts.

drilling contractor. A company that owns and operates drilling rigs.

drilling fluid. The fluid that is circulated down the well during drilling. It is either drilling mud or a gas such as foam. *See also* drilling mud.

drilling fluids engineer. A service or oil company employee who monitors the properties of drilling mud being used on a well. The engineer is responsible for making changes in the mud properties (conditioning) when necessary. (mud man)

drilling line. Wireline made of several strands of braided steel cable wound around a fiber or steel core. It is commonly between 1 and 1⅝ in. in diameter. The line is spooled around a reel in the drawworks on a drilling rig and is used to raise and lower equipment. (hoisting line)

drilling liner. A liner string run in a well as the well is being drilled. It serves the same purpose as a casing string but does not run up to the surface and saves money.

drilling mud. A viscous mixture of clay (usually bentonite) and additives (chemicals) with either (1) freshwater (water-based drilling mud), (2) diesel oil (oil-based drilling mud), (3) synthetic oil (synthetic-based drilling mud), or (4) an emulsion of water with droplets of oil. Mud is circulated on a rotary drilling rig to cool the bit, remove rock chips, and control subsurface fluids. Water-based drilling mud is commonly used on land and synthetic-based drilling mud is used offshore. (DM)

drilling pad. A rectangular area, typically 4 to 6 acres in area, that is bulldozed flat and sometimes covered with gravel to receive a drilling rig. Four to eight horizontal drain holes can be drilled from one pad using deviation drilling; cf. well pad.

drilling recorder. *See* geolograph.

drilling rig. The machinery used to drill a well. Modern drilling rig are rotary drilling rigs. The four major systems on a drilling rig are (1) power, (2) hoisting, (3) rotary, and (4) circulating. Drilling rigs are either mechanical or diesel-electric depending on which power system they use.

drilling spool. A steel cylinder located between the rams on a blowout-preventer stack. The kill and choke lines are attached to it.

drilling-time log. A log showing the rate of drill bit penetration with depth, usually in minutes per foot or meter.

drillpipe. A steel tubular that is 30 ft long and is threaded on both ends. Each section is called a joint. (DP)

drillship. A ship with drilling rig aboard that drills through a hole (moon pool) in its hull. It is kept on position by dynamic positioning. Drillships are used to drill wells in very deep water.

drillsite. The location for a drilling rig.

drillstem test. A test made by running a drillstem made of drillpipe with packers in a well. The packers isolate the zone to be tested. A valve is opened in the drillstem and fluids from the zone can flow into the drillstem. It is a temporary completion of the well. (DST)

drillstring. The kelly, drillpipe, drill collars, subs, and bit that are rotated in the well.

drill to earn. A type of joint operating agreement in which one party is obliged to pay for a certain percentage of the drilling costs to earn a working interest in another party's acreage.

drip, drip gasoline, or **drips.** A common term for condensate that is produced by production and transportation of natural gas. *See* condensate.

drk. derrick.

DRL and **drl.** drill.

drld. drilled.

drlg. drilling.

drlr. driller.

DROI. discounted return on investment.

drop angle. To decrease the deviation of a well; cf. build and maintain angle.

drpt. dropped.

drwks. drawworks.

dry gas. Pure methane (CH_4) gas. It is a gas under both subsurface reservoir and surface conditions; cf. lean gas, rich gas, and wet gas. (DG)

dry hole. A well that was drilled and did not encounter commercial amounts of petroleum; cf. producer. (duster)

dry hole agreement or **contribution.** A type of well support agreement in which a nondrilling party will pay monies to a drilling party that drills a well at a specific location and it is a dry hole. Information from that well is shared with the nondrilling party; cf. bottom-hole contribution and acreage contribution agreements. (DHC)

DS. (1) directional survey and (2) drillstem.

ds. dense.

DSI. drilling suspended indefinitely.

DSO. dead oil show.

DST. drillstem test.

DSU. drilling and spacing unit.

DT. (1) drilling time and (2) drilled tight.

DTD. driller's total depth.

dual completion. A system to keep the production from two zones in a well separate. Usually two tubing strings and two packers are used. There are two surface pumping units or a double-wing production tree on the well; cf. commingle. (DC)

dual gradient drilling. A method for drilling a well in deep water without excessive mud pressure in the well. The well is filled with seawater from the floater down to the mudline and circulating drilling mud from the mudline to the bottom of the well. The circulating drilling mud is pumped from the mudline up to the floater.

dual induction log. An wireline log that gives a medium- and deep-induction resistivity measurement. (DIL)

Dubai. A benchmark crude oil used in the Middle East. It has 31 °API gravity and 2% sulfur.

dune. A hill of sand shaped by blowing wind or flowing water.

duplex pump. A mud pump with two double-acting pistons in cylinders. The mud is pumped on both the forward and backward strokes of the pistons; cf. triplex pump.

duster. A well that did not encounter commercial amounts of petroleum. (dry hole)

DWA. drilling with air.

DWC. daily well cost.

DWM. drilling with mud.

DWT. (1) deadweight ton and (2) deadweight tester.

dynamic positioning. The use of navigational satellites and computers to continuously recalculate a drillship's location and keep the drillship on station with thrusters.

ea. earthy.

E&P. exploration and production.

earth pressure. The pressure on rocks at a specific depth. It is caused by the weight of the overlying rocks; cf. fluid pressure. (lithostatic pressure)

earth science. *See* geology.

ECD. equivalent circulating density.

economic limit. The time in the history of a well in which the revenue from production equals production costs. The well is either stimulated or plugged and abandoned.

edge water. Water located in the reservoir to the side of the oil; cf. bottomwater.

EEB. energy equivalent barrel.

eff. effective.

effective permeability. The permeability of a fluid when it shares the pore space with another fluid.

effective porosity. The percent porosity including only interconnecting pores; cf. total porosity.

EHP. effective horsepower.

EIA. Energy Information Administration (www.eia.gov).

elastomer. rubber-like material.

elec. electric.

electric line. *See* electric wireline.

electrical log. A wireline resistivity log. It is often run with a spontaneous potential or natural gamma ray log. (EL and E-log)

electric or electric-drive rig. *See* diesel-electric rig.

electric submergible pump. An electrical motor that drives a centrifugal pump with rotating blades on a shaft. It is located on the bottom of a tubing string and is used for oil well artificial lift. (sub pump) (ESP)

electric wireline. Wire rope with insulated, electrical wires in the core. It is used to run logging and jet perforating tools in a well. (E-line or electric line)

electrostatic precipitator. A separator that uses charged electrode plates to separate an emulsion.

elephant. An oil field with over 1 billion barrels of recoverable oil.

ELEV and **elev.** elevation.

elevators. A set of metal clamps that are hung from the crown block. They are used to clamp onto tubulars such as drillpipe or casing to raise and lower them in the well.

el gr. elevation of ground.

E-line. *See* electric wireline.

E-log. electrical log. (SP and R)

El/T. electric log tops.

emul. emulsion.

emulsion. Droplets of one liquid suspended in a different liquid such as cream in milk. An oil-in-water emulsion is a common produced fluid from oil wells. (emul)

emulsion mud. Drilling mud made with water containing suspended droplets of oil. The oil improves the lubricating qualities of the mud.

energy equivalent barrel. *See* barrels of oil equivalent. (EEB)

eng. engine.

enhanced oil recovery. The injection of fluids that are not found naturally in a producing reservoir down injection wells into the depleted reservoir to recover more oil (e.g., inert gas injection and steamflood). (EOR) *See also* tertiary recovery.

Eocene. An epoch of geological time from 55.8 to 33.9 million years ago. It is part of the Tertiary period. (Eoc)

eolian. Formed by blowing wind.

EOR. enhanced oil recovery.

epoch. A geological time subdivision of periods such as Miocene epoch. It is subdivided into ages; cf. era.

equip. equipment.

equiv. equivalent.

equivalent circulating density. The density of the circulating drilling mud in a well at the level of a formation in the well. It is slightly above the density of static drilling mud because of friction of the drilling mud in the well during circulation. (ECD)

era. A major time division of earth history such as Paleozoic era. It is subdivided into periods; cf. epoch and age.

ESP. electric submersible pump.

est. (1) estimate, (2) estimated, and (3) estate.

estab. (1) establish and (2) established.

ETH and **eth.** ethane.

ethane. A hydrocarbon composed of C_2H_6. It is a gas under surface conditions and is found in natural gas. (C_2, ETH, and eth)

EUR. estimated or expected ultimate recovery.

eval. (1) evaluate and (2) evaluated.

Evap and **evap.** evaporate.

EW. (1) electric weld and (2) exploratory well.

ex. excellent.

exist. existing.

exp. expense.

expandable casing. Casing made of special steel. It can be run into a well and then have its diameter expanded up to 30% by hydraulically pumping an expander plug through it. Expandable casing is used to make a monobore well.

expander plant. An installation that uses natural gas to drive a turbine to cool the gas and remove natural gas liquids. (cryogenic plant)

expansion-gas drive. A gas field reservoir drive in which the expanding gas produces the energy to force the gas through the reservoir rock. (volumetric drive)

expir. (1) expire, (2) expiring, and (3) expired.

expl. (1) exploration and (2) exploratory.

exploration geologist. A geologist who specialized in the search for petroleum; cf. developmental geologist.

exploratory well. A well drilled to locate new oil and gas reserves. It can be drilled in an area that has no production (new-field exploratory well), or drilled to test a new reservoir rock that has no current production in a producing area (new-pool exploratory well) that is either shallower (shallower pool test) or deeper (deeper pool test) than current production. An exploratory well can also be drilled to significantly extend the limits of a discovered field and to significantly extend the limits of a discovered reservoir. (outpost or extension test, step-out well, wildcat well)

explosive fracturing. To explode nitroglycerin in a torpedo at reservoir depth in a well to fracture the reservoir and stimulate production. (well shooting)

exp plg. expendable plug.

exst. existing.

ext. (1) extended and (2) extension.

extended-reach well. A deviated well with a relatively large horizontal distance between the surface location of the well and the bottom of the well.

extender. An additive to a fluid that reduces its cost (e.g., bentonite in a cement slurry).

extension well. A well that significantly increases the productive area of a field. It is called an outpost well before it is successful.

extr. extremely.

F. (1) factor, (2) flowed, (3) flowing, (4) foam, (5) Fahrenheit, and (6) filtrate.

f. fine.

F/. (1) flowed, (2) flowing, and (4) fractured with.

FAB. faint air blow.

facies. A distinctive part of a rock layer such as a sandstone facies. A lithofacies is based on rock composition such as sandstone facies, whereas a biofacies is based on fossil content.

Fahrenheit. The English system temperature scale. Freshwater freezes at 32° and boils at 212°. To convert Fahrenheit temperature to centigrade or Celsius temperature, subtract 32° and multiply by $5/9$. (F)

fail. failure.

fairway. The area along which a petroleum play occurs. (trend)

farmer's gas. The produced gas that goes free to the lessor for use as part of the lease agreement. It is often used to drive irrigation pumps.

farmin. (1) A lease obtained from another company for drilling in return for a consideration such as a royalty. (2) To receive a farmin; cf. farmout. (FI)

farmout. (1) A lease given to another company for drilling in return for a consideration such as a royalty. (2) To give a farmout; cf. farmin. (FO and F/O)

FARO. flowed or flowing at rate of.

fault. A break in the rocks along which there has been movement of one side relative to the other side. Faults are either dip slip or strike slip; cf. joint. (Flt and flt)

FBH. flowing by heads.

FBHP. flowing bottom-hole pressure.

FBHPF. final bottom-hole pressure flowing.

FBHPSI. final bottom-hole pressure shut-in.

FC. float collar.

FCP. flowing casing pressure.

FCV. flow control valve.

FDC. formation density compensated log.

FDL. formation density log.

fdn. foundation.

fed. federal.

feedstock. A chemical refined from hydrocarbons and used to produce petrochemicals.

fee land. Private land that has both a private surface and mineral rights owner.

FEL. (1) feet from east line and (2) from east line.

F/EL. from the east line.

feldspar. A common group of minerals that are potassium aluminum silicates (potassium feldspar or orthoclase) or sodium-calcium aluminum silicates (plagioclase feldspar). Feldspars are common in igneous and metamorphic rocks and immature sedimentary rocks. They can be white, gray, pink, or pale yellow and can decay to form clay minerals.

fence diagram. A three-dimensional representation of wells and the geological cross sections between them. Cross sections between the wells are called panels that close to form the fence.

FER. fluid energy rate.

FF. (1) fishing for, (2) frac finder (log), and (3) full of fluid.

FFP. final flowing pressure.

FG. full gauge.

FGIH. finish going in hole.

FGIW. finish going in with.

F/GOR. formation gas/oil ratio.

f-gr. fine grained.

FH. full hole.

FHH. final hydrostatic head.

FHP. final hydrostatic pressure.

FI. (1) farmin and (2) flow indicator.

FIC. flow indicating controller.

field. The surface area directly above one or more producing reservoirs on the same trap such as an anticline. (fld)

field print. The original copy of a wireline well log that is made in the logging truck; cf. final print.

field superintendent. An engineer who is in charge of production from a field. The field superintendent gives orders to the production foremen.

FIH. fluid in hole.

fill to spill. A petroleum trap that is filled to capacity. It is filled with oil and/or gas down to the spill point on the trap.

filt. filtrate.

filter cake. A cylinder of clay particles that were plastered against the sides of the well by the drilling mud during drilling. (mudcake)

filtrate. *See* mud filtrate. (F and filt)

fin. (1) final, (2) finish, and (3) finished.

final print. The last copy of a wireline well log that has been cleaned up, computer processed, and printed in the office; cf. field print.

fin drlg. finished drilling.

fireflood. An enhanced oil recovery process in which the subsurface oil is set afire. The heat makes the oil more fluid, and the gases generated by the fire drive the oil to producing wells as air is pumped down injection wells. (in-situ combustion)

fish. (1) A tool or broken pipe that has fallen to the bottom of a well. (junk) (FSH) (2) Fishing.

fishing. A process in which fishing tools are used to retrieve an object (fish) on the bottom of the well. (fish and fsg)

fishing string. A length of tubulars with subs that is run in a well with a fishing tool on the end.

fishing tools. Tools leased from a service company to fish for a fish in a well (e.g., spear, overshoot, and junk basket).

FIT. formation interval tester.

five-spot pattern. A common pattern of injector and producing wells used for a waterflood. Four injecting wells are located at the corners of a square and the producing well is in the center; cf. line drive.

fixed cutter bit. A solid steel bit with no moving parts. Cutters or natural diamonds shear the rock. A PDC bit is a fixed cutter bit.

fixed production platform. A relatively permanent offshore platform with treaters for produced fluids. It has legs that sit on the ocean bottom. Two types are (1) steel jacket and (2) gravity storage; cf. tension leg platform and spar.

FL or **fl.** (1) fluid and (2) fluid level.

FL. (1) floor and (2) flowline.

fl/. (1) flowed and (2) flowing.

flange. A raised edge or projection on the end of a pipe or connection. It can be connected with another flange with bolts or threads or by welding.

flange up. To finish a job.

flare. (1) A flame of burning gas. (2) To burn unwanted natural gas.

flash gas. Wet gas that is separated from crude oil in a low-pressure separator. The separator can be located on a lease or pipeline.

flat spot. A flat seismic reflector in sedimentary rock layers that are not horizontal. It can be off a gas-liquid contact.

fld. (1) failed and (2) field.

flg. flowing.

flint. A very hard sedimentary rock composed of amorphous quartz. (chert)

flo. flow.

float collar. A short length of tubular that is run just above the bottom of a casing string. A one-way valve allows the casing string to float in the drilling mud as it is being run in the well. A constriction in the float collar stops the wiper plug as it is being pumped down the string during a cement job. (FC)

floater. A floating, drilling platform such as a semisubmersible or drillship.

flow-after-flow test. *See* multipoint test.

flow assurance. Any well intervention or production operation that is done to ensure continuous and safe flow of production from the reservoir to the processing facilities.

flow back or **flowback.** The frac fluid and possibly some formation water that flows to the surface after hydraulically fracturing a well. It is treated and them reinjected or disposed.

flow or **flowing by heads.** Intermediate flow of fluids from a well. (FBH)

flowing pressure. Pressure on a fluid as the fluid is flowing; cf. static pressure. (FP)

flowline. A steel, plastic, or fiberglass pipe that conducts (1) produced fluids from the wellhead to the separators, (2) oil from the separators to the stock tanks, or (3) gas to the treaters. (FL)

Flt and **flt.** fault.

flu. fluid.

fluid pound. A problem caused by gas in a downhole, sucker-rod pump.

fluid pressure. The pressure on fluids in the pores of rock at a specific depth. Normal fluid pressure is due to the weight of the overlying waters. (reservoir and formation pressure); cf. earth pressure.

fluor. (1) fluorescence and (2) fluorescent.

fluorescence. The emission of light by a substance under ultraviolet light. Some minerals fluoresce. A mud logger uses the fluorescence of crude oil extracted with a solvent from well cutting to identify oil shows. (fluor)

flw. (1) flowed and (2) flowing.

Flwg. PR. flowing pressure.

Fm and **fm.** formation.

FMS. formation microscanner.

Fm W. formation water.

fn. fine.

FNL. (1) feet from north line and (2) from north line.

F/NL. from the north line.

fnly. finely.

FO and **F/O.** farmout.

FO. (1) faulted out, (2) fuel oil, and (3) full open.

foam drilling. Air drilling with a detergent to form a foam and lift water from the well.

focused log. *See* laterolog.

fold. (1) The number of reflections (traces) off the same subsurface point that are combined in common depth point stacking to form a single reflection. (trace) (2) A bend in sedimentary rock layers such as an anticline or syncline.

footage drilling contract. A drilling contract based on a cost per foot to drill to contract depth. It is commonly used on land; cf. daywork and turnkey drilling contract.

footwall. The side of the fault that protrudes under the other side; cf. hanging wall.

foraminifera. A small, one-cell animal that floats in the ocean or lives on the bottom of the ocean. Many have shells of calcium carbonate. It is a common microfossil. (forams) (Foram and foram)

formation. A mappable, sedimentary rock layer with a sharp top and bottom. The formation is given a two-part name such as Bartlesville Sandstone, Arbuckle Limestone, and Coffeyville Formation. Formations can be subdivided into members and adjacent formations can be combined to form a group. (Fm and fm)

formation damage. A decrease in the permeability of a reservoir rock adjacent to a wellbore. It can be caused by mud filtrate that is forced into the pores during drilling. (skin damage)

formation density log. A radioactive wireline well log used to determine the density and porosity of rocks. It bombards the rocks with high-speed neutrons and measures either the returning slow-speed neutrons or gamma rays. (gamma-gamma log) (FDL)

formation gas/oil ratio. *See* dissolved gas/oil ratio. (F/GOR)

formation pressure. *See* fluid pressure.

formation volume factor. The number of barrels of reservoir oil that shrinks to one stock-tank barrel of oil on the surface after the pressure has decreased and the gas has bubbled out. (FVF and B) *See also* reservoir barrel and stock-tank barrel of oil.

foss. fossiliferous.

fossil. The preserved remains of an ancient plant or animal. Fossils can be either macrofossils or microfossils, depending on their size. (Foss and foss)

fossil assemblage. A group of fossils that identifies a particular geologic time or a rock zone.

fossil fuel. A fuel formed by ancient organic matter (e.g., crude oil, natural gas, and coal).

4-C seismic. A marine seismic survey that uses three vibration detectors and a hydrophone at each location to record both conventional compressional waves (P waves) and shear waves (S waves). It is used to better identify rock types and locate fractures.

4-D and **4D seismic.** The seismic reflection differences between several 3-D seismic surveys run at different times over the same reservoir during production from that oil field. Changes in seismic response such as amplitude from the reservoir can show the flow of fluids through the reservoir and locate undrained areas. (time lapse seismic)

FOT. flowing on test.

fourble. Four tubular joints; cf. single, double, and treble.

FP. (1) flowing pressure, (2) final pressure, and (3) free point.

FPI. free point indicator.

FPSO. floating, production, storage, and offloading.

FPSO vessel. A ship that is stationed above or near an offshore oil field. Produced fluids from subsea wells are brought by flowlines and a riser to the deck of the vessel where they are separated and treated. The purified oil is then transferred to a shuttle tanker.

FR. flow recorder.

fr. (1) fair, (2) from, and (3) front.

Frac. fracture.

Frac and **frac.** fracture.

frac job or **fracing.** *See* hydraulic fracturing.

fracking. The general public term used for hydraulic fracing.

frac pac and **frac/pac.** A hydraulic fracturing technique that uses a viscous gel as the frac fluid with a relatively large amount of proppants. The fractures are relatively short but wide. It is used on high-permeability loose sand reservoirs; cf. slickwater frac.

fractionating. The separation of crude oil by heating and boiling off different components at different temperatures.

frag. fragment.

free gas cap. The uppermost portion of a saturated oil reservoir. The pores of the reservoir rock are occupied by natural gas.

free gas-cap expansion drive. A reservoir drive in which the expanding gas in the free gas cap drives the oil through the reservoir rock. It has a relatively moderate oil recovery efficiency; cf. dissolved gas and free gas cap drive.

free point. The depth in a well that is just above where tubulars are stuck in a well (stuck point). (FP)

free water. Water that readily separates from oil by gravity; cf. emulsion.

free-water knockout. A horizontal or vertical separator that uses gravity to separate gas, oil, and water. (FWKO)

freq. (1) frequent and (2) frequency.

freshwater. Water that contains less than one-part-per-thousand salt; cf. brine.

friable. A rock such as sandstone that easily crumbles or is broken apart.

FRR. final report for rig.

frs. fresh.

FRW. final report for well.

fsg. fishing.

FSH. fish.

FSIP. final shut-in pressure.

FSL. (1) feet from south line and (2) from south line.

F/SL. from the south line.

FSP. flowing surface pressure.

FT. formation test.

F/T. flowline temperature.

ftg. (1) fittings and (2) footage.

ft lb or **ft-lbf.** foot-pound.

ft/min. feet per minute.

FTP. (1) flowing tubing pressure and (2) final tubing pressure.

F Trap. fault trap.

ft/sec. feet per second.

fulcrum assembly. A downhole assembly that uses the sag caused by the weight of a drill collar between two stabilizers to lift the bit and increase the angle (make angle) in a deviated well. *See also* packed hole and pendulum assembly.

full-diameter core. A cylinder of rock that was drilled (cored) from a well. It is between 3½ and 5 in. in diameter; cf. sidewall core.

funnel viscosity. A measure of drilling mud viscosity. It is recorded as the number of seconds that the mud takes to flow through a Marsh funnel. (FV)

FV. funnel viscosity.

FVF. formation volume factor.

FW. freshwater.

FWC. field wildcat.

FWKO. freewater knockout.

FWL. (1) feet from west line and (2) from west line.

F/WL. from the west line.

fx. finely crystalline.

G. (1) gas, (2) gas lift, (3) gallons, and (4) billion.

g. (1) gram or (2) gravity force.

GA. gallons of acid.

ga. gauge.

GAF. gross acre feet.

gage. *See* gauge.

gal. gallons.

gal/min. gallons per minute.

gamma-gamma log. *See* formation density log.

gamma ray log. *See* natural gamma ray log. (GR and GRL)

G&MCO. gas and mud-cut oil.

G&O. gas and oil.

G&OCM. gas and oil-cut mud.

G&W. gas and water.

gas. *See* natural gas. (G)

gas cap. *See* free gas cap.

gas-cap drive. *See* free gas-cap expansion drive.

gas chromatograph. A chemical device used to quantitatively analyze the amount of individual hydrocarbon components in a sample. It is used in mud logging to determine the amount of methane, ethane, propane, butane, and pentane in natural gas from gas-cut mud samples.

gas compressibility factor. *See* Z factor.

gas conditioning. The removal of impurities such as water, acid gases, and solids from natural gas in the field to meet pipeline contract specifications.

gas cut. diluted with gas.

gas detector. An analytical device used in mud logging to separate natural gas from gas-cut mud during drilling a well. It measures the total amount of gas dissolved in the drilling mud.

gas deviation factor. *See* Z factor.

gas effect. A divergence of porosities calculated from the neutron porosity and formation density logs on the same rock. It is caused by gas in the reservoir rock.

gas injection. Injection of natural gas into an oil reservoir or free gas cap to maintain reservoir pressure and produce more oil.

gas in place. The amount of gas in the pores of a subsurface reservoir; cf. recoverable gas. (GIP)

gas jack. A compressor used to increase the pressure on natural gas from a gas well to pipeline pressure.

gas lift. An artificial-lift method for oil wells. An inert gas called lift gas (usually natural gas) is injected into the casing-tubing annulus, through gas-lift valves, and into the tubing to form bubbles that raise the produced liquids. (G and GL)

gas lift valve. A pressure-activated valve in the tubing string of a gas lift well. It allows lift gas, usually natural gas, to flow into the tubing.

gas lock. The failure of a downhole, sucker-rod pump because of gas filling the pump.

gas meter. An instrument used to measure the velocity of gas. Gas volume can be derived from the gas velocity. The most common type used for gas wells is an orifice meter. It measures the pressure drop caused by a change in gas velocity as the gas flows through a restriction (orifice) in the pipe.

gaso. gasoline.

gas/oil contact. The boundary between oil and gas in a reservoir. (GOC)

gas/oil ratio. The number of standard cubic feet of natural gas produced with each barrel of oil. The amount of natural gas in standard cubic feet dissolved in a barrel of oil in the subsurface reservoir is the dissolved, reservoir, or solution gas/oil ratio. The amount of natural gas in standard cubic feet produced with each barrel of oil is the producing gas/oil ratio. A gas well has a producing gas/oil ratio of greater than 20,000, and an oil well has less than 5,000. (GOR)

gas and **gasoline plant.** An installation that removes natural gas liquids from natural gas by cooling or absorption. (natural gas processing plant) (GP)

gas sand. A driller's term for sandstone that contains gas.

gas shale. A black shale source rock that has generated natural gas and crude oil. Although much of the gas and oil has migrated out into overlying reservoirs or leaked onto the surface, the gas shale still contains some gas and sometimes condensate and oil. Wells into a gas shale must be hydraulically fractured with a slickwater frac; e.g., Barnett Shale.

gas show. Natural gas dissolved in drilling mud. It is detected by mud logging while drilling a well. (GS)

gas unit. A measure made by a mud logger of the amount of natural gas dissolved in drilling mud (gas-cut mud). (GU)

gathering line. A flowline that connects a gas well to the gas pipeline. (G/L)

gathering system. A system of flowlines that conducts produced fluids from wells to a central processing unit.

gauge. (1) To measure. (2) The diameter of a bit, wellbore, or tubular. (3) A measuring instrument; e.g., pressure gauge. (ga and gage) (GGE)

gauge hole. A wellbore with a specific minimum diameter.

gauger. A person responsible for measuring the amount and quality of oil in stock tanks.

gauge table. A chart that relates the height of crude oil in a stock tank to the volume of the oil. (tank table)

gauge tape. A metal tape on a reel with a brass weight on the end. It is marked in ⅛-in. increments and is used to measure the height of oil in a stock tank.

GB. (1) good blow and (2) gun barrel.

Gb. billion barrels.

GBDA. gallons of breakdown acid.

GC. gas cut.

GCDM. gas-cut drilling mud.

GCM. gas-cut mud.

GCMS. gas chromatography-mass spectrometry.

GCPD. gallons of condensate per day.

GCR. gas/condensate ratio.

GCSW. gas-cut saltwater.

GCW. gas-cut water.

gd. good.

GDR. gas-distillate ratio.

gel. A homogeneous fluid consisting of dispersed, fine-grained particles that have coalesced to some degree in a liquid such as water. Drilling mud that consists of clay particles and freshwater is a gel.

gel strength. The ability of a fluid such as drilling mud to suspend solids. It is measured on a viscometer or gelometer and reported in lb/100 sq ft.

gen. generally.

geochemist. A geologist who uses chemistry to study source rocks, natural gas, and crude oil. Geochemistry uses soil samples to analyze for traces of hydrocarbons and bacteria that thrive on hydrocarbons to detect surface microseeps.

geographic and **geographical information system.** A computer system of hardware, software, and a database used to display, manipulate, and analyze geographical data at locations on the earth's surface. The information is organized in layers such as culture, topography and well locations. (GIS)

geol. (1) geological, (2) geologist, and (3) geology.

geologic map. A map showing where rock layers, usually formations of sedimentary rocks, crop out on the surface of the earth.

geologist. A scientist who identifies and studies rocks. A petroleum geologist searches for and helps to develop oil and gas deposits. (Geol)

geolograph. A recorder on the floor of a drilling rig that records drilling rate (rate of penetration) on a paper chart. (drilling recorder)

geology or **geoscience.** The study of the earth. It includes the matter that makes up the earth, the processes that act on that matter, and the history of the earth. (earth science)

Geop. (1) geophysical and (2) geophysicist.

geophone. A vibration detector used on land to detect subsurface echoes during a seismic survey. (jug)

geophysicist. A scientist who uses physics and mathematics to study the earth. A geophysicist uses surface methods such as seismic, magnetic, and gravity to image the subsurface and explore for petroleum. (Geop)

geopressure. *See* abnormal high pressure.

geopressured. A reservoir rock with abnormal high pressure.

geosteering. The use of a measurements-while-drilling system, a logging-while-drilling system, and a steerable downhole assembly to drill a horizontal drain well that stays in the pay zone.

geothermal gradient. The rate of temperature increase with depth in the earth.

GFLU. good fluorescence.

GGE. gauge.

GGW. gallons of gelled water.

GI. gas injection.

giant gas field. A gas field that has at least 3 Tcf of recoverable natural gas.

giant oil field. An oil field that has at least 500 million bbl of recoverable oil.

GIDP. gas in drillpipe.

GIH. going in hole.

GIP. gas in pipe.

GIS. geographic information system.

GL. (1) ground level and (2) gas lift.

gl. glassy.

G/L. gathering line.

Global Positioning System. A worldwide navigation method that uses radio signals of accurate time from satellites to obtain precise latitude, longitude, and altitude. It is especially useful in locating seismic lines. (GPS)

GLR. gas/liquid ratio.

gls. glass.

glycol absorber tower. A vertical, metal vessel that causes natural gas to bubble up through a liquid desiccant (glycol) to remove water.

GM. (1) gravity meter and (2) ground measurement (elevation).

GMA. gallons of mud acid.

gn. green.

gneiss. A metamorphic rock with light and dark bands of coarse-grained minerals. It is a basement rock.

GO. (1) gallons of oil and (2) gas odor.

goat pasture. Land that is not worth leasing.

GOC. gas/oil contact.

gone to water. A producing well that has started to produce large amounts of water.

GOR. gas/oil ratio.

gouge zone. The fractured mass of rocks along a fault plane.

GP. (1) gasoline plant and (2) gas pay.

G/P. gun perforate.

GPC. gas purchase contract.

GPM. gallons (natural gas liquids) per 1,000 standard cubic feet (Mcf) (natural gas).

GPS. Global Positioning System.

GR. (1) gamma ray and (2) granite.

Gr and gr. grain.

Gr. group.

gr. (1) grain and (2) ground.

GRA. gallons of regular acid.

graben. The down-dropped block between two normal faults; cf. horst.

grad. (1) grading, (2) gradual, and (3) gradually.

graded bed. A clastic sedimentary rock layer that is coarse grained (sand) on the bottom and grades upward to fine grained (clay) on the top. Turbidity currents can deposit a graded bed.

grainstone. A type of limestone in which the large, sand-sized grains are in contact with each other and fine-grained material is absent. (Grst)

gram. The unit of weight in the metric system. There are 453.59 grams in a pound. (g)

Gran or **gran.** granule.

gran. granite.

granite. The most common igneous rock that crystallizes in the subsurface (plutonic). Granite is coarse grained with a light, speckled texture. It is a common basement rock. (GR and Grt)

granite wash. A sandstone composed of sand grains from weathered granite. It can be a reservoir rock. (gran, Gran W, GW, and G.W.)

Gran W. granite wash.

gr API. °API gravity.

grass hopper. A common term for a sucker-rod pump.

grav. gravity.

gravel pack completion. A well completion used for unconsolidated reservoirs. A large cavity is reamed out in the reservoir. It is then filled with very well sorted, loose sand (gravel pack). A slotted or screen liner is run in the gravel pack. (GVLPK)

gravimeter. *See* gravity meter.

gravity. *See* °API gravity.

gravity-base storage production platform. A fixed production platform that has a large mass of steel and concrete on the bottom to hold it in position. The concrete is hollow, and the cells can be used for storage; cf. steel jacket production platform.

gravity drainage pool. An oil field in which the reservoir drive is gravity pulling the oil down into the wells. It can be very effective over a long period of time; cf. dissolved gas, free gas-cap expansion, and water drives.

gravity meter. An instrument that measures the acceleration of gravity. It detects variations in the density of the earth's crust. Its units of measurement are milligals. (gravimeter) (GM)

greywacke or **graywacke.** A poorly sorted, dark-colored sandstone; cf. orthoquartzite. (gywk)

GRD and **grd.** ground.

grdg. grading.

grd loc. grading location.

greenfield. An oil or gas field that is know but has not been developed; cf. brownfield.

GR log. natural gamma ray log.

grn. green.

ground truthing. To compare surface observations with remote sensing measurements such as satellite photographs.

groundwater. Water that occurs in the subsurface pores of sedimentary rocks.

group. (1) Several geophones that are connected together to record as a single channel. (2) Several adjacent formations that are similar in lithologies and are given a formal name such as the Chase Group. (Gr.)

group shoot. A seismic survey paid for and shared by several different exploration companies; cf. spec survey.

growth fault. A fault that occurs on land where sediments have been rapidly deposited along the margin of a basin. It is parallel to the shoreline and has a curved fault plane that is steepest near the surface and flattens with depth toward the basin. (down-to-the-basin fault) *See also* rollover anticline.

Grst. grainstone.

Grt. (1) granite and (2) grant (land).

grt. grant (land).

Grv. gravel.

gr wt. gross weight.

gry. gray.

GS. gas show.

GSC. gas sales contract.

GSG. good show of gas.

GSI. gas well shut-in.

GSO. good show of oil.

GSW. gallons of saltwater.

GTI. Gas Technology Institute (www.gti.org).

GTS. gas to surface (time).

GTSTM. gas too small to measure.

GU. gas unit.

guard log. *See* laterolog.

guide fossil. A distinctive fossil that represents a particular geologic time.

guide shoe. A short, metal cylinder with a rounded nose having a hole in the end. It is run on the end of a casing string to guide the string into the well.

gun. *See* perforating gun.

gun barrel separator. A steel, settling tank that uses gravity to separate a loose emulsion. (wash tank)

GV. gas volume.

gvl. gravel.

GVLPK. gravel packed.

GVNM. gas volume not measured.

GW and **G.W.** granite wash.

GW. gas well.

GWC. (1) gas-water contact and (2) gas-well gas.

GWI. gross working interest.

Gwke. greywacke.

GY, Gyp, or **gyp.** gypsum.

gypsum. A common salt mineral composed of $CaSO_4.2H_2O$; cf. anhydrite. (GY, Gyp, and gyp)

gypy. gypsiferous.

gywk. graywacke.

h. thickness.

HA. salt.

ha. hectares.

halite. A common salt mineral composed of NaCl. (Hal and hal)

hang. To arrange well logs according to a common, level reference-surface such as sea level going as a straight line through the logs. *See also* structural cross section and stratigraphic cross section.

hanger. A circular device that suspends a casing, tubing, or liner string in a well. It is attached to the top of the tubular by threads or slips.

hanging wall. The side of the fault that protrudes over the opposite side; cf. footwall.

hang rods. To pull sucker-rods out of a well and hang them in the rod hanger of a mast or derrick.

hard rock. Igneous and metamorphic rock. A hard rock geologist explores for ore deposits; cf. soft rock.

HBP. held by production.

HC. hydrocarbons.

HCl. hydrochloric acid.

HCPV. hydrocarbon pore volume.

hd. hard.

hdns. hardness.

hdr. header.

hd sd. hard sand.

header. (1) A large pipe into which several, smaller flowlines feed. (2) The well information at the top of a well log. (3) The seismic information at the side of a seismic record.

heater treater. A separator that uses heat from a fire tube to separate emulsions. (ht)

heave. The horizontal displacement on a fault.

heavy oil. High-density crude oil. It is defined by the U.S. Department of Energy as between 10 and 22.3 °API gravity. Heavy oil is relatively viscous, high in asphalt content, and generally high in sulfur content; cf. light oil.

heavyweight additive. An additive such as galena that is used to increase the density of a fluid such as drilling mud; cf. lightweight additive.

heavyweight drillpipe. Drillpipe that is intermediate in strength and weight, located between drill collars and drillpipe. It has the same outer diameter as drillpipe but a smaller inner diameter and comes in 30½ ft. joints. It is run on the drillstring between the drill collars and drillpipe. (HWDP)

hectare. A metric unit of land area equal to 10,000 sq meters and 2.47 acres. (ha)

held by production. A lease in effect during production on that lease. (HBP)

helirig. A drilling rig that can be broken down in modules and transported by helicopter.

Henry Hub. The pricing location for natural gas futures contracts traded on the New York Mercantile Exchange. It is located in Erath, Louisiana, where 13 natural gas pipelines converge.

hertz. Cycles per second; seismic source frequencies and AC electricity are described in hertz. (Hz)

hetr. heterogeneous.

HEX. heat exchanger.

hex. hexane.

HF. hydrofluoric acid.

HFO. hole full of oil.

HFSW. hole full of saltwater.

HFW. hole full of water.

HGCM. heavily gas-cut mud.

HGCW. heavily gas-cut water.

hgt. height.

HH. hydrostatic head.

HHP. hydraulic horsepower.

hi. high.

high-shrinkage oil. *See* volatile oil.

high-temperature, high-pressure well. A well with a bottom-hole pressure greater than 10,000 psa and a bottom-hole temperature greater than 300°F. (HTHP well)

HO&GCM. heavily oil- and gas-cut mud.

HOCM. heavily oil-cut mud.

HOCW. heavily oil-cut water.

hoisting line. *See* drilling line.

hole opener. A sub that uses roller cones to enlarge a well.

holiday. An area behind casing with no cement.

Holocene. An epoch of geological time from 10,000 years ago until the present. It is part of the Quaternary period. (Recent)

hom. homogeneous.

homocline. Inclined sedimentary rock layers with a constant dip.

hook. A curved steel fastener located below the traveling block on a drilling rig. It is used to suspend the swivel and drillstring in the well.

hor and **horiz.** horizontal.

horizon. a surface.

horizontal drainhole. A relatively short horizontal lateral drilled with a relatively short radius from an existing vertical well.

horizontal well. A well with a highly deviated lateral (70° to 110°) drilled parallel to the pay zone in a reservoir. The well is described by its radius of curvature (short-radius, medium-radius, and long-radius) and reach, the length of its horizontal section. It is drilled with a steerable downhole assembly or rotary steerable motor. The well is commonly completed open-hole or with a perforated liner.

horizontal section. The relatively flat portion of a horizontal drain well. The toe is the furthest point on the horizontal section and the heal is the closest. (lateral)

horizontal slice. A flat, seismic section made at a specific depth in time from 3-D seismic data. It shows where each seismic reflector intersects the slice. (time slice)

horizontal well. *See* horizontal drainhole.

horsehead. A steel plate used on the end of a walking beam to keep the pull on the sucker-rod string vertical.

horsehead pumper. A common term for a sucker-rod pump.

horsepower. A unit of power or rate of doing work that is applied to engines and motors. One horsepower is equal to 33,000 ft-lb/min. (HP and hp)

horst. A ridge between two normal faults; cf. graben.

host company. An oil company owned by a federal government. It operates only in that country.

hot oiler. A service company that removes wax from tubing in wells. A heated, tank truck is used to heat crude oil. The heated crude oil is pumped down the well to dissolve the wax and is then pumped out.

HP. (1) hydrostatic pressure, (2) high pressure, and (3) hydraulic pump.

HP and **hp.** horsepower.

HPF. holes per foot.

HPG. high-pressure gas.

hp-hr and **hp hr.** horsepower-hour.

hr. hours.

HRD. high-resolution dipmeter.

hrs. heirs.

HT and **ht.** high temperature.

ht. (1) heater treater and (2) heat treated.

HTHP. high-temperature, high-pressure.

H₂S. hydrogen sulfide.

huff 'n' puff. An enhanced oil recovery method used for heavy oil. A well is used to first pump steam into the subsurface reservoir to heat the oil and make it more fluid. The same well is then used to pump the heated, heavy oil. (cyclic steam injection)

hvly. heavily.

hvy. heavy.

HWDP. heavyweight drillpipe.

HX. heat exchanger.

HYD. hydraulic.

Hydc. hydrocarbons.

hydrate. A snow-like substance composed of methane gas locked in ice crystals. Hydrate occurs naturally in permafrost on land and below the seafloor in many areas. It can form from water in a flowline as the temperature of natural gas falls.

hydraulic fracturing. A well stimulation method in which a high-pressure frac liquid is pumped down a well to fracture the reservoir rock adjacent to the wellbore. Propping agents such as sand are suspended in the frac liquid and keep the fractures open after pumping has stopped. (frac job, fracing, hydrofrac, hydro-fracing, and sandfrac) *See also* fracpac and slickwater frac job.

hydraulic pump. An artificial lift system. A pump on the surface injects power oil into the well. The power oil drives a pump that is coupled to a sucker-rod pump on the bottom of the tubing string.

hydrocarbon recovery unit. A vessel that uses silica, activated charcoal, or molecular sieves to remove natural gas liquids from natural gas.

hydrocarbons. Molecules formed primarily by carbon and hydrogen atoms. Crude oil and natural gas are composed of hydrocarbon molecules. (Hydc and HC)

hydrochloric acid. A strong acid composed of hydrogen and chlorine. It is commonly used in acid jobs to stimulate a limestone reservoir. (HCl)

hydrofluoric acid. A strong acid composed of hydrogen and fluorine. It is used to dissolve silicate minerals in rocks. (HF) *See also* mud acid.

hydro-fracking. hydraulic fracturing.

hydrostatic head and **pressure.** The pressure due to the weight of the overlying water. It is about 45 psi/100 vertical ft of water. (HH and HP)

hydrofrac. *See* hydraulic fracturing.

hydrogen sulfide. A poisonous and corrosive gas composed of H_2S that can be found by itself or mixed with natural gas that is called sour gas.

hydrogen sulfide embrittlement. The weakening of steel by contact with H_2S.

hydrophone. A vibration detector used at sea to detect subsurface echoes during seismic exploration.

hydrostatic head and pressure. Fluid pressure in subsurface rocks due to the weight of the overlying fluids; cf. abnormal high pressure. (normal pressure) (HH and HP)

Hz. hertz.

IAB. initial air blow.

IADC. International Association of Drilling Contractors (www.iadc.org).

IAGC. International Association of Geophysical Contractors (www.iagc.org).

IB. impression block.

IBHP. initial bottom-hole pressure.

IBHPF. initial bottom-hole pressure flowing.

IBHPSI. initial bottom-hole pressure shut-in.

ice age. *See* Pleistocene.

ICV. inflow-control valve.

ID. inner diameter.

IF. internal flush.

IFP. initial flowing pressure.

IG. in gauge.

igneous rock. A rock formed by cooling and solidifying a hot, molten liquid. A volcanic igneous rock is formed by a liquid (lava) solidifying on the earth's surface whereas a plutonic igneous rock is formed by a liquid solidifying in the subsurface; e.g., granite and basalt (Ig and ig); cf. sedimentary and metamorphic rock.

IHH. initial hydrostatic head.

IHP. (1) indicated horsepower or (2) initial hydrostatic pressure.

ILD. deep induction.

ILM. medium induction.

image log. A well log that uses either conductivity or resistivity to image the rocks along a wellbore.

immature oil. Heavy oil generated at shallow depths in the oil window; cf. mature oil.

immed. immediate.

imp. impression.

impermeable. Rock that does not allow fluids to readily flow through it; cf. permeable. (aquitard)

imperv. impervious.

Imp gal. Imperial gallon.

impression block. A fishing tool used to determine the shape of a fish. It is a weight with wax or lead on the bottom that makes a cast of the fish. (IB)

improved oil recovery. The methods of waterflood and enhanced oil recovery that are used to produce more oil from a depleted reservoir.

inbd. interbedded.

incised valley fill. A long, sinuous body of sandstones and shales deposited during a cycle of sea level fall and rise. During sea level fall, a valley is incised (eroded) by a river. During sea level rise, the valley is filled with sediments from the river and from ocean tides and waves. It can be a petroleum reservoir.

incl. (1) inclusions, (2) include, (3) included, (4) including, and (5) inclusive.

incr. (1) increase and (2) increasing.

ind. indurated.

induction log. A wireline well log that measures the resistivity of the rocks and their fluids with an induced current created by coils in the logging tool. A dual induction log measures both medium and deep induction. (I, IL, and IEL)

indurated sediments. *See* consolidated sediments.

indr. indurated.

inert. A gas that does not burn (e.g., steam, carbon dioxide, and nitrogen).

inert gas injection. An enhanced oil recovery method in which an inert gas such as carbon dioxide or nitrogen is injected into a depleted reservoir to produce more oil.

infill drilling. Drilling between producing wells in a developed field to produce more petroleum at a faster rate.

inflow performance relationship. A plot of drawdown in a well versus production. (IPR)

ingr. intergranular.

inh. inhibit.

inh and **inhib.** inhibitor.

inhibitor. An additive to a fluid to retard a reaction. (inh and inhib); cf. accelerator.

init. initial.

initial potential. The maximum amount a well can potentially produce during the first 24 hours of production.

initial pressure. The original reservoir pressure before any production. (virgin and original pressure)

initial production. The first 24 hours of production from a well. (IP)

inj. (1) inject, (2) injection, and (3) injected.

injection well. A well used to pump fluids down into a producing reservoir for pressure maintenance, waterflood, or enhanced oil recovery; cf. disposal well. (IW)

Inj Pr. injection pressure.

ins. (1) insulate and (2) insulation.

in./sec. inches per second.

insert pump. A common type of oil well downhole pump driven by a sucker-rod string. It is run as a complete unit on the sucker-rod string through the tubing string. (rod pump)

insert tricone bit or **button tricone bit.** A tricone bit in which holes have been drilled into the steel cones and buttons of hard tungsten-steel carbide have been inserted. It is used to drill hard rocks; cf. milled-teeth tricone bit.

in-situ combustion. *See* fireflood.

insol. insoluble.

insp. (1) inspect, (2) inspecting, and (3) inspected.

inst. (1) install, (2) installing, and (3) installed.

instl. installation.

instr. (1) instrument and (2) instrumentation.

insul. insulate.

int. (1) interval, (2) interest, and (3) internal.

intangible drilling costs. Expenditures for drilling and completing a well that cannot be salvaged or recovered. They receive favorable tax consideration. (IDCs)

intbed. interbedded.

intelligent well. A well with downhole sensors for temperature, pressure, and flow velocity. A surface-controlled, downhole adjustable choke is used to regulate flow based on downhole conditions.

interbd. interbedded.

interfinger. A boundary between two rock types in which both form distinctive wedges protruding into each other.

intermediate casing. *See* protection casing.

interval transit time. The sound velocity through a rock measured by a sonic log. The units are in microseconds per foot. (Δt)

INTFP. intermediate flowing pressure.

intrusion or **intrusive body.** A plutonic igneous rock mass that was injected in a molten state into preexisting rock. (Intr or intr)

ints. intersect.

INTSIP. intermediate shut-in pressure.

intv and **intvl.** interval.

invaded zone. The area in a reservoir rock adjacent to the wellbore that has been flushed or diluted with mud filtrate.

I/O. input/output.

I.O.S.A. International Oil Scouts Association (www.oilscouts.com).

IP. (1) initial production, (2) initial potential, and (3) initial pressure.

I.P. in part.

IPAA. Independent Petroleum Association of America (www.ipaa.org).

IPF. intermediate potential flowing.

IPOF. initial production, open flow.

IPP. initial potential pumping.

IPPA. Independent Petroleum Producers of America (www.ippa.org).

IPR. inflow performance relationship.

IR. injection rate.

iron roughnecks. a mechanical device used on the drill floor to make up and break out drillpipe using the rotary table and spinning wrench. The pipe is fed mechanically to the iron roughnecks and is controlled by a driller from the driller's cabin.

irreducible water. Water in the pores of a reservoir rock that will not flow. (residual water)

irreg. irregular.

ISIP. (1) intermediate shut-in pressure (drillstem test) and (2) instantaneous shut-in pressure.

isochron map. A map that uses contours to show the thickness in time (milliseconds) between two seismic horizons. (isotime and time interval map)

isolith map. A map that uses contours to show the thickness of one rock type such as sandstone in a formation.

isomers. Organic compounds that have the same chemical formulas but have slightly different chemical and physical properties. Two isomers of butane (C_4H_{10}) are isobutane and n-butane.

isopach map. A map that uses contours to show the thickness of a subsurface rock layer.

isotime map. *See* isochron map.

ISO. International Organization for Standardization (www.iso.org).

ISP. initial shut-in pressure.

ITD. intention to drill.

IUE. internal upset ends.

IW. injection well.

J. productivity index.

jacket. The legs on an offshore production platform.

jackup rig. An exploratory offshore drilling system with a two hulls and at least three tall legs through the hulls. It is towed into position similar to a barge. The lower hull then rests on the bottom of the ocean and the upper hull is jacked up the legs. The drilling rig is mounted on the upper hull. *See also* cantilevered jackup.

J&A. junked and abandoned.

jar. A tubular run on a drillstring or fishing string that is designed to impart a sharp, upward or downward blow to the string on command.

JB. (1) junction box and (2) junk basket.

JC. job complete.

jct. junction.

jet. An orifice in a tricone drill bit between two cones. Drilling mud flows out the jet. (nozzle)

jet bit. A tricone drilling bit with one large and two small nozzles. The jetting action of the drilling mud from the large nozzle is used to start drilling the well out at an angle.

JINO. joint interest nonoperating (property).

JK. junk.

JKB. junk basket.

jmb. jammed.

jnk. junk.

JOA. joint operating agreement.

joint. (1) A natural fracture in rock along which there has been no movement; cf. fault. (2) A section of a tubular such as drillpipe. (JT, Jt, and jt)

joint interest billing. An accounting procedure that bills each working interest owner for their proportionate share of drilling and lease expenses. (JIB)

joint operating agreement. An agreement between several companies to explore, drill, and develop a common area called the working interest area. The agreement defines how the costs and revenues are to be shared among the parties and which party is the operator. (JOA)

joint venture. A partnership among companies for a common purpose such as exploring and drilling an area. One partner will be the operator. (JV)

JOP. joint operating provisions.

JP. jet perforated.

JP/ft. jet perforations per foot.

JSPF. jet shots per foot.

JT, Jt, and **jt.** joint.

jts. joints.

jug. A vibration detector used on land to detect subsurface echoes during a seismic survey. (geophone)

jug hustler. A seismic crew member who lays cable and plants geophones.

junk. A tool or broken pipe that has fallen to the bottom of a well. (fish) (JK and jnk)

junk basket. A fishing tool. Drilling mud is circulated down along the outside of the tool to pick up pieces of junk on the bottom of the well. The mud then circulates up along the inside of the tool where the junk is caught in a basket; cf. boot sub. (JB and JKB)

junk mill. A fishing tool that is rotated to (1) dress a fish in preparation for another fishing tool or (2) reduce the fish to metal flakes.

Jurassic. A period of geological time from 201.6 to 145.5 million years ago. It is part of the Mesozoic era. (Jur)

JV. joint venture.

K. coefficient.

k. permeability.

karst. A highly dissolved limestone that is or was exposed on the surface of the earth. It exhibits solution features can that range from vugs to caverns.

KB. kelly bushing.

KBM. kelly bushing measurement.

KCl. potassium chloride.

kelly. A strong, four- or six-sided, steel pipe that is located at the top of the drillstring. It runs through the kelly bushing. (KB)

kelly bushing. A device that is fitted on the master bushing and rotating table. The kelly runs through the kelly bushing. (KB)

kelly cock. A valve run on a drillstring just above or below the kelly. A wrench is used to open and close it. It is used to stop fluids from flowing up the drillstring.

kerogen. Insoluble organic matter in sedimentary rocks. It is the part of organic matter in source rock that can be changed into petroleum; cf. bitumen.

keyseat. A section in a well being drilled that has a cross section similar to a key hole. The smaller diameter portion was abraded by the drillpipe and the larger portion by the bit. Drill collars can become stuck in the smaller portion of the keyseat.

kg. kilograms.

kh. permeability thickness.

kHz. kilohertz.

kick. (1) The flow of subsurface fluids into a well. (2) A distinctive deflection on a well-log curve.

kick off. To start drilling a well out at an angle. (KO)

kickoff point. The location in a well where it begins to deviate. (KOP)

kill a well. To stop the flow from a well. Two common methods used to kill a well when drilling are (1) driller's method and (2) wait-and-weigh method. A producing well is killed for a workover by filling the well with a kill fluid.

kill fluid. A liquid used in a well to stop reservoir fluids from flowing into the well in preparation for a workover. Diesel oil or brine is often used.

kill mud. Heavy drilling mud used to stop a kick and control a well.

kilo. thousand.

kilogram. A unit of weight in the metric system. It is equal to 1,000 grams or 2.2 pounds.

kilojoule. The metric system unit of heat used to measure the heat content of natural gas. It is 1,000 joules and is equal to about one British thermal unit in the English system. (kJ)

kilometer. A metric system of length. One kilometer is equal to 1,000 meter or 0.62 miles. (km)

kilopascal. The metric unit for pressure. It is equal to 6.895 psi. (kPa)

Kimmeridgian. An age of geological time from 154 to 151 million years ago. It is part of the Jurassic period.

kitchen. The deep part of a basin where gas and oil are formed. (oven)

kld. killed.

km. kilometers.

K-Monel. A nonmagnetic metal used in some drill collars. It is used on well surveys made with a magnetic compass.

knot. A velocity equal to one nautical mile per hour. It is equal to 1.151 miles per hour.

KO. (1) kick off and (2) kicked off.

KOP. kickoff point.

kPa. kilopascal.

kriging. A statistical method used to estimate a value at a location that has not been sampled. It is based on the variability of that value with distance from locations that have been sampled.

KV. kinematic viscosity.

kv. kilovolt.

KW. killed well.

kw. kilowatt.

L. (1) liter and (2) length.

l. lower.

L/. lower.

LA. load acid.

LACT. lease automatic custody transfer.

LACT unit. *See* lease automatic custody transfer unit.

lag time. The time that it takes the well cuttings to circulate from the bottom of the well, up the well to the screens on the shale shaker.

land. To transfer the weight of something in a well such as a casing string being run into a well to the casing hangers.

landed at. The depth to which a casing string was set in a well.

land farming. Spreading used, freshwater-based drilling mud out on agricultural land to improve the soil; cf. soil farming.

landman. An oil company employee or independent who identifies mineral rights owners and negotiates leases.

landowner royalty. The mineral rights owner's royalty.

Landsat. one of six uncrewed remote-sensing satellites operated by the United States. Landsat pictures of the earth are made in visible light and infrared and are transmitted back to earth.

large-scale map. A map that shows relatively more detail but covers less area than a small-scale map.

LAS file. *See* Log ASCII Standard file.

LAST. logged after short trip.

LAT. logged after trip.

lateral. A general term for a horizontal or near-horizontal wellbore that has been drilled out from an original (mother) well. It can be the horizontal section of a horizontal drainhole or several shorter branches. The toe is the furthest point on the lateral, and the heel is the closest.

laterolog. A wireline electrical log used in conductive muds to measure the true resistivity of the rocks. It uses guard electrodes in the logging tool to focus an electrical current into the rocks. A dual laterolog measures deep and medium resistivity. (focused or guard log) (LL)

lava. Molten rock on the surface of the earth; cf. plutonic rock. (volcanic rock) *See also* basalt.

lay down pipe. To pull drillpipe or tubing from a well and place it horizontally on a pipe rack.

lb and **lbm.** pounds.

lbm/cu ft. pounds per cubic foot.

lbm/gal. pounds per gallon.

LBOS. light brown oil stain.

LC. (1) lost circulation and (2) lease crude.

LCM. lost-circulation material.

LD. (1) land and (2) laid down.

Ld. land.

ld. load.

Ldd. landed.

LDDCs. laid down drill collars.

LDDP. laid down drillpipe.

leakoff test. A test in a well being drilled to determine the strength (fracture pressure) of a formation. The well is shut in and the fluid pressure in the well is increased until the fluid flows into the formation (leaks off). (LOT)

lean gas. (1) Natural gas containing a minor amount of liquid condensate. (2) Natural gas with less than 2.5 gallons of natural gas liquids per thousand standard cubic feet of natural gas; cf. rich gas and dry gas.

lease. (1) A legal document between an oil company (lessee) and a mineral rights owner (lessor) for the purpose of obtaining drilling and production rights on the land under lease. (lse) *See also* primary and secondary term. (2) To obtain a lease for a specific parcel of land. (3) The land under lease.

lease automatic custody transfer unit. A system that uses equipment to measure, sample, test, and transfer oil in the field and to record that transaction. (LACT unit)

left-lateral strike-slip fault. A fault that moves horizontally, with the opposite side of the fault moving toward the left as you face the fault; cf. right-lateral strike-slip fault.

Len. lens.

lent. lenticular.

lessee. The recipient of a lease; cf. lessor.

lessor. The mineral rights owner who grants a lease. The lessor is granted a bonus for signing the lease and is guaranteed a royalty if oil and/or gas is produced; cf. lessee.

LFL. low fluid level.

lg and **lge.** large.

lg. length.

LI. level indicator.

li. lime.

license round. A method used by a country or national oil company to award a license to a company for the exploration of a specific area (license block) during a specific period of time. Competitive, closed bids are submitted. The block is awarded to (1) the highest cash, bonus bidder or (2) the largest financial commitment to explore and drill the block. (bid round)

lift gas. Inert gas, usually natural gas, used for gas lift in a well.

lifting costs. The cost to produce one barrel of oil in the field.

Lig and **lig.** lignite.

light oil. Low-density crude oil of more than 35 °API gravity. It is relatively fluid, has high in gasoline content, and is generally low in sulfur content; cf. heavy oil.

light-sand frac. A frac job using relatively little proppants. A slickwater frac is a light-sand frac.

lightweight additive. An additive used to decrease the density of a fluid; cf. heavyweight additive.

lignite. Soft coal. Bituminous and anthracite are harder coals. (Lig and lig)

LIH. left in hole.

lim. (1) lime and (2) limit.

limb. One side of a fold in sedimentary rocks.

lime. (1) A driller's term for limestone. (2) Calcium oxide. (CaO) (li or lim)

lime mudstone. A type of limestone with a very small percentage of large, sand-sized grains and a considerable amount of fine-grained material.

limestone. A common sedimentary rock composed of $CaCO_3$. It can range from fine- to coarse-grained and can be a reservoir rock. (LS, Ls, and ls)

limy. Containing $CaCO_3$. (lmy)

lin. liner.

linear spread. A geometric pattern of geophone groups that are arranged in a line.

line drive. A common pattern of injector and producing wells used for a waterflood. The injecting and producing wells are located on parallel lines; cf. five-spot pattern.

liner. Relatively thin-walled steel pipe that looks exactly like casing. *See also* liner string.

liner string. A string of tubulars (liners) similar to casing in a well. A liner string, however, does not run all the way up to the surface as a casing string does. A liner string is hung in the well by a liner hanger and may or may not be cemented into the well. A slotted liner has long, narrow, vertical openings (slots) to allow fluids to flow into the liner but exclude sand. A screened, slotted liner has screens wrapped around it to help exclude sand. A prepacked liner has a gravel pack between the liner and the tubing string. Some other types are drilling, production, and scab liner strings; cf. casing string. (lin, LNR, and lnr)

line shooting. A method used to acquire data for a 3-D seismic survey at sea. The seismic is acquired by running the seismic sources and hydrophone streamers in closely spaced, parallel lines. The ship tows at least two arrays of air guns or twin streamers.

liq. liquid.

liquefied natural gas. Methane gas (CH_4) that has been compressed and super cooled into a liquid; cf. liquefied petroleum gas. (LNG)

liquefied petroleum gas. Propane gas in liquid form in the United States. It can be a propane-butane mixture in Europe; cf. liquefied natural gas. (LPG and LP-gas)

liter. Unit of volume in the metric system. It is equal to 1,000 cm^3 or 0.264 U.S. gallon in the English system. (L)

Lith and **lith.** lithology.

lithofacies. One particular rock type such as a sandstone in a rock layer. *See also* facies.

lithofacies map. A subsurface map showing changes in the physical properties of a particular rock layer; e.g., isolith map.

lithologic log. A record of the physical properties of rocks in a well. It includes composition, texture, color, presence of pore spaces, and oil staining. (sample or strip log)

lithology. The composition of a rock such as sandstone or limestone. (Lith or lith)

lithostatic pressure. The pressure on rocks at a specific depth. It is caused by the weight of the overlying rocks; cf. fluid pressure. (earth pressure)

lk. leak.

LLC. liquid level controller.

LLG. liquid level gauge.

lm. lime.

lmy. limy.

Lmy sh. limy shale.

ln. line.

LNG. liquefied natural gas.

LNR and **lnr.** liner.

LO. load oil.

load oil or **water.** Oil or water filling a well to maintain pressure on the bottom of the well. (LO and LW)

load up. To have water fill the bottom of a gas well. It prevents gas from flowing into the well. The water must be removed (unload the well) to have the gas flow again.

loc. (1) locate and (2) location.

loc abnd. location abandoned.

loc gr. location graded.

log. A record or to make a record of rock properties in a well.

Log ASCII Standard file. A digital well log format. Each well log file has well header information and digital values for each well log curve recorded by depth. It is compatible with personal computers. (LAS file)

logged depth or **logger's depth** or **logger's total depth.** *See* measured depth.

logging tool. A metal cylinder, typically 30 to 40 ft long and 4 in. in diameter, that is filled with instruments. It is run in a well on a wireline to make a well log. The instruments sense the electrical, radioactive, and sonic properties of the rocks and their fluids, and the diameter of the wellbore. The tool usually has arms to center it in the hole or against the wellbore. (sonde)

logging-while-drilling. A real-time well log of rock and formation fluid properties made by sensors in the drillstring above the bit. Measurements include gamma ray, resistivity, neutron porosity, formation density, and sonic. The measurements are digitized and transmitted to the surface by pressure pulses in the drilling mud; cf. measurements-while-drilling. (LWD)

long normal resistivity. A wireline resistivity measurement made with electrodes spaced far apart (64 in.); cf. short normal resistivity.

loose. An emulsion that readily separates; cf. tight.

lost circulation. A drilling problem in which relatively large quantities of drilling mud flow into a permeable rock layer (lost circulation zone) in the well. Very little, if any, drilling mud circulates back up the well. (LC)

lost-circulation additive, **control agent**, and **material.** An additive to drilling mud or cement slurry that clogs the pores of a lost-circulation zone. (bridging material) (LCM)

lost-circulation zone. A very permeable rock layer in a well. It takes large amounts of drilling mud during drilling. (thief zone)

lost section. The section of rock that is missing when drilling through a normal fault; cf. double section.

LOT. leakoff test.

low. lower.

low-shrinkage oil. *See* volatile oil.

low-resistivity, **low-contrast pay.** A commercial deposit of oil or gas that appears to be not commercial because it has relatively low resistivity on an induction log and/or low contrast to shale on a natural gamma ray low. It is caused by shale or clay in the pay.

low-temperature separator. An installation that passes natural gas through an expansion choke to cool the gas and separate out natural gas liquids. (LTX)

low-velocity zone. The layer of loose sediments that occurs near the surface of the earth and has a relatively low seismic velocity. Statics correct seismic data for the low-velocity zone. (weathering layer and zone)

LP. low pressure.

LPG and **LP-gas.** liquefied petroleum gas.

LPG drive. An enhanced oil recovery method in which liquefied petroleum gas is injected into a depleted oil reservoir. The LPG is miscible with the oil, which makes it more fluid.

LP sep. low-pressure separator.

LR. (1) level recorder and (2) long radius.

lrg. large.

LS, **Ls**, and **ls.** limestone.

lse. lease.

lss. leases.

LT. (1) lower and (2) light tubing.

lt. light.

LTD. logged total depth.

Ltl and **ltl.** little.

LTS. long tubing string.

LTX. low-temperature separator.

LTX unit. low-temperature extraction unit.

LU. lease use (gas).

lub. (1) lubricant and (2) lubricate.

LV. liquid volume.

lv. leave.

LW. load water.

LWD. logging-while-drilling.

lwr. lower.

m. (1) slope, (2) medium, and (3) meters.

M/. middle.

MA. mud acid.

Ma. mega-annum (millions of years).

Maastrichtian. An age of geological time from 71.3 to 65 million years ago. It is part of the Cretaceous period.

mach. machine.

mag. (1) magnetic or (2) magnetometer.

magnetometer. An instrument that measures the earth's magnetic field intensity. It is able to detect variations in the magnetite content of the rocks. The units of measurement are gauss or nanoteslas. (mag)

maint. maintenance.

maintain angle. To drill a straight section in a deviated well; cf. build and drop angle.

maj. major.

make a connection. To screw a joint of drillpipe to the top of the drillstring below the kelly as the well is drilled deeper.

make up. (verb) (1) to screw togethers such as drillpipe; cf. break out; (2) to mix or prepare.

makeup. (noun or adjective) Something added to a system such as makeup water from another source added to waterflood injection water or drilling mud.

making a connection. Adding another joint of drillpipe to the drillstring.

making a trip. Pulling the drillstring from the well and running it back in.

making hole. Drilling a well.

man. (1) manual and (2) manifold.

M&F. male and female (joint).

M&FP. maximum and final pressure.

mandrel. A device such as a bar, shaft, spindle, or cone that is designed to hold something. A mandrel can be used to lower a tool into a well or be attached in a well to hold equipment such as a sidedoor or sidepocket mandrel used to hold a gas lift valve on a tubing string.

manifold. A tubular with at least one inlet and several outlets. A choke manifold is used on a drilling rig to circulate drilling mud when the blowout preventers are closed. (man and MF)

manometer. An instrument that measures fluid pressure in a well.

man op. manually operated.

mar. marine.

marble. metamorphosed limestone. (Mbl)

marg. marginal.

marginal well. An oil or gas well that due to declining production rate is barely profitable; cf. stripper well.

marine riser. A long length of flexible steel tubular used to connect the blowout-preventer stack on the bottom of the ocean to a floating drilling rig. The drillstring is run down the marine riser.

marker bed. A thin, distinctive, sedimentary rock layer such as volcanic ash used in correlation.

Mark II. An oil well beam-pumping unit that uses levers to balance the weight of the sucker-rod string.

marlstone. A loose term for a rock composed of calcium carbonate and clay. (mrlst)

marn. marine.

Marsh funnel. A funnel used on a drilling rig to measure the viscosity of drilling mud. The time in seconds that it takes the mud to drain through the funnel is related to the mud viscosity.

mass. massive.

massive frac job. A large hydraulic fracturing job on a well. It uses relatively large amounts of frac fluid and proppants and is commonly done on tight gas sands.

mast. (1) A portable steel tower that sits on the drill floor of a drilling or workover rig. (2) The portable steel tower on the bed of a service unit. A mast is assembled horizontally and pivoted vertical as a single unit in contrast to a derrick. Masts are used on land drilling rigs and workover rigs; cf. derrick.

master bushing. A device that attaches to the rotary table. The kelly bushing fits on the master bushing.

materials balance equation. An equation that relates the volume of produced fluids from a reservoir to the change of reservoir pressure to calculate the remaining oil and gas.

materials person. An employee of the operator who is responsible for calculating the amount and ordering the supplies and supervising their timely delivery to a drilling rig.

matl. material.

matrix. The fine-grained particles that bind a poorly sorted sedimentary rock. (Mtrx)

maturation. The chemical alteration of organic matter in sedimentary rocks with burial and increasing temperature and time. It can result in the generation of petroleum.

mature area. An area in which many wells have been drilled.

mature oil. Light oil generated at deep depths in the oil window; cf. immature oil.

MAW. mud acid wash.

max. maximum.

maximum efficient rate. A production rate for a field that balances the economics of rapid production against the waste caused by bypassing of subsurface oil during rapid production. (MER)

maximum potential flow. The maximum rate a well can produce with zero bottom-hole pressure. (absolute open flow)

Mb. member.

Mbl. marble.

Mbr and mbr. member.

MC. mud cut.

MCA. mud cleanout agent.

Mcf. 1,000 cubic feet.

mchsm. mechanism.

m³. cubic meters.

MD. measured depth.

md (or **md.**). millidarcy; millidarcys (pl).

mdl. middle.

MDRT. measured depth from rotary table.

Mdst. mudstone.

md wt. mud weight.

meander. A river channel bend.

meas. (1) measure and (2) measured.

measured depth or **measured total depth.** The depth of a well computed from the number of joints of drillpipe, drill collars, and other parts of the drillstring in the well. (logged depth or logger's depth or logger's total depth) (MD)

measurements-while-drilling. A real-time log of drilling parameters made by sensors in the drillstring above the bit. It measures bit orientation (azimuth and inclination) and downhole temperature and pressure. The measurements are digitized and transmitted to the surface by pressure pulses in the drilling mud. It is used on offshore and deviated wells; cf. logging while drilling. (MWD)

mech. mechanical.

mechanical integrity test. A test used to determine if casing in a well is leaking. A liquid is pumped down the well under pressure. The pumping is stopped, and the liquid pressure is monitored for a period of time. If the pressure drops, this indicates that the casing is leaking. (MIT)

mechanical rig. An older rotary drilling rig that uses just diesel engines as prime movers. The engines are mechanically connected to machinery on the drill floor by a transmission called a compounder; cf. diesel-electric rig.

med and **med.** medium.

med gr. medium grained.

member. A distinctive but local bed that occurs in a formation. It is given a formal, two-part name, similar to a formation name such as the Layton Sandstone Member of the Coffeyville Formation. (Mb., Mbr, and mbr)

MER. maximum efficient rate.

Mesozoic. An era of geological time from 251 to 65.5 million years ago. It is know as the age of reptiles and is divided into the Triassic, Jurassic, and Cretaceous periods. (Meso)

metamorphic rock. A rock that has been altered by heat and/or pressure; e.g., gneiss and marble; cf. sedimentary and igneous rock. (meta)

meteoric water. Fresh subsurface water; cf. conate water.

meter. (1) The unit of length in the metric system. It is equal to 39.37 inches and 3.28 feet. (m) (2) a measuring device. (mtr) *See also* gas meter.

meterhouse. A gas meter shelter. (charthouse)

meter prover. A device that calibrates a meter. It compares the amount of gas or liquid flowing through the meter prover to the meter reading on a meter as the same or equal amount of fluid flows through it. (prover)

METH and **meth.** methane.

methane. A hydrocarbon composed of CH_4. It is a gas under surface conditions and is the most abundant component of natural gas. (C_1, METH, and meth)

metr. metric.

metric ton. The metric system unit for measurement of crude oil weight. A metric ton weighs 2,240 pounds and is the equivalent to 7.5 barrels of average weight oil.

MF. (1) manifold and (2) mud filtrate.

MFP. maximum flowing pressure.

MG. mud gas.

mgal. milligal.

m-gr. medium grained.

MGS. mud-gas separator.

mho. A unit of conductivity recorded on a well log. It is the reciprocal of an ohm.

MHz. megahertz.

MI. (1) moving in and (2) mineral interest.

mi. mile.

MIC and **mic.** mica.

mic. micaceous.

mica. A common mineral that occurs as thin elastic flakes. Two types are white mica (muscovite) and black mica (biotite). (Mic and mic)

micellar-polymer flood. An enhanced oil recovery method in which a surfactant is injected into a depleted oil reservoir to form a microemulsion of the remaining oil. Polymer-thickened water is then injected to drive the oil to producing wells.

micrite. A very-fine-grained limestone. (Micr and micr)

microbial gas. *See* biogenic gas.

microemulsion. An emulsion in which oil occurs as very small droplets suspended in water.

microfossil. The preserved remains of a tiny plant or animal that needs a microscope for identification. They are commonly shells of $CaCO_3$ or SiO_2. Some types of microfossils are foraminifera (forams), radiolaria, coccolithophores, diatoms, spores, and polle; cf. macrofossil. (Microfos or microfos)

micropaleontologist. A person who studies and identifies microfossils. (bug picker)

microresistivity log. A wireline resistivity log that measures resistivity without much penetration into the side of the wellbore.

MICU. moving in completion unit.

Mid. middle.

MIDDU. moving in double drum unit.

MIE. move in equipment.

migration. (1) The vertical and horizontal flow of oil and gas from the source rock to the trap or its ultimate destination. (2) A computer process that moves dipping seismic reflections into more accurate positions on a seismic record.

mill. (1) To grind up or pulverize. (2) A fishing tool with diamond or tungsten-carbide cutting edges used to grind a fish or cut a hole in casing.

milled-teeth tricone bit. A tricone drill bit in which the teeth have been machined out of the steel cones. It is used to drill soft and medium hardness rocks; cf. insert bit. (steel-tooth tricone bit)

milli. $1/1,000$.

millidarcy. $1/1,000$ darcy, a unit of permeability. (md and md.) *See also* darcy.

milligal. The unit of gravity measurement. (mGal)

millisecond. $\frac{1}{1,000}$ second. It is a common depth measurement on a seismic record. (ms)

MIM. moving in materials.

min. (1) minimum and (2) minerals.

min. minute.

mineral. A naturally occurring, relatively pure chemical compound. It can occur as either a crystal or an amorphous grain. Rocks are composed of mineral grains (e.g., quartz and calcite). (Min, min, and mnrl)

mineral interest or rights. The legal ownership of oil and gas below land. The mineral rights owner can explore and drill for gas and oil on that land and can produce the gas and oil. The federal government of most countries owns the mineral rights on land and offshore. The mineral rights of fee land that occurs in the United States is privately owned and can be transferred with a lease; cf. surface rights. (MI)

min P. minimum pressure.

Miocene. An epoch of time from 23 to 5.3 million years ago. It is part of the Tertiary period. (Mio)

MIPU. moving in pulling unit.

MIR. moving in rig.

MIRT. moving in rotary tools.

MIRU. moving in and rigging up.

miscible. The complete mixture of one fluid in another; cf. insoluble.

miscible gas drive. An enhanced oil recovery method in which gases that mix with oil in reservoir, such as carbon dioxide or liquefied petroleum gas, are injected into a depleted reservoir to produce more oil.

MISR. moving in service rig.

Mississippian. A period of geological time from 359 to 318 million years ago. It is part of the Paleozoic era. (Miss)

mist extractor. Wire mesh or vanes that are used to separate liquid droplets from gas in a separator.

mis-tie. A problem in correlating seismic horizons between intersecting seismic lines.

MISU. moving in service unit.

MIT. (1) mechanical integrity test and (2) moving in tools.

mixed-based crude oil. A refiner's term for crude oil that contains both paraffin and asphalt; cf. asphalt-based and paraffin-based crude oil.

ML. mud logger.

ml. milliliter.

mld. milled.

MLT. measured log thickness.

MLU. mud-logging unit.

mly. marly.

mm. (1) millimeter and (2) million.

MMcf. 1,000,000 cubic feet.

MMS. Minerals Management Service (www.mms.gov).

MMscf. million standard cubic feet.

MN. midnight.

mnr. minor.

MO. (1) move and (2) moving out.

MOCU. moving out completion unit.

mod. (1) moderate and moderately.

mode-conversion wave. A wave that has been transformed from one type of wave to another when it was reflected (e.g., a PS wave is a P wave that was converted into an S wave). (c-wave)

MODU. mobile offshore drilling unit.

molecular sieve. A substance, such as the mineral zeolite, that can filter molecules based on size or structure.

Moll. mollusk.

mol wt. molecular weight.

monkeyboard. A small platform located near the top of a derrick or mast on a drilling rig. The derrick operator stands on the monkeyboard when drillpipe is being tripped.

monkeyman. *See* derrick operator.

monobore. A well with a relatively uniform inner casing diameter from the surface to total depth. Expandable casing is used.

moonpool. A reinforced hole in the bottom of a drillship through which the drillstring runs.

MOP. maximum operating pressure.

MOR. moving out rig.

morning report. *See* daily drilling report.

MORT. moving out rotary tools.

motherbore. The original vertical well from which laterals are drilled.

motor-generator rig. *See* diesel-electric rig.

motor operator. The person in charge of maintaining the prime movers on a drilling rig.

mott. mottled.

mottled. A sedimentary rock with spots of different colors. (mott)

mouse hole. A hole in the drill floor used to hold the next joint of drillpipe used to make a connection.

MP. maximum pressure.

MPT. male pipe thread.

MR. (1) marine rig and (2) meter run.

MRG. methane-rich gas.

mrlst. marlstone.

MS. measured depth.

ms. milliseconds.

Mscg/d. thousand standard cubic feet per day.

MSL. mean sea level.

MSP. maximum surface pressure.

MT. measured thickness.

Mt. middle tubing.

MTD. measured total depth.

mtl. material.

MTP. (1) maximum top pressure and (2) maximum tubing pressure.

mtr. meter.

Mtrx. matrix.

MTS. mud to surface.

μ. viscosity.

mud. *See* drilling mud.

mud acid. A mixture of hydrochloric and hydrofluoric acids. It is commonly used in acid jobs. (MA)

mudcake. A cylinder of clay particles that were plastered against the sides of the well by drilling mud during drilling. (filter cake)

mud filtrate. The liquid and fines from drilling mud that are forced into the pores of rocks adjacent to the wellbore (invaded zone) as a well is drilled. (MF)

mud/gas separator. A steel vessel mounted on the mud tanks that is used to separate any gas out of the drilling mud coming from the well. (MGS)

mud hogs. Mud pumps on a drilling rig. (slush pumps)

mud hose. The rubber hose that connects the mud pumps to the swivel on a drilling rig. (rotary hose)

mud-in sample. A drilling mud sample taken from the suction pit on the mud tanks before the mud goes the mud pumps and down the well; cf. mud-out sample. (suction pit sample)

mudline. The bottom of the ocean.

mud log. A record of any natural gas in the drilling mud (gas-cut mud) and any crude oil in the well cuttings (show of oil) made by a service company as a well is being drilled. Either a gas detector is used to determine the total amount of gas or a gas chromatograph is used to determine the amount of methane, ethane, propane, butane, and pentane dissolved in the drilling mud. A rate of penetration (ROP) curve, a gamma ray or spontaneous potential log, and a sample log are also recorded on the mud log.

mud logger. (1) One of two geologists that work 12-hour tours in a mud logging trailer at a drilling site to make a mud log. (2) A service company that makes mud logs. (ML)

mud man. *See* drilling fluids engineer.

mud motor. *See* downhole mud motor.

mud-out sample. A drilling mud sample taken after the drilling mud circulates out of the well and passes through the shale shaker screens; cf. mud-in sample. (shale shaker sample)

mud pit. An earthen excavation near the drilling rig where drilling mud is temporarily stored; cf. mud tanks.

mud report. A daily report of the physical and chemical properties of the drilling fluid used during drilling a well. A mud report is made daily by a drilling fluids engineer.

mudstone. A sedimentary rock composed of silt- and clay-sized particles. (mdst)

mud tanks. Several, rectangular, steel tanks, arranged end-on-end that hold drilling mud on a drilling rig. The tanks are open on the top and connected by pipes. The drilling mud from the shale shakers flows from the shaker tank to the reserve tank to the suction tank where it is pumped back down the well by the mud hogs.

mud up. To increase the density and viscosity of drilling mud by adding dry clay, usually bentonite; cf. water back.

mud weight. The density of drilling mud expressed in pounds per gallon (lbm/gal). Mud weight is commonly 9 to 10 lbm/gal. (md wt, mud wt, and MW)

mud wt. mud weight.

multicomponent seismic. The use of several receivers at each seismic receiver station to record both compressional waves (P) and shear (S) waves. It includes three-component (3-C), four-component (4-C), and nine-component (9-C) seismic. Conventional seismic records only compressional waves.

multilateral well. A well with several smaller branches (laterals) drilled out from the main (mother) well.

multinational. An oil company that operates in several countries.

multiple completion. One well that produces out of two (dual completion) or more reservoirs.

multiple rate flow, **multiple rate**, or **multirate test.** A producing well test that uses pressure recorded during time periods at different well flow rates to (1) evaluate well completion, (2) estimate reservoir parameters such as permeability, (3) satisfy government regulations, and (4) determine well deliverability.

multiple-stage compressors. Several, inline-compressors that increase gas pressure in increments.

multipoint test. A gas well test that measures several flow rates and their bottom-hole pressures to determine the open flow potential of the well. (flow-after-flow test)

multirate test. A type of pressure transient test, such as a four-point test, that measures the flowing bottom-hole pressure at different stabilized flow rates.

multistage cementing. A cement job in which several sections of a casing string are cemented in successive stages.

multistage hydraulic fracturing. The fracking of a lateral in several separate stages. There can be up to 40 stages starting with the toe and ending in the heel.

Musc or **musc.** muscovite.

muscovite. A common mineral formed by white to transparent, thin flakes; cf. biotite. (white mica) (Musc and musc)

mV. millivolts.

MW. mud weight.

MWD. measurements-while-drilling.

MWP. maximum working pressure.

mx. medium crystalline.

mxd. mixed.

m.y. million years.

MYA. millions of years ago.

N. (1) dimensionless number, (2) neutron log, and (3) nitrogen.

NA. (1) not applicable and (2) not available.

NACE. National Association of Corrosion Engineers (www.nace.org).

NaCl. sodium chloride (salt).

NADOA. National Association of Division Order Analysts (www.nadoa.org).

NAG. no appreciable gas.

NALTA. National Association of Lease and Title Analysts (www.nalta.org).

nanotesla. A unit of magnetic measurement. (nT)

NARO. National Association of Royalty Owners.

nat. natural.

national company. *See* host company.

natural diamond bit. A fixed cutter, steel bit with no moving parts. Hundreds of small industrial diamonds are attached to the bottom and sides of the bit in geometric patterns. Watercourses deliver drilling fluid to the face of the bit to remove well cuttings; cf. tricone bit and polycrystalline compact diamond bit. (DB)

natural gamma ray log. A wireline well log that measures the natural radioactivity of rocks. Shales are the only common radioactive rock. It can be run both open and cased hole. (gamma ray log) (GR Log)

natural gas. A naturally occurring gas that is colorless, odorless, and flammable. It is composed of a mixture of hydrocarbon molecules that have one-carbon (methane), two-carbon (ethane), three-carbon (propane), and four-carbon (butane) atoms. Natural gas is measured in thousands of cubic feet (Mcf) and by heat content in British thermal units (Btu) in the English system and thousands of cubic meters and kilojoules in the metric system. Dry gas is pure methane, whereas wet gas also contains ethane, propane, propane, butane, and condensate. Before the gas is sold, an odorant is added to it. (NG)

natural gas liquids. Condensate, butane, propane and ethane that have been removed from natural gas in a natural gas processing plant. (NGL)

natural gasoline. *See* condensate.

natural gas processing plant. An installation that removes natural gas liquids from natural gas by cooling or absorption. (gas plant)

naturally occurring radioactive material. Radioactive material emitting more than 50 microroentgens per hour in oilfield equipment such as tubing. It is primarily from radium in scale that has precipitated out of oilfield brine. (NORM)

nautical mile. A unit of distance used at sea that is ⅟₆₀ degree in latitude. It is equal to 6,080 ft, 1.1516 miles, and 1,852 m. (NMI and nmi)

NB. new bit.

NC. (1) no change and (2) normally closed.

NCT. noncontiguous tract.

ND. (1) not drilling or (2) nipple down.

NDBOPs. nipple, nippling, or nippled down blowout preventers.

NE. nonemulsifying (agent).

NEA. nonemulsion acid.

neg. (1) negative and (2) negligible.

Neogene. A period of geological time from 23 to 2.6 million years ago. It is part of the Cenozoic era and includes the Miocene and Pliocene epochs.

NEP. net effective pay.

net revenue interest. 100% minus all royalties on a well or property. (NRI)

neut. (1) neutral and (2) neutralization.

neutron log or **neutron porosity log.** A radioactive wireline well log that is used to measure porosity. It bombards each rock in the well with a certain number of high-speed neutrons. (N and NL)

NF. (1) natural flow, (2) no fluorescence, and (3) no fluid.

NFD. new field discovery.

NFW. new field wildcat.

NG. (1) no good and (2) natural gas.

NGL. natural gas liquids.

NIC. not in contact.

nine-component seismic. A land seismic survey that uses three vibrator sources and three geophones oriented at right angles to each other at each receiver location to record both conventional compressional waves (P waves) and also shear waves (S waves). It is used to better determine rock types and locate fractures. *See also* 4-C seismic.

NIP and **nip.** (1) nipple and (2) nipple up.

NIPER. National Institute for Petroleum and Energy Research (www.osti. gov/techtran/niper).

nipple. A short pipe with threads or welds on both ends. (NIP and nip)

nipple up. To connect equipment or fittings such as a blowout preventer; cf. nipple down. (NU and NIP)

nipple down. to disconnect equipment or fittings; cf. nipple up. (ND)

NL. neutron log.

NMI and **nmi.** nautical mile.

NMR. nuclear magnetic resonance.

NO. (1) new oil and (2) normally open.

nodding donkey. A beam pumping oil well in England.

No Inc. no increase.

noise. Unwanted seismic energy recorded with the signal. It is everything except direct (primary) reflections that represent the subsurface geology; cf. signal. (N)

noise log. A production log that records sounds with depth in a well.

NOJV. nonoperated joint venture.

nom. nominal.

nominal weight. Calculated weight rather than measured

nonassociated gas. Natural gas that is not dissolved in or in contact with crude oil in the reservoir; cf. associated gas.

nonexclusive. Data shared by several parties; cf. proprietary.

NOP. nonoperating property.

nor. normal.

no rec. no recovery.

no returns. No well cuttings were obtained for that interval in the well. (NR)

NORM. naturally occurring radioactive material.

normal fault. A fault with predominantly vertical movement (dip slip), in which the hanging wall has been lowered in relation to the footwall. It creates a lost section; cf. reverse fault.

normal pressure. Fluid pressure in subsurface rocks due to the weight of the overlying fluids; cf. abnormal high pressure. (hydrostatic pressure)

nose. The lobate surface pattern of an eroded, plunging anticline.

noz. nozzle.

nozzle. An orifice in a tricone drill bit between two cones. Drilling mud jets out the nozzle. (jet) (noz)

NP. (1) no production, (2) not porous, and (3) not pumping.

NPD. new pool discovery.

NPS. nominal pipe size.

NPW. new pool wildcat.

NR. (1) no report, (2) no recovery, (3) no returns, and (4) not reported.

NRI. net revenue interest.

NS and **n/s.** no show.

n.s. no sample.

NSG. no show gas.

NSO. (1) nitrogen, sulfur, and oxygen organic compounds and (2) no show oil.

NSO&G. no show oil and gas.

NSR. no spacing rule.

nT. nanotesla.

NTD. new total depth.

NTS. not to scale.

N/tst. no test.

N$_2$. nitrogen.

NU. nippling up.

NUBOPs. nipple, nippling, or nippled up blowout preventers.

nuclear magnetic resonance log. A wireline well log that uses magnetism to measure porosity and pore sizes. It can be used to calculate permeability and determine types of fluids in the reservoir. (NMR log)

num. numerous.

NVP or **n.v.p.** no visible porosity.

NW. no water.

NYA. not yet available.

NYMEX. New York Mercantile Exchange (www.nymex.com).

O. oil.

OAH. overall height.

OAL. overall length.

O&G. oil and gas.

O&GCM. oil and gas-cut mud.

O&GCSW. oil and gas-cut saltwater.

O&GCW. oil and gas-cut water.

O&GL. oil and gas lease.

O&SW. oil and saltwater.

O&W. oil and water.

OAW. old abandoned well.

OB. off bottom.

OBM. oil-based mud.

OBOC. operated by other company.

observation well. A well drilled to map or monitor subsurface fluids.

obsol. obsolete.

OBW & RS. optimum bit weight and rotary speed.

OC. (1) oil cut, (2) on center, and (3) operations commenced.

occ. occasional.

OCM. oil-cut mud.

OCS. outer continental shelf.

OCSW. oil-cut saltwater.

OCW. oil-cut water.

OD. outer diameter.

od. odor.

OE. (1) open end and (2) oil emulsion.

OEG. oil-equivalent gas.

OEM. oil emulsion mud.

OF. open flow.

offset. (1) The horizontal distance from the seismic source to the receiver. (2) A well location adjacent to a producing well. Each producing well has eight offset locations (four direct and four diagonal offsets).

offshore production platform. *See* production platform.

off structure. Located off the top of a trap; cf. on structure.

OFLU. oil fluorescence.

OFP. open-flow potential.

OH. open hole.

ohm. A unit of electrical resistivity. It is recorded on resistivity well logs as ohm-meter or ohm meter^2/meter.

ohm-m. ohm-meter.

OIH. oil in hole.

oil. *See* crude oil. (O)

oil-base or **oil-based drilling mud.** Drilling mud made with diesel oil. It is designed to prevent formation damage; cf. synthetic-base and water-base drilling mud. (OBM)

oil cut. diluted with oil.

oil-equivalent gas. The number of barrels of oil that is equal in heat content (Btu) to a volume of natural gas. The ratio is about 1 barrel of oil to 6,000 cubic feet of natural gas. It varies slightly from company to company. (barrels-of-oil equivalent or energy-equivalent barrel) (OEG)

oilfield. (adjective) e.g., oilfield equipment.

oil field. A noun for the surface area directly above one or more reservoirs on the same trap; e.g., East Texas oil field.

oilfield brine. Very saline water that is produced with oil.

oil in place. The total amount of oil located in the pores of a subsurface reservoir in an oil field; cf. recoverable oil. (OIP)

oil-in-water emulsion. Droplets of oil suspended in water; cf. water-in-oil emulsion.

oil sand. A common term for sandstone containing crude oil.

oil shale. A fine-grained sedimentary rock containing organic matter called kerogen that, when heated, forms crude oil called shale oil. It is a source rock that is old enough but has never been buried deeply enough to have sufficient heat to generate oil.

oil show. The presence of crude oil in well cuttings. It is detected by mud logging while drilling the well.

oil string. The smallest diameter and longest casing string in a well. (production casing)

oil-water contact. The boundary between crude oil and water in a reservoir. (OWC)

oil wet. A reservoir rock in which water occurs in the center of the pores and oil coats the rock surfaces; cf. water wet.

oil window. The zone in the earth where crude oil is generated from organic matter in source rocks.

OIP. oil in place.

Oligocene. An epoch of time from 33.9 to 23 million years ago. It is part of the Tertiary period. (Olig)

ON. overnight.

on pump. An oil well that uses a sucker-rod pump.

ONRR. Office of Natural Resources Revenue (www.onnr.gov).

on structure. Located on top of a trap; cf. off structure.

OO. oil odor.

OOG. out of gauge.

OOIP. original oil in place.

Ool and **ool.** (1) oolite and (2) oolitic.

oolite. A sand- or silt-sized sphere of calcium carbonate that precipitated from water. (Ool and ool)

oolitic limestone. A limestone composed predominately of oolites.

OP. (1) oil pay and (2) overproduced.

op. opaque.

OPBD. old plug-back depth.

open hole. A wellbore with no casing or liner.

open-hole completion. A well with casing run and cemented down to the level of the reservoir rock, which is left uncased; cf. set-through completion. (barefoot and top-set completion)

open-hole log. A wireline well log such as an electrical log that can only be run in a well without casing. Most wireline logs are open-hole; cf. cased-hole log.

oper. (1) operator and (2) operations.

operator. The company that (1) contracts to drill a well, (2) is responsible for maintaining a producing lease, or (3) is in charge of operations in a working interest area. (oper and opr)

opn. open.

opp. opposite.

opr. operator.

optn to F/O. option to farmout.

Ordovician. A period of geological time from 488 to 444 million years ago. It is part of the Paleozoic era. (Ord)

orf. orifice.

org. organic.

orifice. A hole in a plate through which fluids can flow. An orifice is described by its diameter. (orf)

orifice gas meter. A meter that measures the volume of natural gas flowing through a line bypassing the gas through a specific size orifice. The drop in pressure through the orifice is related to the velocity of the gas.

orig. original.

original pressure. *See* initial pressure.

ORR. overriding royalty.

ORRI. overriding royalty interest.

orthoquartzite. A sandstone composed of well-sorted quartz sand grains. It can be an excellent reservoir rock; cf. graywacke.

OS. (1) oil show, (2) overshot, and (3) operating system.

O/S. out of service.

O, S & F. oil, stain, and fluorescence.

O sd. oil sand.

OSR. oil source rock.

OSTN or **ostn.** oil stain.

OSTOIP. original stock tank oil in place.

Ot. open tubing.

OTC. Offshore Technology Conference (www.otcnet.org).

OTD. old total depth.

otl. outlet.

OTS. oil to surface.

OU. oil unit.

outer continental shelf. The portion of the sea bottom where the federal government owns the mineral rights. It is from the state limit, usually 3 nautical miles from the shoreline, to 200 nautical miles from the shoreline. (OCS)

outpost well. A well drilled to significantly increase the area of a producing field. If the outpost well is successful, it is called an extension well.

OVC. other valuable consideration.

overbalance. The condition in a well in which the pressure of the drilling mud is more than the pressure of the fluids in the surrounding rocks; cf. underbalance. (fluid pressure)

overpull. The amount of upward force exerted on a tubular such as a drillstring in a well that is greater than that tubular's weight in the well. It is recorded in thousands of pounds.

overpressure. *See* abnormal high pressure.

override and **overriding royalty interest.** An interest in production created from a working interest that is free and clear of any costs. (ORRI)

overshot. A fishing tool that is run down and around a pipe (fish) on the bottom of the well. It grips the outside of the pipe to pull the fish out of the well; cf. spear. (OS)

overthrust belt. A zone of thrust faults that moved during the formation of a mountain range. (disturbed belt)

overturned fold. A fold in sedimentary rocks in which the axis is not vertical and the limbs are not symmetrical.

OWC. oil-water contact.

OWDD. oil well drilled deeper.

OWF. oil well flowing.

OWG. oil well gas.

OWPB. oil or old well plugged back.

OWR. oil/water ratio.

OWSI. oil well shut-in.

OWWO. oil or old well worked over.

ox. oxidized.

Oxfordian. An age of geological time from 159 to 154 million years ago. It is part of the Jurassic period.

oz. ounce.

P. (1) compressional wave, (2) pumped, and (3) pumping.

p. pressure.

PA. (1) pooling agreement and (2) pressure alarm.

PAB. (1) per acre bonus and (2) per acre basis.

packed hole assembly. A downhole assembly that uses several stabilizers to make the well to be drilled out straight. *See also* fulcrum and pendulum assembly.

packer. A cylinder of rubber-like material that is run on a tubular string or drillstring and compressed to expand and seal the well at that level. Packers are permanent (nonretrievable) or retrievable. (PRK and pkr)

packer flowmeter. An instrument used to force fluid to flow up the well through an orifice in a packer to measure the flow.

packoff. A sealing tool.

pack off. To seal a space such as the tubing-casing annulus.

packstone. A type of limestone with large, sand-sized grains touching each other and having fine-grained material in between. (Pkst)

pad. (1) The steel plate below the middle of a vibrator truck. During travel, the pad is raised. At the shot point it is lowered onto the ground and used to raise the back wheels of the vibrator truck off the ground. (base plate) (2) *See* drilling pad. (3) *See* well pad.

paid-up lease. A type of lease that does not require delay rental payments to maintain the lease during the primary term; cf. delay rental lease.

paleo. paleontology.

Paleocene. A period of geological time from 65.5 to 55.8 million years ago. It is part of the Tertiary period.

Paleogene. Geological time from 65.5 to 23 million years ago. It is part of the Cenozoic era and includes the Paleocene, Eocene, and Oligocene epochs.

paleogeographic map. An interpretation of the land surface during a certain time of earth's history.

paleontologist. A geologist who studies fossils.

paleontology. A branch of geology that studies of fossils. (paleo)

paleo pick. A horizon in sedimentary rocks defined by fossils.

Paleozoic. An era of geological time from 542 to 251 million years ago. It is divided into the Cambrian, Ordovician, Silurian, Devonian, Mississippian, Pennsylvanian, and Permian periods. (Paleo)

palynologist. A person who studies fossil spores and pollen. (weed and seed person)

P&A. plug and abandoned.

P&F. pump and flow.

P&P. (1) Porosity and permeability and (2) porous and permeable.

PAR. Per-acre rental.

Par and **par.** particle.

par. paraffin.

paraffin. A member of the hydrocarbon series of molecules that are straight chains with single bonds. All hydrocarbon molecules in natural gas and some in crude oil are paraffins. Long paraffin molecules are waxes that are solid at low temperatures. (par)

paraffin-based crude oil. A refiner's term for crude oil with little or no asphalt. It will yield a relatively high percentage of paraffin wax, high-quality lubricating oil, and kerosene when refined; cf. asphalt-based and mixed-base crude oil.

paraffin inhibitor. An additive to crude oil that prevents formation of waxes during production.

paraffin knife or **scratcher.** A tool that use sharp edges to scrape wax (paraffin) out of a tubing string.

parallel-free pump. An oil well hydraulic pump system that uses two tubing strings. One is for the power oil that drives the pump and the other is for the produced fluids; cf. casing-free pump.

patch reef. A small detached reef.

pay. (1) The zone producing gas and/or oil in a well. (2) The vertical thickness of the producing zone. Pay can be measured as either (a) gross pay, including nonproductive zones, or (b) net pay, including only productive zones.

payout. A criterion used to evaluate an investment in an oil or gas well. It is the time necessary for the net production revenues (minus royalties) to equal the costs of drilling, completing, and operating the well up to that time. (PO) *See also* return on investment.

pay sand. A sandstone that produces gas and/or oil.

payt. payment.

pay zone. The vertical portion of a reservoir in a well that produces gas and/or oil.

PB. (1) plug back and (2) plugged back.

PBD. plugged back depth.

PBHL. proposed bottom-hole location.

Pbl and **pbl.** pebble.

PBTD. plugged back total depth.

P_c. capillary pressure.

pct. percent.

PCV. pressure control valve.

PD. (1) proposed depth and (2) per day.

PDC. pressure differential controller.

PDC bit. polycrystalline diamond compact bit.

PDR. pressure differential recorder.

PE. (1) plain end and (2) pumping equipment.

peak oil. A concept that world crude oil production will peak and then decrease at a specific time. It is thought that this will occur when the amount of increased oil production from new, discovered oil fields and new, improved oil recovery applications in older fields is exceeded by the natural decrease in oil production from existing fields.

pen. (1) penetration and (2) penetration test.

pendulum assembly. A downhole assembly that uses the weight of a drill collar below a stabilizer to cause the bit to drop and decrease the angle (drop angle) of a deviated well. *See also* fulcrum and packed hole assembly.

Pennsylvanian. A period of geological time from 318 to 299 million years ago. It is part of the Paleozoic era. (Penn)

percentage map. A map that uses contours to show the percentage of a specific rock type such as sandstone in a formation.

perco. percolation.

PERF. perforated.

perf. (1) perforate, (2) perforated, and (3) perforator.

perf csg. perforated casing.

perforate. To blow holes (perforations) into the casing or lining, cement, and reservoir rock in a well with a perforating gun. (perf)

perforating gun. A tool run on a wireline or tubing string that shoots perforations (holes) in the casing or liner. It uses either steel bullets or, more commonly, shaped-explosive charges (jet perforation). The gun is either expendable or retrievable. (gun)

perforation. A hole (tunnel) shot in casing or liner, cement, and reservoir rock to allow oil and/or gas to flow into the well. Perforations are described by shots per foot (spf) and their angular separation, which is called phase.

period. A subdivision of an era of geological time (e.g., Pennsylvanian period). Periods are subdivided into epochs.

Perm and **perm.** permeability.

perm. (1) permeable and (2) permanent.

permeability. A measure of the ease with which a fluid flows through a rock. The units are millidarcys or darcys. Absolute permeability is the permeability of the rock when only one fluid is in the pores. Effective permeability is the permeability of one fluid in a rock when another fluid also shares the pores. Relative permeability is the ratio of effective permeability to absolute permeability; cf. porosity. (Perm, perm, and k) *See also* darcy.

permeameter. An instrument used to measure the permeability of a rock sample.

Permian. A period of geological time from 299 to 251 million years ago. It is the last period of the Paleozoic era and was characterized by a desert climate. (Perm)

Permian basins. Three tropical-water basins (Midland, Marfa, and Delaware) that were located in west Texas and eastern New Mexico. They are filled with sedimentary rocks and are very productive for gas and oil.

permit person. An employee of a seismic contractor who obtains permission from surface landowners to run seismic exploration across their land.

perp. perpendicular.

Pet and **pet.** petroleum.

petrf. petroliferous.

petrochemicals. Products made from petroleum feedstocks.

petroleum. The strict definition of petroleum includes only crude oil, but by general usage, it also includes natural gas. Petroleum is derived from the Latin words petro (rock) and oleum (oil). (Pet and pet)

petroleum engineer. An engineer who is trained to drill and complete wells and produce petroleum.

petroleum geologist. A geologist who specializes in the search for (exploration geologist) and exploitation (developmental geologist) of petroleum deposits.

petrophysics. The study of the physical and chemical properties of rocks in relation to the pore systems in the rocks and the fluids in the pores.

pf. per foot.

PFT. pumping for test.

PGW. producing gas well.

pH. A scale from 0 to 14 that measures the acidity or alkalinity of a liquid; 7 is neutral, below 7 is acidic, and above 7 is alkaline.

ph. phase.

phase. The angular separation of perforations. e.g., 60°. (ph)

Ø. porosity.

Phos or **phos.** (1) phosphate and (2) phosphatic.

PI. (1) productivity index, (2) production index, and (3) pressure indicator.

pick. (1) An interpretation of where the top or bottom of a subsurface rock layer occurs on a well log. (2) The location of an event such as a seismic horizon of a seismic record.

piercement salt dome. A salt dome that has risen to break through overlying sedimentary rocks.

pill. A batch of a substance or additive such as lost circulation material.

pilot hole. A small-diameter wellbore drilled out from a straight well to kick off the well at an angle.

pin. A male-threaded connection that mates with a box (a female connection); cf. box.

pinch out. (verb) To have a rock progressively narrow to zero thickness in a horizontal direction. (wedge out)

pinch-out. (noun) The termination of a rock as it progressively narrows to zero thickness in a horizontal direction; cf. shale-out. *See also* wedge-out.

pinnacle reef. A small cone-shaped reef.

pipe elevators. Clamplike devices that are attached to the bottom of the traveling block of a drilling rig. They are designed to attach onto the drillpipe.

pipeline oil. Crude oil that is below a maximum basic sediment and water content and meets pipeline transportation specifications. (sales-quality oil)

pipeline-quality gas. Natural gas that has been treated to meet pipeline pressure and chemical standards with a minimum of impurities. (sales-quality gas)

pipe rack. A steel framework on the ground next to a drilling rig. It is used to store horizontal joints of drillpipe.

pipe ramp. A flat, steel incline in the front of a drilling rig. It is used to drag drillpipe and casing up through the V-door and onto the drill floor.

pipe rams. Two large blocks of metal with inserts in a blowout-preventer stack. They are designed to close around drillpipe in a well to close the well; cf. blind rams.

PIT. Pressure integrity test.

pit level. The height of drilling fluid in the mud tanks.

pitman. The steel beam that connects the rotary counterbalance with the walking beam on a beam-pumping unit.

pit volume totalizer. A series of floats in the mud tanks of a drilling rig. They record the volume of mud in the tanks and send an alarm when the volume is decreasing or increasing.

PJ. (1) pump jack and (2) pump job.

pkd. packed.

pkg. packing.

PKR and **pkr.** packer.

Pkst. packstone.

PL. (1) pipeline and (2) property line.

plant. To position a geophone for a seismic survey.

plat. (1) a map. (2) to map.

plate tectonics. A theory in which the crust of the earth is composed of large, moving plates. Each plate originates at a mid-ocean ridge and ends in a subduction zone; cf. seafloor spreading.

play. A proven combination of reservoir rock, caprock, and trap type that contains commercial amounts of petroleum in an area.

pld. pulled.

Pleistocene. An epoch of time, from about 2.6 million years ago to 10,000 years ago, during which glaciers periodically occupied much of the land area. It is part of the Quaternary period. (ice age) (Pleist)

plg. pulling.

plgd. plugged.

Pliocene. An epoch of geological time from 5.3 to 2.6 million years ago. It is part of the Tertiary period. (Plio)

PLO. (1) pipeline oil and (2) pumping load oil.

PLUG. plugged off.

plug. (1) A small cylinder (1 in. diameter) of rock drilled from a core that is used to measure porosity and permeability. (2) To place cement in a well in order to abandon the well or seal off a depleted zone in the well.

plug and abandon and **plug & abandon.** The final stage in any well. Permission to plug and abandon is granted by a government agency and done to specific requirements. A surface cement plug is placed at the surface, and cement plugs are placed at specified depths in the well to prevent any pollution. A steel plate is welded to the top of the casing and covered with dirt. (P&A)

plugback. To plug and abandon one zone and complete in another zone higher in the well. (PB)

plunging anticline. An anticline with an axis oriented at an angle to horizontal.

plutonic rock. An igneous rock that crystallized from a hot, molten liquid below the surface of the earth; e.g., granite; cf. lava. *See also* intrusion.

pm or **pmp.** (1) pump, (2) pumping, or (3) pumped.

pneu. pneumatic.

pneumatic drilling. Drilling with either air or air and water (mist) as the circulating fluid.

pnl. panel.

PO. (1) pay out, (2) pulled out, (3) pumps off, (4) purchase order, and (5) present operation.

POB. (1) plug on bottom and (2) pump on beam.

POE. point of entry.

POGW. producing oil and gas well.

POH. pulled out of hole.

point bar. A sand bar deposited on the inside bend of a river meander.

point of entry. The location where a deviated or horizontal well enters the target formation. (POE)

polished rod. The polished, brass or steel rod that oscillates up and down through the stuffing box of an oil well rod-pumping unit. It is located at the top of the sucker-rod string. (PR)

polycrystalline diamond compact bit. A fixed cutter, steel bit with no moving parts. Synthetic diamonds on blanks on the face of the steel cutters that project out the bottom are designed to shear the rock. Watercourses deliver drilling fluid to the face of the bit to remove well cuttings. PDC bits are known for long life; cf. natural diamond and tricone bit. (PDCB)

polymer. A long-chain, high-weight molecule. When mixed with water, polymers form a thick, viscous fluid called a gel.

pony rod. A shorter than standard sucker rod.

POOH. pull, pulled, or pulling out of hole.

pool or **pooling.** To combine several smaller leases to make a drilling and spacing unit for the purpose of drilling a well.

poorly sorted. A rock or sediments with clastic grains having a large range of sizes; e.g., dirty sands; cf. well sorted.

POP. put or putting on pump.

Por and **por.** porosity.

por. porous.

pore. The space between solid particles in a rock. A primary pore is formed as the sediments are being deposited on the surface such as between grains. A secondary pore forms after the sedimentary rock is buried in the subsurface by solution or fracturing. *See also* porosity.

pore throat. The narrow connection between two pores in a rock.

pore volume. The volume of pores in a rock. (PV and P.V.)

porosimeter. An instrument used to measure porosity in a rock.

porosity. The percent volume of a rock that is pore space. Absolute or total porosity includes all pore spaces in the rock. Effective porosity includes only the interconnected pores; cf. permeability. (Ø, Por, por, and PR) *See also* pore.

porosity cutoff. A minimum porosity value such as 8% for reservoir rock that is used as a guideline in (1) deciding whether to complete a well or (2) making reserve computations.

pos. (1) position and (2) positive.

positive-displacement meter. A meter that measures the volume of a fluid in specific increments of a volume, one at a time.

poss. possible.

possum belly. A closed metal trough at the top of the shale shakers on a drilling rig. It receives the mud and well cuttings from the mud return line and slows them down before they flow onto the shale shaker screens. Mud and well cuttings samples are obtained from the possum belly.

possible reserves. Reserves that exist with at least 10% certainty; cf. proved and probable reserves.

pot. potential.

pot diff. potential difference.

potential test. A test that measures the maximum amount of fluids that a well can produce in 24 hours. (PT)

pound. The English unit of weight. It is equal to 453.59 grams. (pound avoirdupois) (lb and lbm)

pour point. The lowest temperature at which a particular crude oil will still flow. It is an indication of the wax content of the oil.

POW. producing oil well.

power swivel. *See* top drive.

POWF. producing oil well flowing.

POWP. producing oil well pumping.

PP. (1) pulled pipe and (2) production.

PPA. per power of attorney.

PPB and **ppb.** parts per billion.

PPG and **ppg.** pounds per gallon.

PPM and **ppm.** parts per million.

P PRESS. pump pressure.

PPT and **ppt.** parts per thousand.

ppt. precipitate.

PR. (1) poor returns, (2) polished rod, (3) porosity, and (4) pressure recorder.

PR&T. pulled rods and tubing.

prd. period.

Precambrian. An era of geological time from the beginning of the earth (4.5 billion years ago) until 542 million years ago. (Pre Camb)

precipitated. Crystallized from dissolved salts.

pred. predominant.

prelim. preliminary.

Prep and **prep.** (1) prepare and (2) preparing.

prepacked. Production liner or casing that is concentric and double-walled, with the annulus between the walls filled with loose or resin-coated sand. It is used for sand control.

pres. preserved.

present operation. What is currently happening on a well, such as flowing to sales. (PO)

press. pressure.

pressure. Force per unit area such as pounds per square inch (psi). Gauge (or gage) pressure is pressure above atmospheric pressure. Absolute pressure is gauge plus atmospheric pressure. (press)

pressure bomb. An instrument run on a wireline in a well to record pressures. It consists of a pressure sensor, recorder, and clock drive.

pressure buildup curve. A plot of pressure increase after a gas well has been shut in.

pressure integrity test. A test to determine if there is a leak in a tubular, vessel, or cased portion of a well. High pressure is applied, usually with water, and the tubular, vessel, or cased portion of the well is shut in. The pressure is then monitored for a period of time. If there is no pressure decrease, there is no leak. If there is a leak, the pressure decreases. (PIT)

pressure maintenance. A oilfield system in which produced gas is injected into the free gas cap and produced water is injected into the reservoir below the oil-water contact. It is used during primary production to maintain pressure on the remaining oil and increase ultimate production.

pressure transient test. A test that measures changes in pressures with different flow rates in a well. Three types are drawdown, buildup, and falloff tests.

pressure wave. *See* compressional wave.

prestack migration. The migration of seismic data before the data are stacked. *See also* migration.

prev. (1) prevent and (2) preventive.

prim. primary.

primacord. An explosive cord used as a seismic source on land and for a back-off operation on stuck pipe in a well.

primary cementing. A cement job done as the casing is being run; cf. secondary cementing.

primary drive. The original force that causes oil or gas to flow through the reservoir rock and into a well; e.g., water drive and expansion-gas drive.

primary production. The oil or gas that naturally flows into the well due to the reservoir drive. It does not include oil produced during waterflood or enhanced oil recovery.

primary recovery. The amount of oil and gas that is produced from a well or reservoir by its own pressure.

primary stratigraphic trap. A petroleum trap formed by the deposition of a reservoir rock such as a reef that is encased in shale; e.g., a reef and river channel sandstone; cf. secondary stratigraphic trap.

primary term. The time granted in a lease for exploration and drilling; cf. secondary term.

primary wave. *See* compressional wave.

prime movers. The main engines or motors that supply the power to machinery. On a drilling rig, the prime movers are diesel engines, and on a sucker-rod pumping unit, they are electric motors.

prmt. permit.

prncpl lss. principal lessee.

pro. prorated.

prob. (1) probable and (2) problem.

probable reserves. Reserves that exist with at least 50% certainty; cf. proved and possible reserves.

proc. (1) procedure and (2) process.

Prod. production casing.

Prod and **prod.** production.

prod. (1) produce and (2) produced.

prodg. producing.

produced water. Oilfield brine produced from an oil or gas well.

producer. A well that can produce commercial amounts of petroleum; cf. dry hole.

producing gas/oil ratio. The number of standard cubic feet of natural gas that a well produces per barrel of oil.

production casing. The smallest diameter and longest casing string in a well. (oil string)

production foreperson. An employee of the operator of a field who receives orders from the field superintendent and gives orders to the pumpers and work crews.

production index. The downhole pressure drawdown in psi divided by the production in barrels per day from a well. (PI)

production liner. A liner string run on the bottom of a well adjacent to the producing zone. It can be perforated, slotted, or prepacked.

production log. A log run in a producing well to evaluate a problem. Types of logs include the flowmeter, temperature log, manometer, watercut meter, and collar log.

production pad. *See* well pad.

production platform. An offshore platform that treats and separates produced fluids from offshore wells on the deck of the platform. It can have the wellheads on the platform or receive the produced fluids through flowlines from satellite wells or a wellhead platform in deeper water. The oil or gas goes ashore through a submarine pipeline. Two fixed-type platforms that sit on the ocean bottom are steel jacket and gravity storage production platforms. Floating production platforms include tension leg and spar platforms and FPSOs.

production profile. A plot of flow rate per day versus time for a well.

production rig. A mobile well service or workover hoisting unit used for a workover on a well. Two types are workover rig and service unit.

production-sharing contract. A contract between a foreign government and a multinational company (contractor). The company is granted an area of land or ocean bottom (concession) to explore and drill for a specific time (contract time). The company bears the entire cost of exploration and drilling. If no gas or oil is discovered, the contract expires and the cost of exploration and drilling is lost. If gas or oil is discovered, the company is allowed to produce and sell the petroleum. It is reimbursed for exploration and drilling expenditures from the sales of that production (cost oil). After reimbursement, further production is sold (profit oil), and the profit is split by an agreed formula in the contract with the foreign government and the company; cf. tax royalty participation contract.

production tax. State tax on oil and gas produced. (severance tax)

production test. A test that measures the amount of gas, oil, and water that a well contributes to a central processing unit.

production tree. *See* Christmas tree.

productivity index. The flow rate that a well can produce per psi difference between reservoir and bottom-hole pressures. It is an indicator of that well's ability to produce oil. (J and PI)

productivity test. A well test made with portable well test equipment that determines the effect of different flow rates on the reservoir. Fluid pressure is measured with the well shut in and at different stabilized rates. It is used to calculate absolute open flow rate and maximum production rate without reservoir damage.

profit oil. Produced oil that is split between a host company and a multinational company by an agreed formula after the multinational company has been reimbursed for expenditures; cf. cost oil.

prog. progress.

prograde. To deposit sediments out into a basin.

proj. project.

prom. prominent.

PROP. propane.

prop. (1) proposed and (2) proportional.

propane. A hydrocarbon composed of C_3H_8. It is a gas under surface conditions and is found in natural gas. (C_3 and PROP)

proppants and **propping agents.** Small spheres such as well-sorted sands that are suspended in the frac fluid pumped down a well during a frac job. They hold the fractures open. Proppants are described by the screen sizes on which they are caught, such as 20/40 mesh. Ceramic and sintered aluminum pellets are used for high-temperature, high-pressure wells.

proprietary. Kept secret; cf. nonexclusive.

prospect. A location where both geological and economic conditions favor drilling a well.

prot. protection.

protection casing. A casing string with an intermediate length and diameter. It is used to isolate a problem zone such a lost circulation in the well as the well is being drilled. (intermediate casing)

proved reserves. Oil and gas reserves that exist with at least 90% certainty; cf. probable and possible reserves. (proven reserves)

proven reserves. *See* proved reserves.

prover. *See* meter prover.

PRPT. preparing to run potential test.

prtgs. partings.

prtn. partition (land).

PS. pressure switch.

ps. pseudo-.

PSA. packer set at.

PSI and **psi.** pounds per square inch.

PSIA and **psia.** pounds per square inch absolute.

PSIG and **psig.** pounds per square inch gauge.

PSL. public school land.

PT. potential test.

Pt and **pt.** part.

pt. point.

P_{10} and **P10.** At least 90% probability that it exists.

PTG. pulling tubing.

PTR. pulled tubing and rods.

PTTC. Petroleum Technology Transfer Council (www.pttc.org).

PTTF. potential test to follow.

PU. (1) pulling unit, (2) picked up, (3) pulled up., and (4) pumping unit

pull casing. To remove and salvage casing from a well.

pulling unit. A truck-mounted service unit with a winch and mast. It is used to pull and run tubing and sucker rods in a well. The crew usually consists of an operator, derrick operator, and floor person. (PU)

pull rods. To remove the sucker-rod string from a well during a workover.

pulsed neutron log. A type of neutron log that can be used to distinguish gas and oil from water behind casing in a well. It bombards the formation with neutrons and measures the returning gamma rays.

pumpability time. The time a cement slurry remains fluid enough to be pumped. (thickening time)

pumpdown. To pump equipment down a producing well to service the well.

pumper. (1) A well that requires a pump to bring the oil to the surface. (2) The mechanic who is responsible for maintaining producing equipment in the field and receives orders from a production foreman.

pump stroke counter. A device used on a drilling rig to record the number of mud pump strokes per minute (SPM). It is recorded on a mud log and used to estimate lag time.

pump jack. A common term for a sucker-rod pump. (PJ)

push-the bit. A rotary steerable system that uses hydraulically activated steering pads on the drilling unit that expand and contract in the to push the bit of center and drill-deviated wells.

putting the well on pump. Replacing the production tree on a well that has lost pressure and will no longer flow to the surface with a sucker-rod pumping unit.

PV or **P.V.** pore volume.

PV. plastic viscosity.

pvmnt. pavement.

PVT. pressure-volume-temperature.

P wave. compressional wave

PWR. power.

pyls. pyrolysis.

pyr. pyrite.

pyrobit. pyrobitumen.

pyrobitumen. A naturally occurring, dark, hard hydrocarbon. (pyrobit)

pyrolysis. A method of analyzing the composition of a substance by heating the sample in the absence of oxygen. The compositions and temperatures of the gases that are given off as the sample is heated are measured. Source rocks are analyzed by pyrolysis for maturity and organic matter type. (probit)

pyrite. A common, heavy, brassy or bronze-yellow mineral composed of FeS_2. It is commonly called fool's gold. (pyr)

q. rate.

QA. quality assurance.

qty. quantity.

Qtz or **qtz.** quartz.

Qtzt or **qtzt.** quartzite.

quad. quadrangle.

quadrangle. A four-sided tract of land or a map of that land that is bounded by parallels of latitude and meridians of longitude that are 1° apart. (quad)

qual. quality.

quan. quantity.

quartz. A very common and hard mineral composed of SiO_2. Impurities in quartz result in various colors such as milky, rose, and cloudy. Sandstones are composed primarily of quartz sand grains. (Qtz and qtz)

quartzarenite. A sandstone composed of more than 95% quartz sand grains. It can be an excellent reservoir rock.

quartzite. A very hard sandstone composed primarily of quartz sand grains. (Qtzt and qtzt)

Quaternary. A period of geological time from 2.6 million years ago to the present. It is part of the Cenozoic era.

quick-look log. A wellsite, computer-generated log that uses two or more logging measurements to calculate water saturation, porosity, percentages of sandstone, limestone, and shale, and fractures location.

R. (1) resistivity, (2) recovery factor, (3) radioactivity, and (4) range.

r. (1) radius and (2) rare.

RA. radioactive.

R/A. regular acid.

RAD. radius.

radioactivity log. A wireline log that uses a radioactive source in the logging tool to bombard the rocks with either atomic particles or energy to measure porosity; e.g., neutron porosity log and formation density log.

radiolaria. A single-cell animal that floats in the ocean and has a silicon dioxide shell. It is a type of microfossil.

RALOG. running radioactive log.

ramp down and **ramp up.** To gradually and steadily decrease or increase a process or operation.

R&L. road and location.

R&LC. road and location complete.

R&O. rust and oxidation.

R&P. rods and pump.

R&T. rods and tubing.

range. (1) A system of north-south strips six miles wide that are used in land subdivision. (2) The geological time extent that a fossil species existed. (RGE or rng)

rank wildcat. An exploratory well drilled at least two miles away from the nearest production.

raster. Scanned. Raster well logs have been scanned into a computer database.

rate of penetration. The speed with which a drill bit penetrates the rocks at the bottom of a well. It is recorded in minutes per foot (min/ft) on a mud log. (ROP)

rat hole. (1) A hole in the drill floor used to hold the swivel and kelly when tripping out. (2) The lowest portion of the well below the pay zone that is used to accommodate equipment such as a sonde. (anchor hole) (RH)

RB. (1) rock bit and (2) rotary bushing.

RB/D. reservoir barrels per day.

rbds. red beds.

RBM. rotary bushing measurement.

RBP. retrievable bridge plug.

rbr. rubber.

RBSO. rainbow show of oil.

RC. (1) running casing, (2) remote control, and (3) reversed circulation.

RD. (1) rig down, (2) rigged down, (3) rigging down, (4) recorded depth, (5) rotary depth, and (6) random drilling.

R/D. redrilled.

rd. (1) red, (2) road and (3) round.

RDB. rotary drive bushing.

Rd Bds. red beds.

rdd. rounded.

RDRT. rig or rigging down rotary tools.

RDSR. rig or rigging down service rig.

RDSU. rig or rigging down swabbing unit.

RDT. rig or rigging down tools.

rdtr. round trip.

read. reacidize.

ream. To mechanically enlarge or straighten a well or casing string. (RM)

reamer. A sub that uses blades or wheels to ream a wellbore or casing string.

rebar. reinforcing bar.

reboiler. A distillation vessel that heats wet glycol to separate glycol and water.

Rec or **rec.** (1) recover and (2) recovered.

rec. (1) recorder, (2) recovery, and (3) recommended.

Recent. *See* Holocene.

recharge area. An aquifer outcrop where freshwater enters.

recmd. recommend.

recomp. recomplete.

recomplete. To plug and abandon one zone in a well and complete in another. It is done during a workover by either plugging back or drilling deeper. (recomp)

recond. recondition.

recoverable gas or **oil.** The amount of gas or oil that can be produced from a reservoir under current economic conditions. It is a percent of the gas or oil in place; cf. oil in place. See also recovery factor.

recovery factor. The percentage of oil and/or gas in place that can be produced from a reservoir. (R)

recv. receive.

red. (1) reducer and (2) reducing.

red bed. Red-colored sedimentary rocks with an iron oxide coating usually deposited in a desert environment. (rbds and RD Bds)

redrid. redrilled.

reef. A ridge or moundlike structure of wave-resistant, framework-building organisms such as corals. (Rf and rf)

referg. refrigerant.

reflection coefficient. The percentage of seismic energy reflected off a surface.

reg. (1) regular and (2) regulator.

regression. A retreat of the sea from the land; cf. transgression.

regular acid. hydrochloric acid. (R/A)

reinf. reinforce.

reinf conc. reinforced concrete.

rej. reject.

REL. running electric log.

rel. (1) release and (2) released.

relative permeability. The ratio between effective permeability of a fluid at partial saturation to the permeability of that fluid had it been at 100% saturation.

relief well. A well drilled close to a blowout well in order to decrease the pressure on the abnormal high-pressure zone that is causing the blowout. Heavy drilling mud (kill mud) is then pumped into the uncontrolled well to kill the well.

reloc. relocate.

Rem. remains.

rem. (1) remedial and (2) remove.

remotely operated vehicle. An uncrewed submarine propelled by an electrical motor and thruster propellers. It is manipulated from a mother ship through an umbilical that connects the two. The submarine is used to do deep-sea work. (ROV)

rep. (1) replace and (2) report.

repeated section. *See* double section.

repeat formation tester. A wireline tool that samples reservoir fluids and measures reservoir pressures at several levels in a well. (RFT)

reperf. reperforate.

repl and **Repl.** (1) replace and (2) replaced.

repr. repair.

reprocess. The application of new computer-processing methods to older seismic data that was recorded digitally.

req. request.

reqd. required.

reqmt. requirement.

Res and **res.** residue.

res. (1) reservoir, (2) resistance, (3) resistivity, and (4) resistor.

res bbl. reservoir barrels.

reserve pit. An earthen pit, often lined with plastic, located next to a drilling rig. It holds drilling mud that is not being used and well cuttings that flow off the shale shaker.

reserves. The calculated amount of gas and/or oil that is expected to be produced from a well or a field in the future. under current economic and technical conditions. Types of reserves are based on the probability that they exist. Proved or proven reserves have at least 90% probability that they exist. Probable reserves have at least 50% and possible reserves

have at least 10% probability that they exist. Developed reserves can be produced from existing wells, whereas undeveloped reserves cannot be presently produced without either drilling or recompleting a well. *See also* book reserves.

reservoir. The subsurface deposit of oil and/or gas located in the pores of a reservoir rock. Fluids cannot flow from one reservoir to another. (res)

reservoir barrel. One liquid barrel of crude oil in the subsurface reservoir. When the barrel of oil brought to the surface and gas bubbles out, the volume of oil will shrink. (res bbl) *See also* Formation volume factor and stock barrel of oil.

reservoir characterization. The quantification of reservoir properties such as porosity and permeability in an oil and gas reservoir. It is used to make a computer model of the reservoir.

reservoir drive. The source of pressure on subsurface fluids that forces them through the reservoir rock and into the well. It comes from fluid expansion, rock expansion, and gravity. Some types of reservoir drives are solution gas, free gas cap, water, gravity, and expansion gas.

reservoir pressure. The pressure on fluids in the pores of rock at a specific depth. Normal reservoir pressure is due to the weight of the overlying waters. (fluid pressure and formation pressure) *See also* abnormal high pressure.

reservoir rock. A rock that has porosity and permeability. It can hold and transmit fluids. The most common reservoir rocks are sandstones, limestones, and dolomites.

reservoir simulation. The computer modeling of an oil and gas reservoir. The reservoir is divided into a large number of geocells, each with characteristic properties such as porosity and permeability. The flow between each geocell is calculated.

resid. (1) residual and (2) residue.

residual gas. The gas, primarily methane, that exits a natural gas processing plant after the natural gas liquids have been separated. (tail gas)

residual water. Water in the pores of a reservoir rock that will not flow. (irreducible water)

resistivity. The opposition of a substance to the flow of an electrical current through it. Resistivity is a measurement made on an electric and induction wireline log in units of ohm-meter or ohm meter²/meter. It is used to determine the fluid composition in the pores of rocks and oil and water saturation. The inverse of resistivity is conductivity. (R)

resolution. The minimum distance of separation between two features that allows the two features to be distinguished individually. It can be either vertical or horizontal resolution.

resource. A general term for deposits of a valuable gas, liquid, or solid that occur in the world or a geographical areas such as a country. Unlike reserves, resources can also include undiscovered deposits and deposits that cannot be developed with present-day technologies and/or under present-day economics (e.g. crude oil, natural gas, and coal).

restricted basin. A body of water that is separated from the ocean by a shallow sill or bar at the entrance and has limited water circulation.

ret. (1) retain and (2) return.

retained interest. The ownership portion an owner keeps when transferring the remaining ownership.

retarder. An additive that slows a process such as cement setting; cf. accelerator.

retd. returned.

retention time. The time that the produced fluids spend in a separator.

retr. retrieve.

retrograde condensate. The condensate that forms when pressure is dropped on wet gas during production.

retrograde condensate or **retrograde-gas condensate reservoir.** A natural gas reservoir in which condensate forms both in the subsurface reservoir and on the surface during production; cf. dry gas and wet gas reservoir.

retr ret. retrievable retainer.

return on investment. A criterion used to evaluate an investment in an oil or gas well. It is the estimated net production revenue during the life of the well divided by the drilling and completion costs. Discounted return of investment (DROI) uses costs and revenues that have been discounted for the time value of money. (ROI) *See also* payout.

returns. Drilling mud and well cuttings that flow up a well as it is being drilled.

rev. (1) reverse, (2) revise, and (3) revolutions.

reverse circulation. To pump drilling mud down the annulus and back up the tubing string. This method is used to clean out a well. (RC)

reverse fault. A fault with predominantly vertical movement (dip-slip) in which the hanging wall has moved up in relation to the footwall. It creates a double or repeated section; cf. normal fault.

reversionary interest. A well or property interest that becomes effective at a specific time or event in the future.

rev/O. reversed out.

rexlzd. recrystallized.

RF. rig floor.

Rf and **rf.** reef.

rfl. reflection seismograph.

RFT. repeat formation tester.

RGE and **rge.** range.

rgh. rough.

RH. rat hole.

R/H. ran in hole.

ρ. density.

RI. royalty interest.

rich gas. (1) Natural gas that contains a significant amount of condensate. (2) Moderately rich gas that contains between 2.5 and 5 GPM and very rich gas that contains more than 5 GPM cf. lean gas and dry gas.

rift. A large fault with predominantly horizontal movement.

rift valley. A deep, wide fracture.

rig down. To disassemble a drilling rig after drilling; ant. rig up. (RD)

rig floor. The elevated flat steel surface on which the derrick or mast sits and most of the drilling activity occurs. It is supported by the substructure. (derrick or drill floor) (DF)

right-lateral strike-slip fault. A fault that moves horizontally, with the opposite side of the fault moving toward the right as you face the fault; cf. left-lateral strike-slip fault.

rig rel. rig released.

rig up. To assemble a drilling rig to spud a well; ant. rig down. (RU)

RIH. (1) running in hole and (2) ran in hole.

risk. *See* success rate.

rk. rock.

RKB. referenced to kelly bushing (depth measurement).

rky. rocky.

RLN. long normal resistivity.

RM. ream.

rmd. reamed.

rmg. reaming.

rmv. remove.

rnd. rounded.

rng. (1) range and (2) running.

RO. reversed out.

rock. A naturally occurring aggregate of mineral grains. Rocks are classified as igneous, metamorphic, and sedimentary (e.g., granite and shale). (rk)

rod basket. A steel platform with sides that is located near the top of the mast on a well service unit. The derrick operator stands in the rod basket to place the sucker rods in the rod fingers as the rods are pulled from the well.

rod pump. A common type of oil well downhole pump driven by a sucker-rod string. It is run as a complete unit on the sucker-rod string through the tubing string; cf. tubing pump. (insert pump)

rod pumping system. A common artificial lift system for an oil well. A surface beam-pumping unit drives a sucker-rod pump on the bottom of the tubing string. A sucker-rod string that runs down the center of the tubing string connects the walking beam on the surface with the sucker-rod pump.

ROI. return on investment.

ROL. rig on location.

roller-cone bit. A rotary drilling bit that has rotating cones mounted on bearings. A tricone bit with three cones is very common type of roller-cone bit.

rollover anticline. A large fold formed in sedimentary rocks on the basin side of a growth fault. It can be a petroleum trap.

ROP. rate of penetration.

ROS. residual oil saturation.

rot. (1) rotary and (2) rotate.

rotary depth. *See* driller's total depth.

rotary drilling rig. A very common type of drilling rig that rotates a long length of steel pipe with a bit on the bottom to cut the well. Four major systems on the rig are power, hoisting, rotating, and circulating. Rotary drilling rigs are either mechanical or diesel electric depending on the power system used; cf. cable tool rig. (RR)

rotary helper. *See* roughneck.

rotary hose. *See* mud hose.

rotary steerable system. A deviated well drilling system that uses a downhole motor with steering pads on the sides that can be hydraulically expanded and contracted from the surface. The pads push the bit off center in a method called push-the-bit. (RSS)

rotary table. A revolving plate on the drill floor that is driven by the prime movers. The master and kelly bushings are attached to the top of it. It turns the drillstring that runs down through the center of the rotary table. (RT)

rotary total depth. *See* driller's total depth.

roughneck. A drilling crew member who operates and maintains the equipment on the floor of a drilling rig under orders from the driller. (rotary helper)

round trip. A cycle of running a drillstring into the well (tripping in), touching bottom, and putting it back out; cf. short trip. (tripping out).

roustabout. (1) A general helper on producing wells and well service units. (2) A member of the offshore drilling crew who helps bring supplies and equipment aboard under orders from the head roustabout.

ROV. remotely operated vehicle.

roy. royalty.

royalty. A percentage or fraction of the revenue from oil and gas production that is free and clear of production costs. It is paid to the mineral rights owner on fee land and any other royalty owner. Royalty in kind is a share of the production instead of revenue. (roy)

royalty interest. An ownership in production that bears no cost of production. Royalty interest owners receive their share of production revenue before working interest owners; cf. working interest. (RI)

RPM and **rpm.** revolutions per minute.

RPS and **rps.** revolutions per second.

rpt. report.

RR. (1) rotary rig and (2) rig released.

rr. rare.

RR&T. running rods and tubing.

RRB. rerun bit.

RS. rig skidded.

RSN. short normal resistivity.

rsns. resinous.

RSS. rotary steerable system.

RT. rotary table.

Rt. true resistivity.

RTG. running tubing.

rtg. rating.

rthy. earthy.

RTLTM. rate too low to measure.

rtnr. retainer.

RU. (1) rig, rigged, or rigging up and (2) rotary unit.

rub. rubber.

RUM. rig or rigging up machine.

run. (1) The amount of crude oil sold and transferred to a pipeline or tanker truck. (2) To run tubulars or tools into a well. (3) The cycle of lowering (inrun) and raising (outrun) equipment in a well (e.g., a logging run).

run casing or **pipe.** To run and cement casing to complete a well. (set pipe)

run ticket. A form filled out when oil is transferred from stock tanks to a tank truck or pipeline. It lists the quality and quantity of the oil and is used to pay the operator of the wells.

RUP. rig or rigging up pump.

rupt. rupture.

RUR. rig or rigging up rotary (rig).

RURT. rig or rigging up rotary tools.

RUST. rig or rigging up service tools.

RUSU. rig or rigging up service unit.

RUT. rig or rigging up tools.

RVP. Reid vapor pressure.

rvsd. reversed.

RW. reworked.

Rw. resistivity of water.

rwk. rework.

RWTP. returned well to production.

S. (1) saturation, (2) swabbing, (3) sulfur, and (4) signal.

S/. (1) swabbed and (2) show with.

Sa and **sa.** salt.

SAB. strong air blow.

sack. A container for (1) dry cement (94 lb), (2) bentonite clay (100 lb), (3) barite (100 lb), and (4) other dry supplies. (sk or sx)

SAGD. steam-assisted gravity drainage.

sal. salinity.

sales-quality gas. *See* pipeline-quality gas.

sales-quality oil. *See* pipeline oil.

salinity. The weight of all dissolved salts per unit volume in a solution such as oilfield brine. It is often expressed in parts per thousand (ppt), parts per million (ppm), or milligrams per liter (mg/l). (sal)

salt dome. A large mass of salt (salt plug) that is or has been flowing upward through the overlying sedimentary rocks. The salt dome also includes the surrounding and overlying sedimentary rocks that have been deformed.

samp. sample.

sample log. A record of the physical properties of rocks in a well. It includes composition, texture, color, presence of pore spaces, and oil staining. (lithologic or strip log)

Samson post. The steel-beam assembly on which the walking beam pivots on an oil well beam-pumping unit.

sand. A clastic particle between 2 and $\frac{1}{16}$ mm in diameter; cf. silt and clay. (sd)

sand cleanout. A workover in which saltwater or drilling mud is circulated to remove loose sand from the bottom of a well.

sand control problem. Loose sand clogging the bottom of a well.

S&F. swab and flow.

sandface. The surface of the oil or gas reservoir in the wellbore.

sandfrac. *See* hydraulic fracturing.

S&O. stain and odor.

sand/shale ratio map. A map that uses contours to show the ratio of sandstone to shale in a formation.

sandstone. A common sedimentary rock composed primarily of sand grains. It can be a reservoir rock. (SS, ss, and Sst)

Santonian. An age of geological time from 85.8 to 83.5 million years ago. It is part of the Cretaceous period.

sat. (1) saturate, (2) saturation, and (3) saturated.

satellite well. A subsea well in a remote part of an offshore field or in a marginal field with a flowline that conducts produced fluids from the well to a production platform for treating. The well was not drilled from the platform. It was drilled from a jackup rig or floater.

saturated. The condition in which a liquid has dissolved all the gas or salt that it can hold; cf. undersaturated. (Sat and sat)

saturated pool. An oil reservoir with a free gas cap. The crude oil in the reservoir has dissolved all the natural gas that it can hold and is saturated; cf. undersaturated pool.

saturation. The percentage of different fluids such as gas (S_g), oil (S_o), and water (S_w) in the pore space of a rock. (S)

SB. stuffing box.

sb. sub.

SBHP. static bottom-hole pressure.

SC. show of condensate.

scab liner. A liner string run in a well to repair casing.

scale. Salts that have precipitated out of water. Calcium carbonate, barium sulfate, and calcium sulfate are common from oilfield brines.

scale inhibitors. A chemical used to prevent salt formation in a well.

scalped anticline. *See* bald-headed anticline.

scat. scattered.

SCF and **scf.** standard cubic feet.

SCFD and **scf/D.** standard cubic feet per day.

SCO. synthetic crude oil.

scout. An oil company or commercial scouting company employee who gathers information on petroleum-related activities of other companies in a regional area.

scout card and **ticket.** A paper or computer file form completed by an scout on engineering and geological information gathered on a specific well being drilled. It includes well name, location, depth, date completed, major formation tops encountered, well treatments, and initial oil and gas production.

SCR. silicon-controlled rectifier.

scr. (1) scratcher, (2) screen, and (3) screw.

scratchers. Wires on a collar that is attached to the lower part of a casing string being run into a well. It is used to remove mud cake from the well walls.

scrd. screwed.

screened, slotted liner. *See* liner string.

scrub. scrubber.

scrubber. Equipment used to remove liquid from gas. (scrub)

scs. scarce.

SD and **S.D.** shut down.

sd. (1) sand and (2) sandstone.

SDA. shut down for acid.

sd & sh. sand and shale.

SDF. shut down for frac.

SDL. shut down for logging.

SDO. (1) shut down for orders and (2) show of dead oil.

sdoilfrac. sand oil frac.

SDON. shut down overnight.

SDP&A. shut down for plug and abandon.

SDPL. shut down for pipeline.

SDR. shut down for repairs.

Sd SG. sand showing gas.

Sd SO. sand showing oil.

sdtrk. sidetrack.

SDW. shut down for weather.

SDWO. shut down, waiting for orders.

sdy. sandy.

sdy li. sandy lime.

sdy sh. sandy shale.

S/E. screwed end.

seafloor seismic method. Seismic exploration in the ocean with hydrophone streamers positioned on the seabed.

seafloor spreading. A theory in which the earth's crust (seafloor) is formed by basalt volcanoes along the crest of the mid-ocean ridge. The crust is split and spreads to either side of the ridge because of convection currents in the molten interior of the earth. The crust is destroyed in subduction zones. *See also* plate tectonics.

seal. An impermeable rock layer that forms the cap on top of an oil or gas reservoir; e.g., shale. (caprock)

sealing fault. A fault that does not allow fluid flow along or across the fault.

seating nipple. A short pipe that is run on the bottom of a tubing string. It has a constricted inner diameter that stops any tool that falls down the tubing string. It is also used to attach a downhole pump, safety valve, choke, or regulator. (SN)

sec. (1) secondary, (2) section and (3) second.

secondary cementing. A workover cement job on a producing well; cf. primary cementing.

secondary fault. A relatively minor fault oriented parallel to a major fault.

secondary gas cap. A free gas cap that forms from the solution gas that bubbles out of the oil as reservoir pressure drops during production.

secondary recovery. A process of injecting gas or water into an oil reservoir to restore production when the primary drive has been depleted; cf. tertiary recovery.

secondary stratigraphic trap. A petroleum trap formed by an angular unconformity; cf. primary stratigraphic trap.

secondary term. The time granted in a lease for production. It occurs after the primary term and continues as long as commercial amounts of petroleum are being produced; cf. primary term.

section. A surveyed square of land that is one mile on a side; 36 sections make a township. There are 640 acres in a section. (sec)

Sed and **sed.** sediment.

sediment. Loose (unconsolidated), solid particles or salt. Sediments are deposited out of water, air, or ice; e.g., sand grains and mud particles. (Sed or sed)

sedimentary rock. A layered rock composed of sediments that have been solidified (consolidated or lithified). A clastic sedimentary rock is composed of particles formed by weathered rock such as sand grains. A chemical sedimentary rock is composed of salts that have precipitated out of water. An organic sedimentary rock is composed of organic particles such as plant remains. The most common sedimentary rocks are shale, sandstone, and limestone. Sedimentary rocks are drilled to find and produce gas and oil; cf. igneous rock and metamorphic rock.

seep and **seepage.** A natural occurrence of oil and/or gas that has leaked onto the surface.

SEG. Society of Exploration Geophysicists (www.seg.org).

seis. (1) seismic and (2) seismograph.

seismic contractor. A company that maintains and operates seismic equipment.

seismic horizon. A reflection that can be traced on a seismic record.

seismic method. The acquisition, computer processing and display of echoes from subsurface rock layers that are used to image the shape of the rock layers. Seismic energy is put into the earth with a source such as dynamite, vibrator truck, or air gun. The sound energy reflects off subsurface, sedimentary rock layers and is recorded by detectors called geophones or hydrophones on the surface. The data are recorded digitally and processed by computers to make the image. Seismic is 2- or 3-dimensional (2-D and 3-D).

seismic option. A type of mineral rights acquisition. The lessee pays the lessor a bonus for the right to run seismic exploration on the land and to have the option of leasing the land after reviewing the seismic data.

seismic processing. The application of mathematical equations to seismic data by computer to improve the signal/noise ration and increase accuracy and resolution.

seismic record or **section.** A display of seismic reflections recorded off subsurface rock layers similar to a vertical cross section of the earth. Shot points are located along the top of the section. Timelines run horizontally across the section. Zero seconds is always at or near the surface of the earth or at the surface of the ocean. A header with seismic information is located on the record.

seismic stratigraphy. The recognition and use of unconformities on seismic records to correlate and map sedimentary rock packets called sequences. Each sequence was deposited during a major cycle of sea level fall, rise, and fall. Seismic facies (seismic reflection characteristics) are used to identify the depositional environments in each sequence. *See also* sequence stratigraphy.

seismic structure map. A map contoured in units of seismic time (milliseconds) to a specific reflector; cf. structure map.

seismic time-lapse map. A map contoured in milliseconds of vertical distance between two seismic reflectors; cf. isopach map.

seismic tomography. The use of seismic data slices to image subsurface geology.

seismic trace. The response of one seismic recorder to one seismic shot.

seismic wipeout. An area on a seismic record where there are no seismic reflections in contrast to adjacent areas. It is often caused by natural gas in the sedimentary rocks.

selenite. *See* gypsum.

self potential. *See* spontaneous potential.

SEM. scanning electron microscope.

semi and **semisubmersible.** A type of floating, offshore, exploratory drilling rig system anchored above the drillsite. It has large, submerged flotation chambers (pontoons) located on short columns below the drilling platform.

sep. separator.

separator. A long steel tank used to separate produced fluids from oil wells. Separators use gravity, impingement, centrifugal force, filters, and other methods. They can be either horizontal or vertical. (sep)

seq. sequence.

sequence stratigraphy. The use of timelines such as unconformities on well logs, cores, and rock outcrops to map and correlate packets of sedimentary rocks called sequences. A sequence was deposited during an interval of geologic time and can be subdivided in parasequence sets and further into parasequences. *See also* seismic stratigraphy.

sequestering agent. An additive used during acidizing a well to prevent the formation of an iron gel or precipitate.

ser. (1) serial and (2) series.

series. A time-rock division of rocks deposited during an epoch.

service company. A company that supplies services such as logging, mud engineering, or cementing; e.g., Schlumberger and Halliburton.

service unit. A truck with equipment, usually a mast and drawworks with wireline, mounted on it to workover a producing well. A pulling unit is a common type of service unit. (svcu)

set. (1) To position such as set pipe in a well. (2) To harden such as set cement.

set in the dark. To run and cement a string of casing to the top of the reservoir rock in a well without first drilling and testing the reservoir rock.

set pipe. To run and cement casing to complete a well. (run pipe)

set-through drilling. An older drilling method in which the well was drilled into only a short interval on top of the producing reservoir and completed open-hole.

set-through completion. A well completion in which the casing or liner has been cemented into the reservoir. The casing or liner is then perforated; cf. open-hole completion.

severance tax. State tax on oil and gas produced. (production tax)

SF. sand frac.

sfc. surface.

SFL. starting fluid level.

SFLU. slight fluorescence.

SFO. show of free oil.

SFP. surface flowing pressure.

sft. soft.

SG. (1) show of gas, (2) specific gravity, (3) survey gas, and (4) surface geology.

s.g. specific gravity.

S$_g$. gas saturation.

SG&C. show of gas and condensate.

SG&O. show of gas and oil.

SG&W. show of gas and water.

SGCM. slightly gas-cut mud.

SGCW. slightly gas-cut water.

SH, **Sh**, and **sh.** shale.

Sh. share (of land).

shake-out test. A method used to determine the basic sediment and water content of oil by centrifuging a crude oil sample.

shaker. *See* shale shaker.

shale. A very common sedimentary rock composed of clay-sized particles. Most mineral grains in shale are clay minerals. Shales are typically well layered. Black shales are source rocks for petroleum. (SH, Sh, and sh)

shale gas. Natural gas produced from a gas shale. It occurs adsorbed to organic matter (kerogen) and in intergranular micropores.

shale oil. Crude oil obtained by heating (660°F) oil shale.

shale-out. A stratigraphic trap formed by the lateral change (facies change) of a permeable sandstone or limestone into impermeable shale; cf. pinch-out.

shale shaker. A set of vibrating screens in a steel frame on the mud tanks of a drilling rig. Shale shakers are used to separate well cuttings from drilling mud coming from the well. Modern drilling rigs have four or more shale shakers.

shear rams. Two large blocks of metal with chisel edges. They are designed to shear across any drillpipe in the well and close the well. Shear rams are used in a blowout-preventer stack.

shear wave. A wave that causes particles to move up and down as the wave passes through. It is similar to a wave on the ocean. It travels at about half the speed of a compressional wave through rocks and cannot travel across fractures or through a liquid or gas. It is recorded along with compressional waves during multicomponent seismic; cf. compressional wave. (secondary wave) (S wave)

shield. A low-lying, stable area of basement rocks on the surface of the earth.

shld. shoulder.

shls. shells.

shly. shaly.

shock sub. *See* vibration dampener.

shoestring sandstone. A long, narrow, lens-shaped sandstone usually encased in shale and originally deposited as a barrier island, river channel, bar finger, or valley fill.

short normal resistivity. A wireline resistivity measurement made with electrodes spaced close together; cf. long normal resistivity. (16 in.)

short trip. A cycle of running a drillstring in the well (tripping in) without going all the way to the bottom (touching bottom) and putting it back out (tripping out); cf. round trip. (ST)

shot. (1) An explosion used to artificially fracture reservoir rocks in a well to stimulate production. (2) An explosion used as a seismic exploration source to put sound energy into the ground.

shot hole. A shallow hole drilled for an explosive source used for seismic exploration on land. The shot hole directs the explosive energy downward.

shotpoint. The location where a seismic source such as dynamite, vibrator truck, or air gun was activated. (SP)

shotpoint array. The pattern of several seismic sources used simultaneously at a shot point to reduce source noise.

show. Hydrocarbons in an amount above background. (Shw)

show evaluation. A detailed analysis of the composition of hydrocarbons in a show.

shr. shear.

shrinkage factor. The decimal amount to which a barrel of reservoir oil shrinks to on the surface of the ground after the pressure has dropped and the gas has bubbled out of the oil. *See also* stock-tank barrels of oil.

SHT. straight hole test.

shut in. To cease production from a well. The noun and adjective form is shut-in. (SI)

shut-in pressure. Pressure on a fluid that is not moving. (static pressure)

Shw. show.

SI. shut in.

SIBHP or **SIBP.** shut-in bottom-hole pressure.

SICP. shut-in casing pressure.

sidetrack. (1) A new wellbore branch drilled out from an existing well (often the deviated portion of a well drilled around a fish). (2) To drill a sidetrack. (sdtrk and ST)

sidewall core. A 1-in. diameter core from the sides of a well. It is obtained by (1) an explosive-propelled tube (percussion sidewall coring) or (2) drilling (rotary sidewall coring); cf. full-diameter core. (SWC and S.W.C.)

sieve. (1) A screen used to sort particles by size. (2) To use sieves to sort particles by size.

signal. The desired seismic energy (direct or primary reflections) received from the subsurface; cf. noise. (S)

SIGW. shut-in gas well.

Sil, sil, and **silic.** siliceous.

siliceous. A rock containing silica. (Sil, sil, and silic)

silicon-controlled rectifier. A device that converts alternating electric current to direct electric current. It is used on a diesel-electrical drilling rig to convert the AC electric current from the prime movers to DC electric current used by the motors on the drill floor. (SCR)

sill. An igneous rock that was injected as a molten liquid between sedimentary rock layers; cf. dike.

silt. A clastic particle between $\frac{1}{16}$ and $\frac{1}{256}$ mm in diameter. (Slt or slt)

silt. siltstone.

siltstone. A sedimentary rock composed primarily of silt-sized particles that are intermediate between sand- and clay-sized particles. (silt, Sltst, and sltst)

Silurian. A period of geological time that occurred from 444 to 416 million years ago. It is part of the Paleozoic era. (Sil)

sim. similar.

simulfrac, simultaneous frac job, or **simultaneous hydraulic fracturing.** The hydraulic fracturing of two or more, closely spaced (500 to 1,000 ft), parallel horizontal drainholes at the same time. It intensifies the reservoir fracturing between the well. It is used in gas shales.

single. One tubular joint; cf. double, treble, and fourble.

single-pole unit. A well-servicing unit used for shallow wells that has only one telescoping steel tube for a mast that can be set to several different heights. It has one or two drums and can be run by an operator and floor person; cf. double-pole mast.

SIO. shut-in oil.

SIOW. shut-in oil well.

SIP. shut-in pressure.

SITP. shut-in tubing pressure.

SIWHP. shut-in working head pressure.

SIWOP. shut-in, waiting on potential.

SK. skimming plant.

sk. sacks.

skeletal sands. Sands formed by fragments of shells.

skim. skimmer.

skin damage. *See* formation damage

skt. socket.

SL. (1) sea level, (2) section line, and (3) south line.

sl. (1) sleeve and (2) slight.

slab. (1) a core that has been cut lengthwise with a diamond saw to better view the rocks. (2) to cut a slab.

slant rig. A drilling rig with a mast or derrick that is or can be adjusted to be at an angle (usually 30 to 45°) to vertical. The well is spudded at an angle to drill a deviated or horizontal drainhole to a relatively shallow drilling target.

SLAR. side-looking airborne radar.

sli. (1) slight and (2) slightly.

slick assembly. A downhole assembly that has no stabilizers. It is used to drill a straight hole.

slick line. A single strand of wire that is used to raise and lower equipment in a well; cf. wireline.

slickwater or **slickwater frac.** A hydraulic frac job that uses freshwater with a friction reducer additive and relatively little proppants. It produces relatively long, narrow fractures and is used in tight reservoirs and gas shales. A slickwater frac is less expensive than a normal frac job; cf. frac pac. (waterfrac)

slim hole. A well with a small-diameter wellbore (6¾ to 4¾ in.). Slim holes are less expensive to drill and are used for exploration.

slip logs. To move the well log from one well up and down vertically to correlate the formations in that well with the formations on another well log from another well that is held stationary.

slips. A steel wedge with teeth used in the bowl of a rotary table to grip and prevent the drillstring from falling down the well.

sli SO. slight show of oil.

SLM. steel line measurement.

slnd. solenoid.

slotted liner. *See* liner string.

sloughing shale. Shale along the walls of a well that absorbs water and expands.

Slt and **slt.** silt.

Sltst or **sltst.** siltstone.

SLT WT. saltwater.

slty. (1) silty and (2) salty.

sluff. The collapse of well walls into the hole. (cave)

slug. A batch of water and/or chemicals that is injected into a well or reservoir.

slur. slurry.

slurry. A mixture of a liquid and suspended, fine-grained, insoluble particles. Cement is a slurry as it is being pumped into a well during a cement job; cf. solution. (slur)

slush pumps. *See* mud hogs.

sly. slightly.

SM. surface measurement.

sm and **sml.** small.

small scale. A map that shows relatively less detail but covers more area than a large-scale map.

smls. seamless.

smth. smooth.

smwt. somewhat.

SN. seating nipple.

S/N. signal-to-noise ratio.

sniffers. A chemical devise towed behind a ship to detect hydrocarbons in ocean water.

snub. to run tools or pipe into a high-pressure well that is still flowing.

snubbing unit. A production rig designed to workover wells under high pressure.

SO. (1) show of oil, (2) shake out, and (3) side opening.

S$_o$. oil saturation.

SO&G. show of oil and gas.

SO&GCM. slightly oil and gas-cut mud.

SO&W. show of oil and water.

SOCM. slightly oil-cut mud.

SOCW. slightly oil-cut water.

soft rock. sedimentary rock; cf. hard rock.

soil. A surface layer of weathered rock particles containing organic matter.

soil farming. The mixing of drilling fluid and cuttings with soil for disposal; cf. land farming.

soil investigation. *See* subsea site investigation.

SOL. percent solids.

sol. (1) solenoid and (2) solids.

solids control system. The shale shakers, desanders, desilters, and settling tanks on a drilling rig that remove the well cuttings from the drilling mud circulation out of a well being drilled.

soln. solution.

solution. A homogeneous liquid formed by dissolving a gas or solid in the liquid; cf. slurry.

solution gas. The dissolved natural gas that bubbles out of crude oil on the surface when the pressure drops during production.

solution gas drive. A reservoir drive in which the drop in reservoir pressure during production causes dissolved gas to bubble out of the oil and force the oil through the reservoir rock. It has a relatively low oil recovery efficiency. (dissolved gas and depletion drive)

solution gas/oil ratio. The standard cubic feet of natural gas dissolved in one barrel of oil in the reservoir. (formation and dissolved gas/oil ratio)

solv. solvent.

sonde. *See* logging tool.

sonic amplitude log. A wireline well log that measures the attenuation of sound through rocks to detect fractures.

sonic log. A wireline well log that measures sound velocity through the rocks in microseconds per foot (μsec/ft). The porosity of the rock can be calculated from the sound velocity of the rock. (acoustic velocity log) (SL and SONL)

SOP. standard operating procedure.

sorting. A measure of the range of different sized particles in a clastic rock. Sedimentary rocks can be well (narrow range) or poorly (wide range) sorted.

sour. Gas or oil with a high sulfur content. Sour oil generally contains more than 1% sulfur; cf. sweet.

source rock. A sedimentary rock rich in organic matter that can or has been transformed under certain geological conditions into natural gas and/or crude oil. Black shales are common source rocks.

SP. (1) spontaneous or self potential, (2) shotpoint, (3) set plug, (4) slightly porous, and (5) surface pressure.

sp. spare.

spacing. *See* drilling and spacing unit.

spar. A type of offshore floating production platform. The above-water production equipment is located on decks on top of a long, vertical, closed, floating cylinder or cylinders held in position by a mooring system to anchors. Flowlines from subsea completion wells bring produced fluids to the spar where they are separated and treated.

spcl. special.

spcr. spacer.

SPD and **spd.** spud.

SP-DST. straddle packer, drillstem test.

SPE. Society of Petroleum Engineers (www.spe.org).

spear. A fishing tool that is run into pipe (fish) on the bottom of the well. It grips the inside of the pipe as the fish is being pulled out of the well.

specific gravity. The ratio between the weight of a solid or liquid and the weight of an equal volume of water. Quartz, a common mineral, has a specific gravity of 2.65. (SG, s.g., and sp gr)

speck. speckled.

spec survey. A seismic survey paid for and run by a seismic contractor. Various exploration companies can pay to view the nonexclusive data; cf. group shoot.

spf. shots (perforations) per foot.

sp gr. specific gravity.

spiking. To add condensate to crude oil in the field to lighten the density of the oil and make the oil more valuable.

spill point. The lowest elevation down to which a trap can be filled with gas and oil. If the structure is filled down to the spill point (fill to spill), the addition of more gas or oil will cause the oil to flow out at this point.

spinning chain. A chain used on the floor of a drilling rig to wrap around drillpipe to start screwing together or finish unscrewing the pipe.

spinning wrench. A pneumatic- or hydraulic-operated wrench that is suspended above the drill floor by a cable. It is used to grip and turn the drillpipe when screwing together and unscrewing the pipe.

spiral-grooved drill collar. A drill collar with three spiraling grooves cut into the outer wall. It is used to reduce the drill collar's surface area in contact with the well walls to prevent stuck pipe.

spkt. sprocket.

Spl and **spl.** sample.

SPM or **spm.** strokes per minute (mud pumps).

spontaneous potential. A wireline measurement of the electrical current caused by the contact of mud filtrate in the pores of a reservoir rock with the natural waters in the rock. It is plotted in track 1 and used to identify reservoir rocks. (self potential) (SP)

SPOT. One of two uncrewed, remote-sensing satellites operated by France. They take pictures of the earth in visible light and infrared.

spot. To place.

spot a well. To locate and put a well on a base map.

spot price. The short-term delivery price for a barrel of oil or 1,000 cubic feet of natural gas traded on the spot market. The spot market is a commodities market that is very sensitive to supply and demand. The prices are constantly changing.

spotting fluid. A liquid lubricant such as diesel or mineral oil that is put in a well (spotted) at the stuck point to loosen stuck pipe.

spread. The geometric pattern of geophone groups in relation to the seismic source. It is described by names such as split spread, cross, end-on, and in-line offset.

sps. sparse.

spsly. sparsely.

SPT. shallower pool test.

spt. spot.

spud and **spud in.** Starting to drill a well. It can be done either (1) when any work is done on preparing the site such as digging the cellar or (2) when the rig that is capable of drilling down to contract depth starts drilling. (SPD and spd)

spud date. The day a well is started.

squeeze cementing. To pump cement under pressure down a cased well to force the cement through casing perforations.

SPWLA. Society of Petrophysicists and Well Log Analysts (www.spwla.org).

sq. (1) square and (2) squeezed.

sq pkr. squeeze packer.

sqz. squeeze.

SQZD. squeezed.

SR. short radius.

srt. sorted.

srtg. sorting.

SS. (1) subsea, (2) slow set (cement), (3) small show, (4) stainless steel, (5) string shot, and (6) subsurface.

SS and **ss.** sandstone.

SSG. slight show of gas.

SSO. slight show of oil.

SSO&G. slight show of oil and gas.

Sst. sandstone.

SSU. Saybolt seconds universal.

ST. (1) sidetrack and (2) short trip.

S/T. (1) sample top and (2) suction temperature.

st. (1) state and (2) stand.

stab. To guide the end of a pipe such as casing into a coupling or tool joint to make a connection.

stab. (1) stabilizer and (2) stabilized.

stabilized. steady and unchanging. (stab)

stabilizer. A sub with blades running along the length of it. It is designed to keep the downhole assembly in the center of the well. (stab)

stack. (1) The number of seismic reflections used in stacking to make a common-mid-point stack. It can be expressed either as a number or a percentage with 100% equal to one reflection. (2) To deactivate and store a drilling rig.

stacking. The combining of several different seismic reflections off the same point in the subsurface. It reduces noise and amplifies weak reflections.

stage. (1) A time-rock subdivision of rocks deposited during an age. (2) A portion of the horizontal section (lateral) in a horizontal drainhole that is pressure isolated and hydraulically fractured. Horizontal drainholes are fracked in several stages starting with the toe and ending with the heal.

stage separation. The use of two or more decreasing-pressure separators in line to treat oil and retain more of the lighter fractions in the liquid.

stake a well. To survey the exact location and elevation of a proposed well and make a map (plat) of the site.

stand. (1) Several connected lengths of tubulars such as drillpipe that are raised, stacked, and/or run as a unit. They can be doubles (2), trebles (3), or fourbles (4). (St, STD, or std) (2) To set a tubular such as tubing on end.

standard cubic foot. The English system unit of natural gas volume measurement under standard temperature and pressure (STP) that is defined by law. It is often a surface temperature of 60°F and a surface pressure of 14.65 psia (1 atmosphere). (SCF and scf)

standard tools. A cable tool drilling rig.

standing valve. One of two valves in a downhole pump on the bottom of a tubing string driven by a sucker-rod string. The standing valve does not move up and down; cf. traveling valve.

stat. stationary.

static pressure. Pressure on a fluid that is not moving; cf. pressure. (shut-in pressure)

statics. Corrections applied to seismic data for elevation and the thickness and velocity of the loose sediments near the surface in the low-velocity zone.

STB. stock-tank barrels.

STB/D. stock-tank barrels per day.

STBOIP. stock-tank barrels of oil in place.

STD. (1) salinity-temperature-depth and (2) stand.

std. (1) standard and (2) stand.

stds. stands.

stdy. steady.

steam-assisted gravity drainage. A method used to produce heavy oil. Two parallel horizontal laterals are drilled, one on top of the other.

Steam is pumped into the upper lateral to heat the heavy oil that drains to the lower lateral where it is produced. (SAGD)

steamdrive or **steamflood.** An enhanced oil recovery method used on heavy oil reservoirs. Very hot steam is pumped down injection wells to heat the heavy oil and make it more fluid. The steam condenses into hot water that drives the heated, heavy oil to producing wells.

steam injection. *See* cyclic steam injection.

steel jacket. The legs on an offshore fixed production platform.

steel-jacket production platform. An offshore production platform that is held in place by piles driven into the ocean bottom. They are bolted, welded, or cemented to the legs that are called the steel jacket. The production equipment is located on a deck(s) on the jacket; cf. gravity storage production platform.

steel-tooth tricone bit. A tricone drill bit in which the teeth have been machined out of the steel cones. It is used to drill soft and medium hardness rocks; cf. insert tricone bit. (milled-teeth tricone bit)

steerable downhole assembly. Made up of a bent sub, stabilizers, a downhole turbine motor, and a diamond bit and run on the bottom of a drillstring. It is used in the rotating mode to maintain angle and in the sliding mode to drop or build angle in deviated wells.

step out well. A well drilled to the side of a discovery well to determine the extent of the new field. (appraisal and delineation well)

stepping out. Drilling to the sides of a discovery well to determine the limits of the reservoir.

STH. side tracked hole.

stg. sidetracking.

stging. straightening.

stk. (1) staked , (2) streaks, (3) streaked, and (4) stuck.

stl. steel.

STM. steel tape measurement.

Stn and **stn.** stain.

stochastic. A process or method in which the outcome cannot be predicted with certainty. The variables are random. Statistics and probabilities are used to estimate the outcome; cf. deterministic.

stock tank. A large bolted or welded steel tank that holds oil in the field. It has a thief hatch on the top for sampling and an oil sales outlet near the bottom for transferring the oil. Several stock tanks are connected together to form a tank battery.

stock-tank barrel. One stabilized barrel of oil on the surface after the gas has bubbled out. (STB) *See also* reservoir barrel.

STOIP. stock tank oil in place.

stor. storage.

STP. (1) standard temperature and pressure and (2) shut-in tubing pressure.

stp. stopper.

stpd. stopped.

straddle packer. One of two packers on a drillstem. They are expanded above and below a zone to be tested to isolate that zone. (SP and STRD)

straddle plant. An installation on a gas pipeline that removes condensate from natural gas.

straight hole. See vertical well.

strain. The deformation of an object by stress.

strapping. To measure the height and volume of oil in a specific tank to prepare a tank table.

strat. stratigraphic.

strata. layers of rocks. (Strat and strat)

stratigraphic column. A column showing the vertical succession of rock layers in an area. It is drawn as a cliff with rocks shown as they would weather. Weaker rocks (shales) are indented and stronger rocks (limestones and sandstones) protrude out.

stratigraphic cross section. A cross section made by correlating well logs that have been hung from a common marker bed or horizon in each well; cf. structural cross section.

stratigraphic test well. A well drilled primarily to determine the characteristics of the subsurface rocks. (strat test)

stratigraphic trap. A petroleum trap formed during the deposition of the reservoir rock such as a limestone reef (primary stratigraphic trap) or by erosion of the reservoir rock such as an angular unconformity (secondary stratigraphic trap); cf. structural trap. (Strat Trap)

strat test. *See* stratigraphic test well.

Strat Trap. stratigraphic trap.

STRD. straddle packer.

strd. straddle.

streamer. A long plastic tube containing hydrophones and a cable connecting them. It is towed behind a boat or left on the ocean bottom for seismic exploration at sea.

stress. Force acting on an object, cf. strain.

strg. (1) storage, (2) stringer, and (3) strong.

strike. The horizontal, compass direction of a plane such as a sedimentary rock layer or fault; e.g., North 10° East; cf. dip.

strike-slip fault. A break in the rocks accompanied with horizontal movement of one side with respect to the other. It can be either a right or left lateral strike- slip fault; cf. dip-slip fault.

string. A long length of tubulars such as casing (casing string), tubing (tubing string), or drillpipe (drillstring) made by screwing together joints.

strip. To remove a liquid from a gas.

strip log. *See* sample log.

stripper well. A well that is barely profitable. In the United States a stripper well produces less than 10 bbl of oil or 60 Mcf of gas per day.

Strk and **strk.** (1) streak and (2) streaking.

strt. straight.

structural casing. *See* conductor casing.

structural cross section. A cross section made by correlating well logs that have been hung by modern sea level in each well; cf. stratigraphic cross section.

structural map. A map that uses contours called structural contours to show the elevation of the top of a subsurface rock layer. It is made from well data. (structure-contour map)

structural trap. A petroleum trap formed by the deformation of the reservoir rock such as a fold or fault; cf. stratigraphic trap.

structure-contour map. *See* structural map.

STS. short tubing string.

STTD. sidetracked total depth.

stuck pipe. A drillstring stuck along the sides of a well. It is caused by either differential wall pipe sticking or a keyseat and can be loosened with a jar or treated with a spotting fluid.

stuck point. The depth in a well at the top of a section of stuck pipe. The free point is just above it.

stuffing box. The steel container on the wellhead of a rod pumping unit on an oil well. It contains packing that seals around the polished rod which oscillates up and down through it. (SB)

stwy. stairway.

Su and **su.** sulfur.

sub. A short section of pipe run on the drillstring between or below the drill collars; e.g., stabilizer sub and bumper sub. (sb)

subcrop map. A geologic map of rock layers cropping out under an angular unconformity.

subduction zone. An area described in the seafloor-spreading theory as the place two opposite-moving seafloors collide. It is seen as a deep ocean trench and/or a mountain range. *See also* seafloor spreading.

submarine fan. A large wedge of sediments deposited in deep water at the base of a submarine canyon.

submersible electrical pump. *See* electric submergible pump.

sub pump. *See* electric submergible pump.

subsalt. Sedimentary rock structures located below a layer of salt.

subsea. Measured from the bottom of the ocean. (ss)

subsea completion or **well.** A well with the wellhead equipment such as the production tree or gas lift located on the bottom of the ocean. It can be either wet or dry. It is drilled from a jackup or floater rig and is tied to a production platform, semi, or FPSO vessel by flowline.

subsea site investigation. A survey of the ocean bottom to determine slope, composition, and load-bearing capacity for a drilling rig or platform. (soil investigation)

substructure. The steel framework on a rotary drilling rig used to elevate the drill floor above the ground.

subsurface safety valve. A valve run in a tubing string in a well located in the ocean. The valve closes when pressure drops below a specific level.

subsurface trespass. To illegally drill a well under land without permission from the mineral rights owner.

success rate. The number of wells completed as producers divided by the number of wells drilled. It is expressed as a decimal or percent. (chance of success or risk)

sucker rod. A narrow-diameter solid-steel rod (usually 25 ft long) with threaded ends. A sucker-rod string is run in a well down the tubing to connect a walking beam on a surface, rod pumping unit with a downhole pump on the bottom of the tubing. Sucker rods in 37½-ft lengths are also made with fiberglass that is lighter than steel.

sucker-rod pumping system. *See* rod pumping system.

suct. suction.

suction pit sample. *See* mud-in sample.

sul. sulfur.

SULW and **sul wtr.** sulfur water.

supercompressibility factor. *See* Z factor.

supply company. A company that provides materials such as casing.

support agreement. An agreement between parties to encourage and support drilling a well. Three types are dry hole agreement, bottom-hole agreement, and acreage contribution agreement. (contribution agreement)

sur. survey.

Surf. surface casing.

surf. surface.

surface casing. The largest diameter and shortest casing string in a well. It is used to protect freshwater aquifers and prevent the sides of the well from caving.

surface rights. The legal ownership of the surface of fee land. The surface rights owner can build, ranch, or farm on that land; cf. mineral rights.

surfactant. A detergent-like chemical used in enhanced oil recovery to reduce the surface tension of oil and wash it from the rock surfaces and out of small pores.

surp. (1) suspended and (2) surplus.

suspended well. A well that has been producing but is shut-in. It eventually will have to be put on production again or plugged and abandoned.

SURV and **surv.** survey.

SUS. Saybolt universal seconds.

susp. suspended.

svcu. service unit.

SVG. survey gas.

SW or **S.W.** saltwater.

S_w. water saturation. Gas or oil saturation is equal to 100% minus S_w.

swab. To remove liquids from a well with a swabbing tool. (swb)

swage. A tapered tool that is run on a workstring to reopen collapsed casing in a well. (swg)

swath shooting. A method used to acquire data for 3-D seismic exploration on land. The receiver cables are laid out in parallel lines. The shot points are run perpendicular to the receiver lines.

S wave. shear wave.

swb. (1) swab, (2) swabbing, and (3) swabbed.

swbd. swabbed.

swbg. swabbing.

SWC and **S.W.C.** sidewall core.

SWCM. saltwater-cut mud.

SWD. saltwater disposal.

SWDS. saltwater disposal system.

SWDW. saltwater disposal well.

sweep. The frequency range that is injected into the subsurface by a vibrator truck at a shot point for seismic exploration.

sweep efficiency. The ratio of pore volume contacted by an injected fluid to total pore volume in a reservoir during waterflood or enhanced oil recovery.

sweep length. The time during which a vibrator truck shakes the ground at a shot point for seismic exploration.

sweet. Gas or oil with a low sulfur content. Sweet oil generally has less than 1% sulfur; cf. sour.

sweetening. Removal of acid gases such as hydrogen sulfide and carbon dioxide from natural gas.

sweet spot. An area in a reservoir that has relatively high permeability and produces gas and oil at a high rate. It is often an area where natural fractures are concentrated.

swet. sweetening.

swg. (1) swage and (2) swaged.

SWI. saltwater injection.

swivel. A device on a drilling rig which allows the drillstring to rotate while being suspended from the derrick. It is located at the top of the kelly and hangs from the hook on the traveling block.

SWS. side wall sample.

SWTR. saltwater.

SWTS. saltwater to surface.

SWU. swabbing unit.

sx. sacks.

syn. (1) synthetic and (2) synchronous.

syncline. A large, long fold of sedimentary rocks that are bent downward; cf. anticline.

Syncrude. A company (Syncrude Canada, Ltd.) in Alberta, Canada, that makes synthetic crude oil from the Athabaska tar sands.

synthetic-base and **synthetic-based drilling mud.** Drilling mud made with a synthetic oil. It is commonly used offshore; cf. oil-base and water-base drilling mud.

synthetic crude oil. A combination of naphtha, distillate, and gas oil produced from upgrading bitumen extracted from tar sands. It is generally 32 °API gravity with less than 0.2% sulfur. It is unlike natural crude oil in that synthetic crude oil contains no residuum. About 1.16 barrels of bitumen are processed to make 1 barrel of synthetic crude oil. Syncrude Sweet Blend (SSB) is the name for a light, sweet, synthetic crude oil that has no residual bottoms and is gold-colored. (SCO)

synthetic seismogram. An artificial, computer-generated seismic record made from the acoustical impedance differences of subsurface rock layer contacts.

sys. system.

system. A time-rock division of rocks deposited during a period of geological time.

sz. size

T. (1) temperature, (2) ton, and (3) township.

t. time.

T/. top of.

TA. (1) temporarily abandoned and (2) turn around.

tab. tabular.

tadpole plot. A diagram which shows the dip of subsurface rock layers in a well as determined by a dipmeter.

TAI. Thermal alteration index.

tail gas. The gas, primarily methane, that exits a natural gas processing plant after the natural gas liquids have been separated. (residual gas)

tally. (1) A record of a repetitive event count such as a pipe tally that shows the number of drillpipe joints used in a drillstring. (2) To measure and record the length of tubulars such as casing or pipe. A drilling tally sheet is a record of drillstring components measured in $1/100$ foot.

T&B. top and bottom.

T&BC. top and bottom chokes.

T&G. tongue and groove (joint).

T&R. tubing and rods.

tank battery. Two or more stock tanks connected by a flowline. The tank battery is connected by flowline to a separator. First one stock tank and then another is filled with oil. (TB)

tank table. A table that relates the height of oil in a stock tank to the volume of the oil. (gauge table)

tar. A viscous material composed of very heavy, high-molecular weight hydrocarbons.

target. (1) The potential reservoir rock to which a well is drilled. (2) The proposed bottom-hole location for a deviated well.

tar sands. Very thick, heavy crude oil bitumen mixed with sand and water. *See also* Athabaska tar sands.

tax royalty participation contract. A contract between a foreign government and a multinational company. The multinational company receives an exclusive concession and bears the entire cost and risk of exploration, drilling, and production. The host government is paid bonuses, taxes, and royalties from production; cf. production sharing contract. (concession agreement)

TB. (1) tank battery and (2) thin bedded.

tb. tube.

t.b. thin-bedded.

tbg. tubing.

tbg ch or **tbg chk.** tubing choke.

tbg press. tubing pressure.

TC. (1) temperature controller, (2) tool closed, (3) top choke, and (4) tubing choke.

Tcf. trillion cubic feet.

TCV. temperature control valve.

TD. total depth.

TDed. To drill a well to (1) total depth (TD), (2) contract depth, or (3) the drilling target.

TDI. temperature differential indicator.

TDS. total dissolved solids.

tech. (1) technical and (2) technician.

Temp and **temp.** temperature.

temp. temporary.

temperature log. A production log that records fluid temperatures at various levels in a well. It can detect gas flowing into the well and where cement is setting behind casing. (TL)

temporarily abandoned. A producing well that has been shut in. Eventually the well is either put back on production or plugged and abandoned. (TA)

tendons. Long steel tubes about 2 ft in diameter that connect a tension-leg platform to the anchor weights on the bottom of the ocean.

tension. Forces that pull apart; cf. compression.

tension fault. A fault formed by tension; e.g., normal fault; cf. compressional fault.

tension-leg platform. A floating wellhead and production platform held in place by large weights on the bottom of the ocean; cf. tension-leg well platform. (TLP)

tension-leg well platform. A floating wellhead platform. It is similar to a tension-leg platform except that production is sent by submarine pipeline to a processing platform in shallow water for treating; cf. tension leg platform. (TLWP)

tent. tentative.

ter. terrigenous.

term. terminal.

termin. terminate.

Tertiary. A period of geological time from 65.5 to 2.6 million years ago. It is part of the Cenozoic era. (Ter)

tertiary recovery. A process used after the primary drive mechanism has been depleted and secondary recovery has been completed on an oil reservoir. Either (1) chemicals or steam is injected into a reservoir or (2) the subsurface oil is set afire; cf. secondary recovery. *See also* enhanced oil recovery.

Tex and **tex.** texture.

TG. trip gas.

tgh. tough.

TH. tight hole.

thd. thread.

thermogenic gas. Natural gas formed by subsurface heat on organic matter or by the thermal cracking of oil. It can be either dry or wet gas; cf. biogenic gas.

therst. thermostat.

THF. tubinghead flange.

THFP. top hole flow pressure.

thickening time. The time a cement slurry remains fluid enough to be pumped. (pumpability time)

thief. (1) A brass or glass container that is used to obtain an oil sample from a stock tank. (2) To obtain an oil sample.

thief hatch. The hatch on the roof of a stock tank. It is used to gain access to the tank to measure the height of the oil and obtain a sample.

thief zone. (1) A highly permeable zone in reservoir rock through which waterflood or enhanced oil recovery fluids flow, bypassing oil in other parts of the reservoir. (2) A very permeable rock layer in a well that takes large amounts of drilling mud during drilling. (lost-circulation zone)

thin-section. A paper-thin slice of rock mounted on a glass slide.

thk. (1) thick and (2) thickness.

thn. thin.

thread protector. A plastic or metal cap screwed to the ends of tubulars such as casing and drillpipe to prevent damage to the threads.

3-C seismic. Land seismic acquisition using three geophones at right angles to each other at each receiver station to record both compressional and shear waves.

3-D and **3D seismic method.** The acquisition and computer processing of seismic data to make a three-dimensional image of the subsurface rock layers. The data are often shown on a cube display along with slices of the subsurface. A slice (1) at a specific depth is a time or horizontal slice, (2) along a vertical plane is a vertical slice, (3) along a seismic reflector is a horizon slice, and (4) along a fault is a fault slice. *See also* visualization center.

thribble. *See* treble.

thrling. throttling.

thrm. thermal.

throw. The vertical displacement on a fault.

thru. (1) through and (2) throughout.

thrust fault. A reverse fault with a dip of less than 45° from horizontal. The hanging wall has been thrust over the footwall.

TI. temperature indicator.

ti. tight.

tie back and **tie in.** (1) To connect something such as a subsea well by flowline to a production platform. (2) To run seismic lines together. (3) To run a seismic line through a well.

tight. (1) An emulsion that resists separation; cf. loose. (2) A rock with very low (<0.1 md) permeability. (ti)

tight hole. A well being drilled in which the results are being kept secret. (TH)

tightness. The degree to which an emulsion resists separation.

tight gas sand. A natural gas sandstone reservoir with less than 0.1 md permeability. A tight gas sand well is usually stimulated by hydraulic fracturing.

tight sands. A general term for any reservoir rock with very low permeability.

TIH. trip in hole.

time interval map. A map that uses contours to show the span in time (milliseconds) between two seismic horizons. (isotime and isochron map)

time-lapse seismic. The seismic differences between several 3-D seismic surveys run at different times over the same reservoir during production from that oil field. Changes in seismic responses from the reservoir such as amplitude can show the flow of fluids through the reservoir. (4-D seismic)

time slice. A flat horizontal section made at a specific depth in time from 3-D seismic data. It shows where each seismic reflector intersects the slice. (horizontal slice)

time structure map. A contoured map that shows depth in time (milliseconds) to a seismic horizon.

time-to-depth conversion. A seismic process in which the vertical scale on a seismic record is converted from time in milliseconds to depth in feet or meters.

title opinion. A legal history of mineral rights ownerships on a parcel of land.

TJPF. tubing jets per foot.

tk. tank.

tkg. tankage.

tl. (1) tool and (2) tools.

TLH. top of liner hanger.

TMC. total mud cost.

tn. tan.

TNS. tight no show.

TO. tool open.

T.O. tool open.

TOC. (1) top of cement and (2) total organic carbon.

TOCP. top of cement plug.

TOF. top of fish.

TOH. trip out of hole.

TOL. top of liner.

tol. tolerance.

ton. A weight in the English system equal to 2,240 pounds or 1.016 tonnes in the metric system; cf. tonne. (long ton)

tongs. A wrench-like device that is suspended by a cable above the drill floor on a drilling rig. It is used to grip and hold the drillpipe as it is being screwed together and unscrewed by the spinning wrench.

tonne. A weight in the metric system equal to 1,000 kg or 0.9842 long tons in the English system; cf. ton. (short ton)

TOOH. trip out of hole.

tool face. The direction the drill bit is facing.

tool joint. A short steel cylinder with female-threads. It is used to connect joints of pipe. (collar and coupling)

tool pusher. A drilling company employee at the drillsite who is ultimate in charge of the drilling crews and the drilling rig. (TP)

TOP. testing on pump.

top. The depth to the top of a formation or zone in a well.

top drive. (1) an AC or DC electrical motor that rotates the drillstring from the swivel. It replaces the rotary table and kelly bushing. (power swivel) (2) a drilling rig with a top drive motor.

topo. (1) topographic and (2) topography.

topographic map. A map that uses contours to show the elevation of the surface of the ground.

tops. *See* top.

top-set completion. *See* open-hole completion.

torque. A twisting force that can cause rotation.

TORT. tearing out rotary tools.

tot. total.

total acid number. A measure of the acidity and corrosiveness of a crude oil. It is reported in units of mg KOH/g. Higher numbers are more corrosive.

total depth. The depth of a well to the bottom measured by number of drillpipe joints in the well (driller's or rotary depth) or by a wireline (logged or measured depth); cf. true vertical depth. (TD)

total mud cost. The cumulative cost of drilling fluids on a rig up to the date of the daily drilling report. (TMC)

total organic carbon. The percent by weight of organic carbon in a sample. It is a measure of a source rock's ability to generate and expel hydrocarbons. Organic carbon values below 0.5 to 1.0% are considered too low to generate hydrocarbons.

total well cost. The cumulative drilling expenses until to date of the daily drilling report. (TWC)

to the right. Clockwise; the rotary table on a drilling rig turns to the right.

touch bottom. The contact of a tool or drillstring with the bottom of a well.

tour. A crew shift on a drilling rig. There are usually three 8-hour tours on a land rig and two 12-hour tours on an offshore rig. On a land rig the graveyard or morning tour is from midnight to 8 a.m., the day tour is from 8 a.m. to 4 p.m., and the evening tour is from 4 p.m. to midnight.

tour pusher. An assistant tool pusher. The tour pusher can relieve the tool pusher, often during a night tour.

tour sheet or **report.** A report made by the driller on the drilling activities during that tour. It is used to make the daily drilling report. (driller's report)

township. A surveyed square of land 6 miles on a side. Townships are divided into 36 sections. (T and twp)

TP. (1) tubing pressure and (2) tool pusher.

Tp. top.

T/pay. top of pay.

TPC. tubing pressure closed.

TPF. tubing pressure flowing.

TPSI. tubing pressure shut in.

TR and **tr.** trace.

TR. temperature recorder.

tr. tract.

trace. The response of a single seismic detector to a single seismic shot. It is recorded as a vertical line with peaks and troughs to the right and left sides that represent recorded seismic energy. (wiggle trace)

trace fossil. Indirect remains of a plant or animal in a sedimentary rock. Tracks, burrows, root casts, and trails are examples.

tracer log. A log that uses a radioactive tracer and detectors to measure fluid flow characteristics in a well.

tractor. A device used to pull tools down highly deviated and horizontal wells. It is run down the well on an electric line. When electronically activated, hydraulic-driven wells are deployed from the tractor body. It is controlled electronically.

trans. (1) transformer and (2) transmission.

transgression. The advance of seas onto the land; cf. regression.

transp. transparent.

transportable gas. Natural gas that has had minimal field processing so that it can be transported to a final processing plant. *See also* pipeline-quality gas.

trap. A high area on the subsurface reservoir rock such as a dome or reef where oil and/or gas can accumulate. It is overlain by a caprock that is a seal. *See also* structural trap and stratigraphic trap.

traveling block. A steel frame with steel wheels on a horizontal shaft. It is suspended in the derrick or mast of a drilling or workover rig by the hoisting line. A hook is attached to the bottom of the block. It travels up and down in the derrick or mast as equipment and tools are raised and lowered in the well.

traveling valve. One of two valves in a downhole pump on the bottom of a tubing string driven by a sucker-rod string. The traveling valve moves up and down with the sucker-rod string; cf. standing valve.

treater. A vessel used to separate an emulsion. A heater treater uses heat, and an electrostatic treater uses high-voltage electric grids. Chemicals, called emulsion breakers, can also be used.

treble. Three tubular joints; cf. single, double, and fourble. (thribble)

tree. *See* production tree.

trend. The area along which a petroleum play occurs. (fairway)

Triassic. A period of geological time from 251 to 201.6 million years ago. It is part of the Mesozoic era. (Tri)

tribble. *See* treble.

tricone bit. A common type of roller cone drill bit with three rotating cones on the bottom. Two types are milled-teeth and insert tricone bits. The bit is designed to chip or crush the rocks on the bottom of the well to produce well cuttings; cf. diamond bit.

trip. A cycle of pulling (tripping out) and running (tripping in) a drillstring in a well. A round trip involves tripping in, touching bottom with the drillstring, and tripping out. A short trip does not touch bottom. A wiper trip uses the drill bit on the drillstring to ream out the wellbore.

trip. (1) tripped and (2) tripping.

triplex pump. A mud pump with three single-acting pistons in cylinders. The mud is pumped only on the forward stroke of the pistons; cf. duplex pump.

tripping in. Running the drillstring into the hole; cf. tripping out.

tripping out. Pulling the drillstring out of the hole; cf. tripping in.

trip tank. A small tank, usually on the floor of a drilling rig, that holds drilling mud that is added to a well during tripping out.

trmt. treatment.

trnsl. translucent.

trnsp. transparent.

TRRC. Texas Railroad Commission.

trt. treat.

trtd. treated.

trtg. treating.

trtr. treater.

true vertical depth. The depth of a well measured straight down; cf. total depth. (TVD)

truncated. The lateral termination of rocks, usually either by erosion or faulting.

TS. tensile strength.

T/S. top of salt.

TSD. temporarily shut down.

T/sd. top of sand.

TSI. temporarily shut in.

TSITC. temperature survey indicated top of cement.

TST. true stratigraphic thickness.

tst. test.

tstd. tested.

tste. taste.

tstg. testing.

TSTM. too small to measure.

tstr. tester.

TT. through tubing.

TTF. test to follow.

TTL. total time lost.

tubing. A small-diameter (¾ to 4½ in.) steel tubular string that runs down the center of a well to conduct the produced fluids up the well. (tbg)

tubing anchor. A device that grips the casing to secure the bottom of the tubing string.

tubinghead. The forged- or cast-steel fitting on the top part of the wellhead. It contains the tubing hangers that suspend the tubing string in the well; cf. casinghead.

tubingless completion. An oil well completion in which the production is brought up the casing without using a tubing string. It is used in a very high-capacity oil well.

tubing packer. A packer run on the bottom of the tubing string to seal the space (annulus) between the tubing and casing. (completion packer)

tubing pressure. Pressure on the fluid in the tubing string. It can be either flowing or static; cf. casing pressure.

tubing pump. A sucker-rod pump that is run on the tubing string; cf. rod pump.

tubing swage. A tool with a cylindrical body that tapers toward the bottom. It is run on a wireline to open collapsed tubing.

tubular. A general term for any long steel cylinder such as a joint of drillpipe, casing, tubing, or drill collar.

tungsten carbide. An extremely hard alloy (W_2C) used in granular form to hard-face drilling tools.

turbidite. A layer of sedimentary rocks deposited by a turbidity current coming to rest on the ocean bottom. It can be graded with the coarsest grains such as sand on the bottom and the finest on the top or it can be relatively uniform sand deposited by several turbidity currents.

turbidity current. A dense mixture of water and sediment flowing down a submarine slope. *See also* turbidite.

turbine. A motor driven by fluid flowing through revolving vanes on a shaft.

turbine meter. A gas or liquid meter that measures the volume by the turns per unit time on the turbine shaft.

turbine motor. *See* downhole turbine motor.

turnkey drilling contract. A drilling contract based on a fixed fee to drill to contract depth. It can also have the obligation to complete and equip the well based on well tests; cf. daywork and footage drilling contract.

turning to the right. To drill a well.

Turonian. An age of geological time from 93.5 to 89 million years ago. It is part of the Cretaceous period.

TVD. true vertical depth.

TVT. true vertical thickness.

TWC. total well cost.

twin. To drill a well adjacent to an existing well.

two-way travel time. The recorded time in milliseconds from the seismic source to the reflector and back to the surface detector.

twp. township.

twst off. twist off.

TWTM. too weak to measure.

typ. typical.

u. (1) upper and (2) unit.

U/. upper.

UC. unconformity.

U/C. under construction.

UG. (1) under gauge and (2) underground.

U/L. upper and lower.

ult. ultimate.

ultimate oil production and **recovery.** The oil in a reservoir can be commercially recovered by primary production, waterflood, and enhanced oil recovery.

ultrasonic testing. The use of very high frequency (0.1 to 25 MHz) sonic waves to test tubulars such as joints of tubing and drillpipe for wear and defects.

un. unit.

Unconf and **unconf.** unconformity.

unconformity. An ancient erosional surface. (UC, Unconf, and unconf) *See also* disconformity and angular unconformity.

uncons. unconsolidated.

unconsolidated sediments. Loose sediments; cf. consolidated sediments. (uncons)

unconventional resources. Crude oil and natural gas that occur in a reservoir that is not a conventional porous and permeable sandstone or carbonate; e.g., gas shale, tar sand, tight gas sand, and oil shale.

underbalance. The condition in a well in which the pressure on the drilling mud is less than the pressure on fluids in the surrounding rocks; cf. overbalance.

underbalanced drilling. Drilling a well while circulating relatively lightweight drilling fluid. This prevents formation damage and lost circulation and increases the rate of drilling. It does not have pressure control and fluids will flow out of the rocks and up the well. A rotating head on top of the well acts as a seal and diverts the produced fluid to a separator.

undercompacted. Sediments that have not compressed as much as would be expected at that depth.

underream. To enlarge the diameter of a wellbore to a diameter larger than that of the last string of casing in the well. An underreamer is used.

underreamer. A tool run on a drillstring and rotated to enlarge the bottom of a well using cones on arms that expand.

undersaturated. A liquid that can dissolve more of a gas or a salt; cf. saturated.

undersaturated pool. An oil reservoir without a free gas cap. There is some natural gas dissolved in the oil but it can hold more and is unsaturated; cf. saturated pool.

undershoot. To acquire a 3-D image of the subsurface of an area without the seismic equipment ever being on that land. The geophones are positioned on one side and the source(s) on the other side of the land.

undiff. undifferentiated.

undly. underlying.

unf. unfinished.

uni. uniform.

unitize. The coordination of all operators in a unit or field to increase ultimate oil production with a common pressure maintenance, waterflood, or enhanced oil recovery project. A unit operator is declared to manage the project. The costs and additional production are shared proportionally to each participant's acreage or reserves in the unit. Unitization can be either voluntary or compulsory (mandated by a government agency).

unit operator. The company in charge of drilling and production on a unitized field.

unload. To remove liquid from a well. Gas wells are unloaded when water on the bottom of the well prevents gas from flowing into the well (load up).

updip. A direction or location up the slope or angle (dip) of a plane such as the top of a rock layer or fault; cf. downdip.

uplift. An area of the earth that has been forced upward.

upset. A thicker wall section on a tubular such as drillpipe. It is used to strengthen the pipe where it has been threaded on each end.

upstream. Petroleum exploration, drilling, and production. Downstream involves transportation, refining, chemicals, and marketing.

upthrown. The side of a dip-slip fault that moved up. cf. downthrown

UR. (1) underreaming and (2) ultimate recovery.

USDOE. United States Department of Energy (www.doe.gov).

USGS. United States Geological Survey (www.usgs.gov).

UT. upper tubing.

U/W. used with.

V. (1) volt, (2) volume, and (3) mud viscosity in API seconds.

v. velocity, volt, very.

VA. volt-ampere

vac. (1) vacuum and (2) vacant

valve. A gate used to open, close, and regulate flow. (vlv)

vap. vapor.

var. (1) variable or (2) various.

vari. variegated.

variable area wiggle trace. A method of displaying seismic data using vertical lines with wiggles to the left and right. Wiggles to the right are reflections and are often colored black; cf. variable density display.

variable-bore rams. Two large blocks of steel with inserts cut into them. They are designed to fit around a range of different-sized drillpipes and drill collars to close a well. Variable bore rams are used in a blowout-preventer stack.

variable density display. A method of showing seismic data using shade of gray to show reflections. Darker colors are stronger reflections; cf. variable area wiggle trace.

variegated. A pattern of irregular spots and patches in a sedimentary rock. (vari and vgt)

v.c. very common.

VD. variable density.

V-door. An inverted, V-shaped opening in the front of a mast or derrick. It allows the joints of drillpipe and casing to be run up the ramp onto the drill floor.

vel. velocity.

vent. ventilator.

vert. vertical.

vertical seismic profiling. A method used to measure seismic velocities of rock layers in a well. The seismic source is located on the surface next to the well. A borehole geophone is raised in the well to measure seismic velocities at various depths. It is similar to a check shot, but the geophone stations are closer together. (VSP)

vertical well. A well drilled almost straight down with a maximum tolerange in degrees per 100 ft for any well section and within a cone of specific degrees as defined in the drilling contract. (straight hole) cf. crooked hole and deviated well.

v-f-gr. very fine grained.

vgt. variegated.

VHF. very high frequency.

v-HOCM. very heavily oil-cut mud.

VI. viscosity index.

vibration dampener. A sub used in a downhole assembly to reduce vibrations. (shock sub)

vibrator. A truck with hydraulic motors on the bed of the truck and a steel plate (base plate or pad) below the motors. It is used as a seismic source to shake the ground by vibroseis.

vibroseis. A seismic method in which the source energy is put into the ground with a vibrator truck or trucks that shake the ground.

virgin pressure. The original reservoir pressure before any production. (initial or original pressure)

vis. (1) viscosity and (2) visible.

viscometer. A laboratory instrument used to measure the viscosity and gel strength of drilling mud.

viscosity. The resistance of a fluid or slurry such as drilling mud to flow. The units of dynamic viscosity are centipoises (cp). Kinematic viscosity is viscosity divided by fluid density and is measured in centistokes (cs). (μ, V, and vis)

vis room. *See* visualization center.

visualization center or **station.** A room used to project and display 3-D seismic data on the walls and the floor. It is also called a visionarium or decisionarium. (vis room)

vitrinite reflectance. A method that uses a microscope to measure the amount of light reflected off a type of plant organic matter (vitrinite) in shale to determine if the shale has generated crude oil or natural gas.

V/L. vapor to liquid ratio.

VLAC. very light amber cut.

vlv. valve.

v.n. very noticeable.

vol. volume.

volat. volatile.

volatile oil. Crude oil with relatively more intermediate-sized molecules and less longer-sized molecules than black oil; cf. black oil.

Volc and **volc.** (1) volcanic and (2) volcanic rock.

volcanic rock. *See* lava.

vol %. volume percent.

volumetric drive. A gas field reservoir drive in which the expanding gas produces the energy to force the gas through the reservoir rock. (gas-expansion drive)

VP. vapor pressure.

VPS and **V.P.S.** very poor sample.

v.r. very rare.

vrtl. vertical.

vs. versus.

V/S. velocity survey.

VSGCM. very slightly gas-cut mud.

v-sli. very slightly.

VSP. (1) vertical seismic profile or profiling and (2) very slightly porous.

VSSG. very slight show of gas.

VSSO. very slight show of oil.

vug. A roughly spherical, relatively small solution pore in limestone.

vug. (1) vuggy and (2) vugular.

vuggy. A limestone or dolomite containing vug pores. (vug)

W. (1) watt, (2) water, (3) weight, and (4) mud weight in ppg.

w. watt.

w/. with.

WAB. weak air blow.

wackestone. A type of limestone with significant amounts of large, sand-sized particles supported in fine-grained material. (Wkst)

WAG. water-alternating-gas.

wait-and-weigh method. A method used to control a well with a kick. After the blowout preventers have been thrown, kill mud is prepared, and the kick-diluted mud is replaced by kill mud during one circulation; cf. driller's method.

waiting on cement. The time that operations on a well are shut down as the cement sets behind a casing string. It can be hours to several days. (WOC)

walking beam. The steel beam that pivots up and down on the Samson post of an oil well beam-pumping unit.

walk to the right. The tendency of a bit to whirl clockwise and the well to be drilled down in a clockwise, corkscrew pattern because the bit is being turned clockwise (to the right). Anti-whirl bits are designed to counter this effect.

wall scraper and **scratcher.** A ring with protruding, metal wires that is attached to the outside of a casing string. It is rotated or reciprocated to scratch the mud cake off the well walls.

wash job. Acidizing a well to remedy skin damage on the wellbore.

washout. (1) Excessive erosion and enlargement of the wellbore by drilling mud. (2) Damage on a drillstring caused by fluids flowing through the walls of a tubular.

washover pipe and **washpipe.** A fishing tool that consists of a section of casing. It is run onto the fish. Drilling mud is pumped out the bottom of the washover pipe to clear debris from around the fish. Another tool is then used to retrieve the fish. (WP)

wash tank. A settling tank that uses gravity to separate a loose emulsion. (gun barrel separator)

water-alternating-gas. An enhanced oil recovery method in which slugs of inert gas and water are alternately injected into a depleted oil reservoir. The water helps prevent fingering of the gas and early breakthrough. (WAG)

water back. To dilute water-based drilling mud with water to make it less viscous and dense; cf. mud up.

water-base or **water-based drilling mud.** Drilling mud made with freshwater. It is commonly used on land; cf. oil-base and synthetic-base drilling mud. (WBM)

water cushion. *See* cushion. (WC)

water cut. The percentage of water that an oil well produces.

water-cut. Adjective for diluted with water.

water-cut meter. An instrument used to measure the water content of fluids in a well.

water drive. A reservoir drive in which the expansion of water beneath or beside an oil or gas reservoir forces the oil or gas through the rocks. It has a high oil recovery efficiency but only a moderate gas recovery efficiency.

water encroachment. Water flowing into the oil-producing part of a reservoir.

waterflood. A method used to produce more oil from a depleted reservoir. Water is pumped down injection wells into the reservoir in order to force oil through the reservoir to producing wells. Waterfloods are described by their pattern of water injection and producing wells such as 5-spot and line drive. (WF)

waterfrac. A frac job using water with very little additives as the frac fluid. A slickwater frac is a type of waterfrac.

water hauler. A service company that uses a tank truck to pick up and take oilfield brine to a disposal well.

water-in-oil emulsion. Droplets of water suspended in oil. It is the most common emulsion produced from an oil well; cf. oil-in-water emulsion.

water table. The subsurface level below which the pores in the soil or rock are saturated (filled) with water.

water washing. A process in which water flowing by crude oil removes the lighter fractions by solution.

water wet. A reservoir rock in which oil or gas occupies the center of the pores, and water coats the rock surfaces; cf. oil wet.

WAX. wax plant.

waxy crude. Crude oil with a high wax (paraffin) content. It has a high pour point. A waxy crude is liquid under high temperature in the subsurface reservoir and solid under surface temperature.

WB. (1) water base and (2) water blanket.

WBIH. went back in hole.

WBM. water-based drilling mud.

WC. (1) water cushion (drillstem test), (2) wildcat, and (3) water cut.

WCM. water-cut mud.

WCO. water-cut oil.

WCTS. water cushion to surface.

WD. water depth.

WDW. water disposal well.

weathering. The physical and chemical breakdown of rock. Weathering produces sediments.

weathering layer and **zone.** *See* low-velocity zone. (WZ)

wedge out. To have a rock progressively narrow to zero thickness in a horizontal direction. (pinch out)

wedge-out. The line or edge of a rock that pinches out.

weed and seed person. *See* palynologist.

weighting agent and **material.** A drilling mud additive such as barite, siderite, or hematite used to increase the density of the mud.

well. A relatively narrow-diameter hole drilled through subsurface rocks to either produce fluids or inject fluids. Type of well drilled in the petroleum industry are (1) wildcat, (2) appraisal, (3) production, (4) injection, (5) disposal and (6) observation. The well can be either straight or deviated.

wellbore. The hole made by a drilling rig.

well completion report. A form required by a government agency after a well has been completed. It lists well, formation, and test information and is certified as accurate.

well control. The methods used to prevent a blowout on a well.

well cuttings. Rock flakes made by the drill bit. (cuttings)

wellhead. The forged or cast steel fitting on the top of a well. It consists of one or more casingheads located on the bottom and a tubinghead on the top. It is bolted or welded to the top of the surface casing. (WH)

wellhead equipment. Equipment attached to the top of the tubing and casing strings in a well. It includes the casingheads, tubinghead, Christmas tree, stuffing box, and pressure gauges.

well intervention. A general term for work on a producing well. It includes repairing, replacing, and installing equipment and well stimulation.

well log. A continuous record of rock properties measured in a well. Some types are sample, mud, and wireline logs.

well log library. A place where copies of well logs from a region are on file.

well pad. A drilling pad than now is the location for a wellhead on a producing well. It is often surrounded by a fence. There can be four to eight horizontal drainhole wellheads on the same pad; cf. drilling pad.

well production profile. A plot of flow rate per day versus time for a well.

well service unit. *See* well-servicing unit.

well servicing. Maintenance on a well to maintain or improve production rates. It includes swabbing, removal of scale or paraffin, and running or retrieving tools and equipment such as sucker rods, tubing, and downhole pump.

well-servicing unit. Truck-mounted equipment used to service a producing well. It has either a single- or double-pole telescoping mast, one or two drums, and is operated by a two- to four-person crew. (well service unit)

well shooting. To explode nitroglycerin in a torpedo at reservoir depth in a well to fracture the reservoir and stimulate production. (explosive fracturing)

wellsite geologist or **well sitter.** A geologist at the drillsite who is responsible for sampling and testing.

well sorted. A rock or sediments composed of clastic grains that are all about the same size. Clean sands are an example; cf. poorly sorted.

well spotting. To locate wells on a base map.

well stimulation. Any engineering method used to increase the permeability of a reservoir around the wellbore to increase production. It includes acidizing and hydraulic fracturing.

West Texas Intermediate. The benchmark crude oil for the United States. It has 38 to 40 °API gravity and 0.3% S. (WTI)

wet. (1) A reservoir that produces only water. (2) A well that encountered no commercial hydrocarbons.

wet combustion. A fireflood in which hot steam and water are alternately injected into the reservoir. The steam generated from the water helps drive the oil to producing wells. (combination of forward combustion and waterflooding)

wet gas. A hydrocarbon that occurs as a gas under both initial reservoir conditions and during production as the pressure decreases in the reservoir. A liquid condensate separates from the gas after production under surface conditions but not in the reservoir; cf. dry gas.

wetting fluid. The fluid such as oil or water that coats the rock surfaces of pores in a reservoir. A water-wet oil reservoir has water located on the outside of the pores and oil in the center.

WF. (1) waterflood and (2) wide flange.

WFD. wildcat field discovery.

wgt. weight.

WH. (1) wash and (2) wellhead.

wh. white.

whip. whipstock.

whipstock. A long metal wedge used in a well to bend the drillstring and kick off a deviated well. (whip and WS)

white gas. *See* condensate.

white mica. *See* muscovite.

white oil. *See* condensate.

WHOF. wellhead open flow.

WHP. wellhead pressure.

Wh Sd. white sand.

WI and **W.I.** working interest.

WI. water injection.

wiggle trace. *See* trace.

WIH. (1) went in hole and (2) water in hole.

wildcat well. *See* exploratory well.

wild well. A well that is blowing out of control.

WIN. water injection. (well)

wind gas. nitrogen.

window. A hole cut in casing to kick off a deviated well.

wing. The fittings and tubular on the side of a production tree used to direct the produced fluids to a flowline. It can be either a single- or double-wing tree.

wiper plug. A cylinder of aluminum and rubber used during a cement job. It is pumped down the casing to remove the cement slurry from the casing and to separate fluids in the casing. A top and a bottom plug are used.

wiper trip. The tripping in and out of a well with a drillstring to use the drill bit to ream the wellbore.

wireline. (1) Wire rope used to run tools in a well. It is made of strands of wire wrapped around a fiber or steel core. Each strand is made of steel wires wrapped in a helical pattern. The wireline used in logging has insulated, electrical wires in the core and is called electric wireline. (2) Slick line made of a single wire used to run tools in a well. (WL)

wireline well log. A record of rock properties and their fluids that are measured by an instrument (logging tool or sonde) raised up the well on a wireline. The diameter of the wellbore can also be measured. The properties are recorded as curves on a long strip of paper called a well log. Recent wireline well logs are also recorded digitally and transmitted by radio telemetry; e.g., electrical, gamma ray, and neutron porosity logs.

wireline spear. A fishing tool that uses barbs to engage and remove a wireline fish from a well.

wire rope. A general term for cable made with braided, steel wires forming several twisted strands wound around a steel core. Wireline is a type of wire rope.

wk. weak.

wko. workover.

wkor. workover rig.

Wkst. wackestone.

WL. (1) wireline and (2) water loss.

Wl and **wl.** well.

W/L. water load.

WLC. wireline coring.

wld. (1) welded and (2) welding.

WLT. wireline test.

WLTD. wireline total depth.

WNSO. water not shut off.

WO. (1) waiting on, (2) workover, (3) wash over, and (4) workover or work over.

w/o. without.

WOA. waiting on acid.

WOB and **W.O.B.** weight on bit.

WOB. waiting on battery.

WOC. waiting on cement.

WOCR. waiting on completion rig.

WOCT. waiting on completion tools.

WODP. without drillpipe.

WOG. water, oil, or gas.

WOO. waiting on orders.

WOP. (1) waiting on pipe, (2) waiting on pipeline, (3) waiting on permit, and (4) waiting on pump.

WOPE. waiting on production equipment.

WOPL. waiting on pipeline.

WOPT. waiting on potential test.

WOPU. waiting on pumping unit.

WOR. (1) waiting on rig, (2) waiting on rotary, and (3) water/oil ratio.

working interest. An ownership in a well that bears 100% of the cost of production. Working interest owners receive their share of the production revenue after the royalty owners have taken their share and after expenses have been deducted; cf. royalty interest. (billing interest) (WI and W.I.)

working pressure. The maximum pressure that equipment is designed to operate under. (WP)

work over. A verb for to have a service company do work (a workover) such as pull rods or sand cleanout on a producing well. A production rig, either a workover rig or a smaller service or pulling unit, is used. (wko and WO)

workover. A noun or adjective for the methods used to maintain, restore, or improve production from a well. (wko and WO)

workover rig. A portable rig with a mast and hoisting system used in the workover of a well. A workover rig can drill and circulate. It is a type of production rig; cf. service unit. (wkor)

workstring. A tubing string used to convey a treatment or run tools in a well during a workover. It can be either coiled or jointed (straight) tubing.

worm. An inexperienced worker.

WORT. waiting on rotary tools.

WOS. washover string.

WOSP. waiting on state potential.

WOST. waiting on standard tools.

WOT. (1) waiting on test and (2) weighting on tools.

WOT&C. waiting on tank and connections.

WOW. waiting on weather.

WP. (1) working pressure and (2) wash pipe.

WS. (1) whipstock and (2) water supply.

WSD. whipstock depth.

wshd. washed.

wshg. washing.

WSO. water shut off.

WSONG. water shut off no good.

WSO OK. water shut off OK.

WSRT. well sorted.

W/SSO. water with slight show of oil.

W/sulf O. water with sulfur odor.

WSW. water supply well.

WT. wall thickness (pipe).

wt. weight.

wtg. waiting.

wthd. weathered.

wthr. weather.

WTI. West Texas Intermediate.

wt%. weight percent.

wtr. water.

WTS. water to surface.

W/2. west half.

WW. (1) wash water and (2) water well.

wxy. waxy.

WZ. weathered zone.

XBD, **X-bd**, **X-bdd**, and **x-bdd.** crossbedded.

X-hvy. extra heavy.

Xln and **xln.** crystalline.

X-R. x-ray.

X-ray diffraction. A method used to identify minerals such as clay minerals in rocks. A sample of the rock is bombarded with x-rays to measure the crystal structure of each mineral grain. (XRD)

XRD. x-ray diffraction.

X-stg. extra strong.

XTAL and **xtal.** crystal.

Xtree. Christmas tree.

yel and **yell.** yellow.

yield. (1) To permanently deform. (2) The volume of cement slurry with a specific density formed by the mixture of one sack of dry cement with water and additives. It is expressed as cubic feet per sack (cu ft/sk). (3) The minimum yield strength of steel used to make tubulars. (4) The oil to gas ratio expressed in bbl/Mcf. It is the inverse of the gas-to-oil ratio.

yield point. The minimum stress that causes permanent deformation. (YP and yp)

yield stress. The stress (force) that must be applied to something to make if flow.

YP or **yp.** yield point.

yr. year.

YTD. year to date.

Z. (1) elevation, (2) Z factor, and (3) zone.

Z/A. The ratio of atomic number to atomic weight.

zeolite. A group of naturally occurring aluminum and silica minerals that can also be manufactured. These minerals are used for water softening and treating crude oil and natural gas. (zeol)

Z factor. A number, usually between 1.2 and 0.7, that compensates for natural gas not being an ideal gas under high pressure and temperature in the reservoir. It is used in gas reserve equations and can be found in tables of gas composition, temperature and pressure. (gas deviation factor, gas compressibility factor, and supercompressibility factor) (Z)

zip collar. A drill collar with a recess in the box end for the attachment of elevators.

ZN, **Zn**, and **zn.** zone.

zonal isolation. The use of packers to isolate a producing zone in a well. (zone isolation)

zone. (1) a rock layer identified by a specific lithology, age, or porosity. (Z, ZN, Zn, and zn) (2) a rock layer identified by a characteristic microfossil species or species such as the Siphonia davisii zone. (biostratigraphic zone) (3) *See* pay zone.

zone isolation. *See* zonal isolation.

Figure References

All line drawings in this book are original. Many have been modified, especially simplified, from the sources listed below:

Arabian American Oil Company Staff, 1959, Ghawar Oil Field, Saudi Arabia. *AAPG Bulletin*, vol. 43, 434–454.

Baker, R., 1979, *A Primer of Oil-Well Drilling*. Petroleum Extension Service, University of Texas at Austin, 94 p.

Baker, R., 1983, *The Production Story*. Petroleum Extension Service, University of Texas at Austin, 81 p.

Barlow, J. A. Jr., and J. D. Haun, 1970, Salt Creek Field, Wyoming, in M. T. Halbouty, ed., *Geology of Giant Petroleum Field* (AAPG Memoir vol. 14), 147–157. Tulsa: American Association of Petroleum Geologists.

Baugh, J. E., 1951, Leduc D-3 Zone Pool. *World Oil*, vol. 132, 210–214.

Beebe, B. W., 1961, Drilling the Exploratory Well, in G. B. Moody, ed., *Petroleum Exploration Handbook*, 17-1–17-36. New York: McGraw-Hill.

Berg, R. R., 1968, Point-Bar Origin of Fall River Sandstone Reservoirs, Northeastern Wyoming. *AAPG Bulletin*, vol. 52, 2116–2122.

Beydown, Z. R., 1991, Arabian Plate Hydrocarbons Geology and Potential—A Plate Tectonics Approach. *AAPG Studies in Geology*, vol. 33, 77.

Bishop, W. F., 1968, Petrology of the Upper Smackover Limestone in northern Haynesville Field, Clairborne Parish, Louisiana. *AAPG Bulletin*, vol. 52, 92–128.

Bruce, L. G., and G. W. Schmidt, 1994, Hydrocarbon Fingerprinting for Application in Forensic Geology; Review With Case Histories. *AAPG Bulletin*, vol. 78, 1692–1710.

Burke, D. C. B., 1972, Longshore Drift, Submarine Canyons and Submarine Fans. *AAPG Bulletin*, vol. 56, 1975–1983.

Busch, D. A., 1974, Stratigraphic Traps in Sandstones—Exploration Techniques (AAPG Memoir vol. 21), 174. Tulsa: American Association of Petroleum Geologists.

Charles, H. H., 1941, Bush City oil field, Anderson County, Kansas, in A. I. Levorsen, ed., *Stratigraphic Type Oil Fields*, 43–56. Symposium, AAPG, Tulsa.

Collingwood, D. M., and R. E. Retter, 1926, Lytton Springs Oil Field, Texas. *AAPG Bulletin*, vol. 10, 953–975.

Colter, V. S., and D. J. Harvard, 1981, *Petroleum Geology of the Continental Shelf of North-West Europe*. London: Institute of Petroleum, 521 p.

Craig, D. H.,1988, Caves and Other Features of Permian Karst in San Andres Dolomite, Yates Field Reservoir, West Texas, in N. P. James and P. W. Choquette, eds., *Paleokarst*, 342–365. New York: Springer Verlag.

Dyni, J. R., 2006, Geology and Resources of Some World Oil-Shale Deposits, *USGS Scientific Investigations Report* 2005-5293, 47 p.

Exploration Logging Inc., 1979, *Field Geologist's Training Guide: An Introduction to Oil Field Geology, Mud Logging and Formation Evaluation*. Sacramento: Baker International.

Frey, M. G., and W. H. Grimes, 1970, Bay Marchand-Timbalier Bay-Caillou Island Salt Complex, Louisiana, in M. T. Halbouty, ed., *Geology of Giant Petroleum Fields* (AAPG Memoir vol. 14), 277–291. Tulsa: American Association of Petroleum Geologists.

Galloway, W. E., T. E. Ewing, C. M. Garrett, N. Tyler, and D. G. Bebout, 1983, Austin/Buda Fractured Chalk, in *Atlas of Major Texas Oil Reserves*, 41–42. Austin: Bureau of Economic Geology.

Gallup, W. B., 1982, A Brief History of the Turner Valley Oil and Gas Field. *Canadian Society of Petroleum Geology Guidebook*, n. 8, 81–86.

Gatewood, L. E., 1970, Oklahoma City Field—Anatomy of a Giant, in M. T. Halbouty, ed., *Geology of Giant Petroleum Fields* (AAPG Memoir vol. 14), 223–254. Tulsa: American Association of Petroleum Geologists.

Gearhart-Owen Industries, *Chart Book*, Fort Worth, 38 p.

Gerding, M., 1986, *Fundamentals of Petroleum*. Petroleum Extension Service, University of Texas at Austin, 452 p.

Gill, D., 1985, Depositional Facies of Middle Silurian (Niagaran) Pinnacle Reefs, Belle River Mills Gas Field, Michigan Basin, Southeastern Michigan, in P. D. Roehl and P. W. Choquette, eds., *Carbonate Petroleum Reservoirs*, 121–140. New York: Springer Verlag.

Graversen, O., 2005, *The Jurassic-Cretaceous North Sea Rift Dome and Associated Basin Evolution*. AAPG Annual Convention, Calgary, Alberta, abstract.

Halbouty, M. T., 1979, *Salt Domes, Gulf Region, United States and Mexico*. Houston: Gulf Publishing, 61 p.

Halbouty, M. T., 1991, East Texas Field-USA., in E. A. Beaumont and N. H. Foster, eds., *Stratigraphic Traps II, Treatise of Petroleum Geology Atlas*, 189–206. Tulsa: American Association of Petroleum Geologists.

Handford, C. R., 1981, Sedimentology and Genetic Stratigraphy of Dean and Spraberry Formations (Permian), Midland Basin, Texas. *AAPG Bulletin*, vol. 65, n. 9, 1602–1616.

Harris, P. M., and S. D. Walker, 1990, McElroy Field—USA, in E. A. Beaumont and N. H. Foster, eds., *Stratigraphic Traps I, Treatise of Petroleum Geology Atlas of Oil and Gas Fields*, 195–227. Tulsa: American Association of Petroleum Geologists.

Heritier, F. E., P. Lossel, and E. Wathne, 1980, Frigg Field: Large Submarine-Fan Trap in Lower Eocene Rocks of the Viking Graben, North Sea, in M. T. Halbouty, ed., *Giant Oil and Gas Fields of the Decade 1968–1978* (AAPG Memoir vol. 30), 59–80. Tulsa: American Association of Petroleum Geologists.

Hilyard, J., ed., 2008, *International Petroleum Encyclopedia*. Tulsa: PennWell, 46–55.

Hull, C. E., and H. R. Warman, 1970, Asmari Oil Field of Iran, in M. T. Halbouty, ed., *Geology of Giant Petroleum Fields* (AAPG Memoir vol. 14), 428–437. Tulsa: American Association of Petroleum Geologists.

Hurley, N. F., and R. Budros, 1990, Albion-Scipio and Stoney Point Fields—USA, in E. A. Beaumont and N. H. Foster, eds., *Stratigraphic Traps I, Treatise of Petroleum Geology Atlas of Oil and Gas Fields*, 1–38. Tulsa: American Association of Petroleum Geologists.

Hyne, N. J., 1991, *Dictionary of Petroleum Exploration, Drilling and Production*, Tulsa: PennWell Books, 624 p.

Hyne, N. J., 1995, Sequence Stratigraphy: A New Look at Old Rocks, in N.J. Hyne, ed., *Sequence Stratigraphy of the Mid-Continent*, n. 4, 5–17. Tulsa Geological Society Special Publication.

Jamison, H. C., L. D. Brockett, and R. A. McIntosh, 1980, Prudhoe Bay: A 10-Year Perspective, in M. T. Halbouty, ed., *Giant Oil and Gas Fields of the Decade 1968–1978* (AAPG Memoir vol. 30), 289–314. Tulsa: American Association of Petroleum Geologists.

Jardine, D., 1974, Cretaceous oil sands of western Canada, Oil Sands: Fuel of the Future. *Canadian Society of Petroleum Geologists Memoir*, vol. 3, 50–67.

Jardine, D., D. P. Andrews, J. W. Wishart, and J. W. Young, 1977, Distribution and Continuity of Carbonate Reservoirs. *Journal of Petroleum Technology*, vol. 29, 873–885.

Koeberl, C., W. V. Reinold, and R. A. Powell, 2994, Ames Structure, Oklahoma: An Economically Important Impact Crater, Twenty-Fifth Lunar and Planetary Congernce: Abstract of Papers, 721–722. Part 2, Houston.

Kious, W. J., and R. I. Twilling, 2996, *This Dynamic Earth: The Story of Plate Tectonics*, Washington, D.C.: U.S. Government Printing Office, 80 p.

Kirk, R. H., 1980, Statfjord Field: A North Sea Giant, in M. T. Halbouty, ed., *Giant Oil and Gas Fields of the Decade 1968–1979* (AAPG Memoir vol. 30), 95–116. Tulsa: American Association of Petroleum Geologists.

Kolb, C. R., and J. R. Van Lopik Jr., 1966, Depositional Environments of the Mississippi River Delta Plain-Southeastern Louisiana, in M. L. Shirley, ed., *Deltas in Their Geological Framework*, 17–62. Houston: Houston Geological Society.

Kuykendall, M. D., and T. E. Matson, 1992, Glenn Pool Field—USA, Northeastern Oklahoma Platform, in N. H. Foster and E. A. Beaumont, eds., *Stratigraphic Traps III, Treatise of Petroleum Geology Atlas of Oil and Gas Fields*, 155–185. Tulsa: American Association of Petroleum Geologists.

Lamb, C. F., 1980, Painter Reservoir Field, in M. T. Halbouty, ed., *Giant Oil and Gas Fields of the Decade 1968–1978* (AAPG Memoir vol. 30), 281–288. Tulsa: American Association of Petroleum Geologists.

Landes, K. K., 1970, *Petroleum Geology of the United States*. New York: Wiley, 571 p.

LeMay, W. J., 1972, Empire Abo Field, Southeast New Mexico, in R. E. King, ed., *Stratigraphic Oil and Gas Fields-Classification of Exploration Methods and Case Histories* (AAPG Memoir vol. 16), 472–480. Tulsa: American Association of Petroleum Geologists.

Levorsen, A. I., 1967, *Geology of Petroleum*. New York: Freeman, 724 p.

MacKay, A. H., and A. J. Tankard, 1990, Hibernia Oil Field-Canada, in E. A. Beaumont and N. H. Foster, eds., *Structural Traps III, Treatise of Petroleum Geology Atlas of Oil and Gas Fields*, 145–176. Tulsa: American Association of Petroleum Geologists.

Martinez, A. R., 1970, Giant fields of Venezuela, in M. T. Halbouty, ed., *Geology of Giant Petroleum Fields* (AAPG Memoir vol. 14), 326–336. Tulsa: American Association of Petroleum Geologists.

Mayuga, M. N., 1970, Geology and Development of California's Giant Wilmington Field, in M. T. Halbouty, ed., *Geology of Giant Petroleum Fields* (AAPG Memoir vol. 14), 158–184. Tulsa: American Association of Petroleum Geologists.

McGreger, A. A., and C. A. Biggs, 1970, Bell Creek Oil Field, Montana, a Rich Stratigraphic Trap, in M. T. Halbouty, ed., *Geology of Giant Petroleum Fields* (AAPG Memoir vol. 14), 128–146. Tulsa: American Association of Petroleum Geologists.

Mero, W. E., 1991, Point Arguello Field–USA, in E. A. Beaumont and N. H. Foster, eds., *Structural Traps V, Treatise of Petroleum Geology Atlas of Oil and Gas Fields*, 27–58. Tulsa: American Association of Petroleum Geologists.

Mills, H. G., 1970, Geology of Tom O'Conner Field, Refugio Co., Texas, in M. T. Halbouty, ed., *Geology of Giant Petroleum Fields* (AAPG Memoir vol. 14), 292–300. Tulsa: American Association of Petroleum Geologists.

Morgridge, D. L., and W. B. Smith Jr., 1972, Geology and Discovery of Prudhoe Bay Field, Eastern Arctic Slope, Alaska, in R. E. King, ed., *Stratigraphic Oil and Gas Fields*. SEG Special Publication n. 10, 489–501.

Murphy, J. A., 1952, Good Field Data Are Necessary for Good Reservoir Engineering. *Petroleum Engineer*, vol. 24, B-91–B-99.

Ottman, R. D., P. L. Keyes, and M. A. Ziegler, 1976, Jay Field, Florida—A Jurassic Stratigraphic Trap, in J. Braunstein, ed., *North American Oil and Gas Fields* (AAPG Memoir vol. 24), 276–286. Tulsa: American Association of Petroleum Geologists.

Pippen, L., 1970, Panhandle-Hugoton Field, Texas-Oklahoma-Kansas— The First Fifty Years, in M. T. Halbouty, ed., *Geology of Giant Petroleum Fields* (AAPG Memoir vol. 14), 204–222. Tulsa: American Association of Petroleum Geologists.

Proctor, R. M., G. C. Taylor, and J. A. Wade, 1983, Oil and Natural Gas Resources of Canada. *Canadian Geological Survey Paper 83-31*, 59 p.

Rajasingam, D. T. and T. P. Freckelton, 2004, Subservice Development Changes in the Ultra Deepwater Na Kika Development, *Offshore Technology Conference Paper* 16699, 3 p.

Schlumberger Educational Services, 1987, *Log Interpretation Principles/ Applications*. Houston: Schlumberger Educational Services, 198 p.

Short, K. C., and A. J. Stauble, 1967, Outline of Geology of Niger Delta. *AAPG Bulletin*, 51, n. 5, 761–779.

Shumard, C. B., 1991, Stockholm Southwest Field—USA, in E. A. Beaumont and N. H. Foster, eds., *Stratigraphic Traps II, Treatise of Petroleum Geology Atlas of Oil and Gas Fields*, 269–304. Tulsa: American Association of Petroleum Geologists.

Sieverding, J. L., and F. Royse Jr., 1990, Whitney Canyon-Carter Creek Fields—USA, in E. A. Beaumont and N. H. Foster, eds., *Structural Traps III, Treatise of Petroleum Geology Atlas of Oil and Gas Fields*, 1–30. Tulsa: American Association of Petroleum Geologists.

Sills, S. R., 1992, Drive mechanisms and recovery, in D. Morton-Thompson and A. M. Woods, eds., *Development Geology Reference Manual*, n. 10, 518–522. AAPG Methods in Exploration Series. Tulsa: American Association of Petroleum Geologists

Stafford, P. T., 1957, Scurry Field, Scurry, Kent and Borden Counties, Texas, in F. A. Herald, ed., *Occurrence of Oil and Gas in West Texas*. Austin: University of Texas Bureau of Economic Geology, 295–302.

Stafford, P. T., 1959, Geology of Part of the Horseshoe Atoll in Scurry and Kent counties, Texas. *U. S. Geological Survey Professional Paper* 315-A.

Stanley, T. B. Jr., 1970, Vicksburg Fault Zone, in M. T. Halbouty, ed., *Geology of Giant Petroleum Fields* (AAPG Memoir vol. 14), 301–308. Tulsa: American Association of Petroleum Geologists.

State of Ohio, 1966, *Oil & Gas Fields Map of Ohio*. Department of Natural Resources, Division of Geological Survey, Columbis, Ohio.

Thorne, J. A., and A. B. Watts, 1989, Quantitative Analysis of North Sea Subsidence, *AAPG Bulletin*, vol. 37, 88–116.

Viniegra, F., and C. Castillo-Tejero, 1970, Golden Lane Fields, Veracruz, Mexico, in M. T. Halbouty, ed., *Geology of Giant Petroleum Fields* (AAPG Memoir vol. 14), 307–325. Tulsa: American Association of Petroleum Geologists.

Unit Conversions

English to Metric

1 inch = 2.54 centimeters

1 foot = 0.3048 meter

1 mile = 1.609 kilometers

1 pound = 0.4536 kilogram

1 gallon (U.S.) = 3.7854 liters

°F − 32 = 5/9 °C

1 barrel = 0.1590 cubic meter

1 cubic foot = 0.0283 cubic meter

1 British thermal unit = 1.0542 kilojoules

Metric to English

1 centimeter = 0.394 inch

1 meter = 3.2808 feet

1 kilometer = 0.6214 mile

1 kilogram = 2.2046 pounds

1 liter = 0.2642 gallon (U.S.)

$9/5\,°C + 32 = °F$

1 cubic meter = 6.2899 barrels

1 cubic meter = 35.3144 cubic feet

1 kilojoule = 0.9842 British thermal unit

Oil and Gas Records

First oil wells in North America

1858: Oilsprings, Ontario, Canada. The well was dug to 60 ft by James
M. Williams.

1859: Titusville, Pennsylvania. The Drake well was drilled to 69.5 ft by
William "Uncle Billy" Smith.

First gas well in North America

1825: Fredonia, New York. The well was drilled by William Aaron Hart to
27 ft. The gas reservoir rock was the fractured, Devonian age Dunkirk
black shale.

First dry hole

1859: Grandin well, Warren County, Pennsylvania, 134 ft total depth

First recorded lease

July 4, 1853: Cherrytree township, Venango Co., Pennsylvania, with a
5-year term

First use of a rotary drilling rig for oil

Between 1895 and 1900 by M. C. and C. E. Baker in Corsicana, Texas

First casing

1808: Great Buffalo Lick, West Virginia, using wood strips in a
saltwater well

First use of drilling mud

1900: Spindletop, Texas. It is reported that cattle were driven through a
pond to produce the mud.

First wireline well log

September 5, 1927, a resistivity log in the Pechelbronn oil field, France, by
Marcel and Conrad Schlumberger

First well stimulation

1865: Ladies well, Titusville, Pennsylvania, using a tin torpedo with a
gunpowder primer and a fuse

First hydraulic frac job

1947: Hugoton field, Oklahoma

First offshore well

1897: Summerland, California, on a pier

First offshore well drilled out of sight of land

1947: Gulf of Mexico southeast of Morgan City, Louisiana, in 18 ft of
water by Kerr-McGee Company

First subsea well

1943: Lake Erie, United States

First refraction seismic discovery

1924: Orchard salt dome, Texas

First reflection seismic discovery

1928: Maud pool (Seminole field), Oklahoma

Deepest well drilled for gas or oil

31,441 ft.: Bertha Rogers No. 1. Washita County, Oklahoma, in 1974 – it
was a dry hole.

Deepest well ever drilled

40,230 ft.: The Kola Superdeep Borehole, a scientific well drilled by Russia
with a turbine motor in the Kola Peninsula between 1970 and 1989.

Most oil from a single well

57,000,000 bbl from Cerro Azul No. 4, Mexico, in the early 1900s

Most oil from a well in a single day

260,858 bbl from Cerro Azul No. 4, Mexico, on February 19, 1916

Largest conventional oil field

Ghawar, Saudi Arabia, 82 billion bbl recoverable oil

Largest gas field

North Dome in Qatar and South Pars in Iran with 1,200 trillion cu ft of
natural gas

First reported waterflood

1880: Pithole City, Pennsylvania

Longest continuously producing oil well

McClintock Well No. 1, Rouseville, PA, since 1861. The 200-ft-deep well
initially produced 175 b/d but is currently producing 1/10th b/d.

Index

A

abandonment pressure, 456
abnormal high pressure, 286–289
Abo reef, 142, 143
El Abra reef, 150
absolute age dating, 33–36
absorption tower, 381
AC motors, 276–277
accumulation (petroleum), 117–119
   in fractured basement rocks, 157
accumulator, 273
acid gases, 380
acid job, xxiii, 440
acid rain, 4
acidizing, 440
acoustic impedance, 228
acoustic velocity log, 333–334, 335
acquisition (seismic data), 216–222
   cable-free, 223
acreage contribution agreement, 241
additives (chemical), 268–269
adjustable bent subs, 296
adjustable choke, 359
aerial photography, 196

aeromagnetic survey, 211, 212
Agbada Formation, 166
age
   dating, 33–41, 137
   petroleum, 120
air drilling, 303–304
air gun, 218
air-balanced beam-pumping unit, 367
Akata Formation, 166
alarm call-out, 387
Alaska
   Mukluk, 289
   North Slope, 174
   Prudhoe Bay field, 48, 131, 173, 174, 175
Alberta
   Athabaska tar sands, 483
   Devonian reefs, 147
   Leduc oil and gas field, 147, 178
   Pembina oil field, 181, 182
   Redwater oil field, 147, 148
   Turner Valley field, 428, 429
Aleutian Islands, 76
Aleutian Trench, 76
alkane molecule, 2, 5
allowable amount, 244

B

C

cable tool rigs, 246–248
cable-free seismic acquisition, 223
Caillou Island salt domes, Louisiana, 188
calcite, 16, 27
California
    Kern River steamflood, 465, 466
    Los Angeles basin, 53, 54, 70, 88, 89
    Point Arguello oil field, 156, 157
    Potrero oil field, 66
    San Andreas Fault, 66, 67, 80
    Southern California oil basins, 88, 89
    Wilmington oil field, 67, 119, 156, 161, 162
caliper log, 331–333
Cambrian Period, 43
Canadian shield, 21
Cantarell field, Mexico, 45
cantilevered jackup rig, 392–393
cantilevered mast rig, 251, 252
Capitan reef, 141–142, 143
caprock, xv, 172, 187. See also specific caprocks
    production, 190
carbon atoms, 2
    dating, 36
carbon dioxide, 9–10, 286
    flood, 462–463
carbon-14 age dating, 137
carbonate, 140
    reservoir rocks, 140
Carboniferous Period, 42
Cardium Formation, 181
carrier, 366
    unit, 413
cased-hole log, 316, 434
casing (well), xx, 343, 346–351, 353. See also perforated casing completion
    cemented into well, xxi
    conductor, 243, 395–396
    crew, 344–345
    elevators, 345
    expandable, 356, 418
    hangers, 345, 358
    intermediate, 352
    oil string, 352

    patch, 418
    pressure, 431
    program, xx, 351
    pump, 362
    repair, 418
    roller, 418
    string, xx, xxii, 344–345
    tongs, 345
    types, 352
casing-free pump, 370
casingheads, 358
Caspian basin, 22
catheads, 254
cathodic protection, 447
catshaft, 254
catwalk, 383, 384
caustic flooding, 462
cellar, 243
    deck, 398
cement, xxi
    additives, 347
    bond log, 419
    slurry, 347, 348–351
    squeeze job, 419
cement job, 343, 344, 347, 349, 350
    pressure-tested, 353
    waiting on cement, 351
cementation, natural, 16, 17
cementing (well), xxi
    multistage, 351
    primary, 419
    secondary, 419–420
Cenozoic Era, 45
Central basin platform, Texas, 149, 150
central processing unit, 373, 374
centralizer, 345
centrifugal compressors, 381
chalk, 152, 158, 191. See also Austin Chalk
chance of success, 289
channel (geophones), 220
charcoal test, 385–386
charthouse, 386
checkshot survey, 231
chemical
    additives, 268–269
    composition of crude oil, 1–2
    cutter, 283, 284, 418
    flood, 463–464
chemistry (petroleum), 1–2

chert. *See* flint
Chicxulub crater, Mexico, 45
chokeline, 273
chokes, 359
  manifold, 273
Christmas tree, xxii, xxiii, 359, 360,
    370, 371
  gas conditioning and, 380
circle wrench, 416
circulating system, xvii–xviii, 266–273
clastic grain size, 18
clastic sedimentary rocks, 17
clastic sediments, 15
clean oil, 383
clean sands, 126
cleanout (well), 415–416
cleats, 478
cleavage, 25–26
Clinton Sandstone gas fields, Ohio,
    128, 129
closed hydraulic pump, 370
closed loop drilling, 278
closeology principle, 194
closure (trap), 199, 200
cloud point, 5
coal, 30, 113–114
  bed, 478–480
  seam/seam gas, 113, 478–480
coastal mountain range, 77
coastal plains, 91–92, 128, 129
coccoliths, 41
Cognac (platform), 296–297
coiled tubing unit, 413–414
collar, 344, 356
  drill, 259
  float, 350
  log, 435
  spiral-grooved drill, 283, 285
color (mineral), 25
Colorado, 178, 179
colored seismic displays, 225
columnar legs, 392
column-stabilized semisubmersible
    rig, 393–394
combination contract, 241
combination traps
  bald-headed structures, 182–186
  salt domes, 186–191
commercial land ownership maps,
    236

common-depth-point, 222
common-mid-point, 222
  stacking, 232
company representative, 274
compensated log, 316
completing (well). *See* well
    completion
completion
  barefoot, 353
  bottom-hole, 353–355
  cards, 206
  dual, 370–371
  fluid, 356
  gravel pack, 353, 354, 355
  horizontal well, 355
  multiple, 370–371
  open-hole, xx–xxi, 353, 354
  packer, 356
  perforated casing, xx–xxi, xxii,
    355
  perforated liner, 355
  rig, 344
  set-through, 354
  top set, 353
  tubingless, 358
compliant platform, 400–401
composite sample, 306
compressibility factor, 456
compression, 55
  ratio, 381
  test, 385–386
compressional wave, 234
computer-generated log, 338
concentric tubing workover, 414
concession, 237
  agreement, 238
condensate, xiii, xxiii, 10–11, 381
conditioned (hole), 313
conductor
  casing, 243, 395–396
  hole, 243
  pipe, 351
conglomerate, 29
coning, 438–439
consolidated sediments, 16
constructive delta, 133, 134, 136
  delta switching on, 137
continental drift, 74
continental margins, 67–72
Continental Oil Company, 154
continental shelf, 67–68

G

J

K

M

P

S

T

U

Y

Z